Thin Film Deposition Techniques

Tahir Iqbal Awan · Sumera Afsheen · Sabah Kausar

Thin Film Deposition Techniques

Thin Film Deposition Techniques and Its Applications in Different Fields

Springer

Tahir Iqbal Awan
Department of Physics
University of Gujrat
Gujrat, Pakistan

Sumera Afsheen
Department of Zoology
University of Gujrat
Gujrat, Pakistan

Sabah Kausar
Department of Physics
University of Gujrat
Gujrat, Pakistan

ISBN 978-981-96-1363-2 ISBN 978-981-96-1364-9 (eBook)
https://doi.org/10.1007/978-981-96-1364-9

This Springer imprint is published by the registered company Springer Nature Singapore Pte Ltd.
The registered company address is: 152 Beach Road, #21-01/04 Gateway East, Singapore 189721, Singapore

Contents

Chapter 1
Fundamentals of Thin Film

Abstract In this introductory section, the elementary history and beginning of the shrill film have been described in detail. Nanomaterials play a crucial role in the formation of the shrill film as the growth of the shrill film is greatly affected by the different properties of the nanomaterials. Shrill film is deposited by the CVD and PVD methods, and certain types of these processes affect the growing formation of the shrill film. The main properties of the shrill film such as mechanical, electrical, and optical also affect the growth rate and application aspects of different shrill films. We also tried to speechless the challenges and difficulties that occurred during the formation of the shrill film. The significance of the statement techniques is required to describe the statement processes of the shrill film. Some challenges, such as adhesion issues, film uniformity, and film thickness, are also explained, along with the process through which these challenges can be overcome and film uniformity enhanced. The applications of shrill-film statement techniques can be seen in numerous fields such as optoelectronics and photonics, biocompatible shrill-film coatings, photovoltaic cells, and also shrill-film requests in the field of medical and muscle manufacturing.

1.1 Introduction and Overview

A shrill film is defined as a coating of a solid whose thickness ranges from less than a nanometer to numerous microns. In particular, thin film is a coating deposited on a material atom by atom. If we consider bulk materials that have fixed properties, they must have limited applications due to this condition. If we require more properties of such materials, then we must have to reduce the size of bulk materials. So, when the thickness is reduced beyond a certain limit, then these properties change dramatically. As a result, we can easily fabricate new devices by decreasing their size. In our everyday lives, we see many examples of shrill film like as follows:

- Soap bubbles.
- Antireflection covering on eyeglasses.
- When we add oil drops to water, oil slicks water to show the presence of a thin film.

T. I. Awan et al., *Thin Film Deposition Techniques*,
https://doi.org/10.1007/978-981-96-1364-9_1

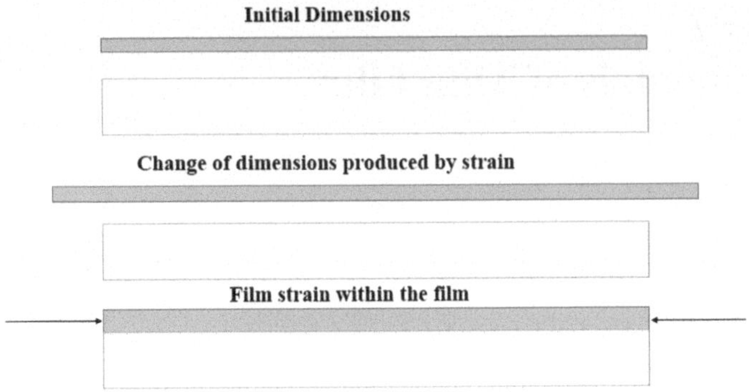

Fig. 1.1 Overview of shrill film

Developing novel materials has led to many revolutionary technological advancements during the past century. In addition, a number of the current global issues about energy, health, and security will necessitate the development of new material classes, such as multicomponent alloys and composites or artificially structured materials that are known as advanced materials [1]. It is becoming progressively obvious that new materials will be essential to promoting economic growth and making the industrial sector more competitive. Several US funding agencies are involved in the multistakeholder Materials Genome Initiative as a response to this challenge. By eliminating the time needed to introduce novel materials to the market from the current 10–20 years, it aims to speed up the rate of advanced material discovery. Although the Initiative is founded on a robust computational and modeling methodology, it is acknowledged that comprehensive and precise sets of data on material properties are the only way to develop useful models of material behavior. The Initiative states that "Researchers must have access to the largest possible dataset upon discovery [1]," upon which to build their models, to offer a more comprehensive representation of a material's properties [2]. Figure 1.1 demonstrates the general overview of shrill film.

1.2 Historical Development of Shrill-Film Technology

1.2.1 Early Discoveries and Experiments

Modern shrill films are ordinary and have more than 5000 years of known history. On the additional pointer, shrill-film growth by the sputter statement is a considerably more recent phenomenon, requiring the creation of space pumps and electric power in the 1600s and 1700s. The splutter statement was first documented in the early 1800s, and by 1880, it was the dominant technique in the optical coating business. The 1800s described the description of multitarget sputtering, favored spatter of compounds, spatter of fluids, and visual spectroscopy for procedure classification. In the late 1800s, beginning dynamism and harvests were measured, and in the 1930s, harvests that were reasonably consistent with current statistics were published. The mid-1930s observed the introduction of roll-to-roll splutter covering on elastic substrate, and the first prolonged self-sputtering (i.e., spatter devoid of gas) demonstration was carried out in 1970. The earliest magnetron sputtering experiment results were published in the late 1930s, while the term magnetron primary looked in print in 1921 [3]. Rotatable magnetrons first surfaced in the early 1980s, while tunable "unbalanced" magnetron spatter was invented in 1992. The initial imageries of a parallel-plate magnetron were given in an obvious application filed in 1962 [4]. Ionized-magnetron spatter and high-octane impulse magnetron spatter, the final of which is currently offered in multiple variations, are two further types of magnetron spatter that developed throughout the 1990s to effectively ionize sputter-ejected metal atoms. Mid-frequency-ac and pulsed-dc sputtering were made possible by a patent application in 1975 after modern capacitive-coupled RF sputtering systems were invented and designed in the early 1960s. Although the phrase "reactive sputtering" was not used in the literature until 1953, the deliberate manufacture of metal-oxide films dates back at least to 1907, and it resulted in initial metal-oxide and nitrate spatter researches in 1933. In 1940, a study was conducted on the impact of board corrosion on secondary electron harvests and spatter charges. Reactive sputtering's earliest kinetic models were created in the 1960s, and in the early 1980s, partial-pressure control-based high-rate reactive sputtering was developed [5].

1.2.2 Evolution of Thin-Film Deposition Techniques

Several papers were published in the mid to late 1800s on the use of visual release to investigate the transition from a continuous glow discharge to an arc, the electrical structure and configuration of DC glow discharge, and the nature of metal atoms sputter ejected from the cathode (target). An English physicist, Michael Faraday (1791–1867), reported on electrical and optical characterization. In 1838, there were glow discharges from the surface of the cathode. His glass discharge tube contained brass electrodes and was powered by air, nitrogen, and electricity [6]. At 4.4 inches of pressure, oxygen, hydrogen, and other gases are present (112 Torr). While Faraday deposited films throughout these tests, this was not the goal. Later work on the optical characteristics of vacuum-arc-deposited shrill metal films was described in his fifth Bakerian Lecture in 1857 and published that year [7].

In 1775, Henry Baker (1698–1774) donated the Bakerian Medal and Lecture to the Royal Society of London. It is given out once a year to someone working in the domains of "natural history or experimental philosophy" (i.e., the physical sciences). Turner, Senior Assistant Curator of the Museum of Science at Oxford, described Baker as "a typical polymath in the eighteenth-century manner." Although he made no important contributions to scientific research, he did make valuable contributions to the diffusion of scientific knowledge, particularly in the field of microscopy, and was an eager participant in London's scientific and literary life.

Michael Faraday, well renowned for his work in electromagnetics and electrochemistry, is regarded by science historians as one of the most influential scientists and the best experimentalists in history, despite having little formal training beyond elementary school. Between 1829 and 1857, Faraday was asked to give five Bakerian Lectures to the Royal Society [3].

1.2.3 Development of Nanomaterials

In the discipline of thin-film growth, nanomaterials have become revolutionary participants that present intriguing opportunities for customizing the characteristics and functions of these films. Due to their special size-dependent characteristics, which include a large surface area, an electrical structure that can be adjusted, and increased chemical reactivity, they are the best options for solving problems and making advances in thin-film technology [8].

There are some factors through which the thin-film growth is altered by nanomaterials.

1.2.3.1 Patterns and Structures

When it comes to controlling the formation of thin films with distinct morphologies and crystal structures, nanoparticles can serve as patterns and structures. As a result, complex nanostructured films with the required electrical, optical, and catalytic capabilities can be produced. To generate effective light-emitting diodes, for instance, semiconductor nanowire growth can be patterned using gold nanoparticles [9].

1.2.3.2 Enhanced Growth and Nucleation

Some nanomaterials, like zeolites and metal oxides, decrease the energy barriers and provide active sites for nucleation and development of thin films. This leads to greater mastery over the layer thickness and morphology, faster deposition rates, and increased homogeneity of the film. For example, good electrical characteristics may be achieved in the crystalline oxide thin films grown from the titanium dioxide nanoparticles used as seeds [10].

1.2.3.3 Property Tailoring and Functional Doping

The thin film introduced with nanoparticles of different compositions allows its magnetic behavior, optical absorption, and electrical conductivity to be accurately established and controlled. This would make it possible to develop thin films for specific purposes, differentiated by unique capabilities. This ability will result in the growth of thin films for transparent electrodes in solar cells through the doping of zinc oxide thin films with silver nanoparticles to be more electrically conductive [7].

1.2.3.4 Hybrid and Multipurpose Thin Films

Hybridization of nanoparticles with other materials, like organic molecules or polymers, can provide hybrid thin films that express combinations of properties unique to each of the involved materials. Hybrid films potentially hold a wide range of features such as photovoltaic activity from light converting to electric energy conductivity, self-healing, and stimuli-responsive activity. Therefore, the hybridization of graphene with metal nanoparticles could actually produce transparent, flexible, and conductive thin films for possible use in display touchscreens [10].

1.2.3.5 Sustainable and Expandable Methods

The researchers are currently working on ecologically acceptable, scalable, and cost-effective techniques for growing thin films based on nanomaterials. Mass production of high-quality thin films requires applying bioinspired methodologies, roll-to-roll processes, and solution-based procedures. For instance, a scalable and environmentally friendly method of producing flexible and biodegradable electronics is the use of nanocellulose templates in thin-film deposition [11]. The synthesis of nanomaterials for thin-film growth is an immensely promising but quickly developing matter. We can generate thin films with previously unattainable functions by utilizing the special qualities of nanomaterials, providing the door to breakthroughs in a diversity of fields as well as electronics, photonics, energy gathering, and sustainable technology [12].

Epitaxial growth of thin film and nanodots on the surface of substrates with variable lattice oddity is essential for the development of advanced and better application of thin-film technology. As the size of structures shrinks, a particular dislodgment will be able to govern scheme presentation. When the lattice eccentricity is minimal, films raise pseudomorphically up and around to a "critical thickness," anywhere it develops actively beneficial for the film to include disruptions. Because dislocations of the film externally float to the boundary, the Burgers vectors and planes of the displacements are determined by the slip vectors and glide planes of the film's quartz erection. Depending on structures, these disruptions reflect misplaced or additional planes [13]. For example, it has been discovered that during the 3D development of germanium on silicon, 90-degree disruptions with a/2110 Burgers vectors are formed at the control of germanium atolls and invention in (001) thin film–substrate boundary. According to unoriginal thinking, LME throughout thin-film evolution is conceivable extended as the lattice eccentric among the film and the substrate is less than 7–8%. Minor lattice crank results in lower interfacial energy and the formation of coherent epitaxy. Overhead this discrepancy, it was predicted that film would become surfaced or mostly polycrystalline. The underlying principle of lattice reduction including lost or additional planes powerfully recommended that during thin-film growth, we should emphasize plane matching rather than lattice constants. In some situations, the film grains may organize themselves end-to-end the disorders of atoms that are a portion of the corresponding planes of the substrate to reduce the system's free energy [14].

The film in the DME paradigm has an immovable or continuous location about the substrate, which may become the same, especially under low misfit conditions. TiN can be produced epitaxially on a silicon substrate with almost 22% lattice misfit via DME, where four TiN lattice planes are matched with roughly three silicon planes across the film–substrate interface. According to a simple theoretical demonstration, the DME increased total energy. Many systems with a substantial mismatch have been cultivated with amazing success since the initial DME report. The atomistic specifics of the reduction process, as well as the environment of displacements in such schemes, will determine future advancement in this crucial sector [14].

1.2.4 Need for Nanomaterials for the Development of Thin Film

In numerous essential and high-priority research fields, such as health and the production of energy from renewable sources, nanomaterials in the usage of nanoparticles and nanometric thin films are significant. As materials with particular optical, electrical, or mechanical capabilities, they are widely used. Nanoparticles have been used in silicon and perovskite solar cells as phosphor layers and as semiconductor layers in dye-sensitized solar cells. Conversely, thin films are commonly utilized in solar cells of various generations as transparent conductive, antireflection, or passivating materials. A significant area in the rapidly expanding fields of science and business is nanomaterials [14]. A significant challenge in large-scale deployment is the production and deposition process, wherein appropriate processes for nanomaterials and thin films need to be identified. It is, therefore, of utmost necessity to develop scalable industrial processes and research the relationship between production and deposition parameters for the properties of thin films and nanomaterials produced. The reproducibility of the carried-out synthesis and deposition procedures is therefore very important. When doing research in the field of product lifecycles of adopted new materials and technologies, it is also crucial to consider the economic and environmental factors [15]. The application of thin and nanostructured solid films on exteriors allows for significant new applications in grounds of vitality change and storing today. Shortly, these films, or thousands of films, will be integrated into energy systems for the production of useful heat and electricity and for energy-saving applications—specifically for buildings. We review work in this area. Even now, when we are facing the threats of energy deficit and greenhouse gas emissions as a result of fossil fuel consumption through the intensive use of these resources, we understand how much the consequences and future of solar energy systems have been greatly overrated. Let us review the study on solar thermal systems and energy-saving solutions [16]. Effective application of solar thermal technology is vital for producing a significant percentage of ecologically harmless renewable energy sources. Besides the design and geometry of sophisticated systems capable of reaching high temperatures, new and nanostructured materials have been proposed and integrated collecting and storage systems. The latest technological and material developments have been having great impacts on the quality and cost of thin films. This has promoted new thermal absorbers and appraisal methodologies. We also elaborate on low-e coatings and electrochromic thin efficiency films, which are critical for thermal energy savings in buildings and increasing devices [17].

1.3 Properties of Thin Film

There are numerous properties of thin films that present an important role in the applications of modern thin films. Popular the general process of manufacturing thin films, we must take some precautions and observe certain conditions, thin-film application is not affected by these conditions.

When we deposit thin films on a substrate, the most crucial characteristic is the technique we are using for deposition which will be responsible for the long-term use and performance by enhancing the properties of the materials. The ability of the discipline of engineering and material science to imagine efficient and unique materials with astonishing combinations with various qualities has revolutionized the present world. Technological advancement is accelerating. Thin films are compulsory in recent technology for a diversity of resolutions. Thin-film expertise is the foundation of astonishing progress in solid-state-based microelectronics [18]. The use of metal films' optical features, as well as scientific interest in the performance of 2D solids, has contributed to the vast curiosity in thin-film research and technology. The extraordinary increase in thin-film research is undoubtedly owing to their numerous uses in several fields such as space science, electronics and optics, aircraft, defense, and further productions [19]. These inquiries have resulted in frequent developments such as piezoelectric procedures, micro-miniaturization of influence supplies, refinement and intensification, instrument basics, solar energy storing and alteration to other forms, magnetic memories, superconduction films, interference filters, reflecting and antireflection coverings, and numerous others. Thin-film materials are critical components of ongoing technical breakthroughs in optoelectronic, photonic, and magnetic devices. The dispensation of materials into thin films enables simple assimilation into a variety of devices.

- Thin film exhibits low density.
- Thin film has properties completely different from bulk materials.
- Thin film may be under stress.
- Thin films are usually two-dimensional materials [20].

1.3.1 Electrical Properties

These can be broadly classified into three types of electron materials: metal, semiconductor, and dielectric. The nature of electrical transport is different for all of them, so it becomes troublesome to define them in a general sense. But here, scaling effects start dominating the electrical characteristics of the materials similarly even in the thin-film state. The character of electrical transport depends upon various parameters like film thickness, lattice dimensions, purity, surface roughness, and defective level of the layer [19].

Electrical properties of thin films cover a broad field if we consider the variation in resulting microstructure they can be in the form of metallic films, insulators, or semiconductors in films, and also can be the type of substrate on which the thin film is dumped. Besides this, an extensive choice of thin films possesses some morphological properties that are common in the global usage of the conductivity of thin films. The main factor that affects the electrical conductivity of thin films in comparison with bulk materials is the size factor. When the size of materials transform from bulk phase to thin film, the result of that thing average path of electrons also reduces because the phenomenon of additional scattering occurs. When we decrease the size of materials, it causes an increase in the properties of materials like size effect whether we are working with amorphous, polycrystalline, or epitaxial structures keeping all other components of the conductivity constant [21]. When the thin film is formed by semiconductor material where a large number of grain restrictions exist and due to the presence of semiconductor either electrons or holes as majority carriers, so in this case the mechanism of transport becomes very complicated because of scattering. In this type of mechanism, due to higher charge carrier mobility, a larger volume fraction of the crystalline is offered. When the mobile carriers are scattered in between the crystallites, then the grain restrictions are considered disordered regions. For example, the electrical conductivity of the bulk single crystalline is larger than the electrical conductivity of nanocrystalline and polycrystalline. The presence of microstructural defects in grain boundaries also causes of reduction of electrical conductivity. These defects limit the motion of charge carriers forming a potential energy barrier of doped semiconductors. Also, electrical conductivity is reduced if the number of free carriers that are available for conduction is reduced [22].

1.3.2 Optical Properties

The optical properties of thin films can be influenced by the confession parameters that result from microstructures. To explain the optical properties with the microstructural factors, there is a dire need to define the relationships of optical coefficients. These optical coefficients can be explained based on transmission and reflection coefficients. The dependence of these coefficients is on the thickness of the film. For some metals like silver, gold varies from bulk materials to thin-film materials, and when the width of the film rises, the optical coefficients of materials increase which are very low for bulk materials [23]. These optical coefficients also depend upon the incident wavelength because to wavelength of the incident beam dominated evolutions in the observable or the infrared area. The mechanisms of the surface smattering increase when the thickness of the film is reduced, and as a consequence of this, the scattering time and mean free path for electrons also decrease which affects the optical conductivity contribution. Hence, a decrease in electrical conductivity causes an increase in resistivity. The reflection and transmission in the thin film hang on the roughness of materials [22, 24]. Optical experiments provide a good method to characterize semiconductors. Measuring the absorption coefficient for varying energies in particular provides information about the bandgaps of material. Knowing about the bandgaps is crucial in understanding the electrical properties of semiconductors and therefore of immense practical value [19].

1.3.3 Mechanical Properties

The mechanical properties of thin films may often be different from their counterparts in bulk materials. This can be partly explained by the nanostructure of thin films as well as by the fact that thin films are bonded onto a substrate. Under normal circumstances, the yield strengths of thin films are extremely high. For these reasons, very high residual stresses are no problem for thin films. The residual stress can be dissipated later in the processing or during the operation of the device by plastic deformation, thin-film fracture, or interfacial delamination. Elastic and plastic characteristics are both important in the characterization of thin films. Elastic and plastic characteristics are both important in the characterization of thin films [25]. Thin films may be measured for mechanical properties using the tensile testing of freestanding films and the microbeam cantilever deflection technique, but nanoindentation is the most convenient method. It does not require special sample preparation and allows for fast and inexpensive tests [19].

1.3.4 Morphological Structures

The morphological erection of the thin film is affected by the influence of deposition parameters because the influence of various factors influences the growth phases. However, there also exist some typical structures of thin film that we can consider in a broad variety of thin films. The most common feature of morphological structures for thin films is probably grown by techniques that are vapor-based where grown directions don't need to be perpendicular to the surface of the substrate. The final morphological structure of thin film is determined by how the atoms are arranged in the formation process. These structures increase in groups when more atoms are arranged layer by layer on the surface of the substrate during the diffusion process.

This flux of the atoms creates the thickness of the film and more atoms means more flux, and this will directly affect the thickness of thin film. This growth in the thickness of the thin film forms a columnar structure that looks like a cauliflower. It was earlier reported that if we want to form a columnar structure from a rough surface, then the necessary conditions that we must have to fulfill are the limited surface mobility of the atoms that we are adding in the growth of thin film and the flux of the arriving vapor must be non-normal to the thin film. Morphological structures of the thin film also vary from the technique of deposition because there are different techniques for the development of thin films on the substrate surface and every technique has a different mechanism of formation and different thickness of the thin film [26].

1.4 Challenges and Problems

Thin-film deposition is an important procedure in the manufacture of numerous electronic and optoelectronic devices. However, it has several challenges and problems that must be addressed to produce high-quality, reproducible, and scalable thin films. These difficulties can be divided into three categories.

1.4.1 Process Control and Material

- **Uniformity and Thickness Control**: It is acute for any device performance to attain constant thickness and composition through wide areas. Uniformity can be artificial by factors such as substrate research, salutation parameters, and ambient variables.
- **Surface Morphology**: Regulatory thin-film surface morphology is necessary for optical, electrical, and mechanical things. Device disasters can be caused by faults such as pinholes, roughness, and crashes [27].

Understanding and directing thin-film nucleation and progression mechanisms are critical for altering their microstructure and physical appearance. Optimization of deposition constraints and substrate interactions is required.

1.4.2 Impurities and Flaws

- **Impurity Incorporation**: Impurities from the base material, deposition space, or substrate can have a main impact on thin film possessions. It is critical to classify and reduce impurity bases.

- **Point Defects**: Point defects, for example, vacancies and interstitials, can influence thin-film electrical conductivity, charge carrier transportation, and optical characteristics. Controlling the meditations of point defects is acute [28].
- **Extended Defects**: Extended defects, for instance, particle boundaries, dislocations, and stacking faults, can have an effect on mechanical characteristics, carrier mobility, and longevity of the device. These faults can be condensed by optimizing conditions of deposition.

1.4.3 Reproducibility and Scalability

- **Wide-Area Deposition**: Due to subjects such as non-uniformity, stability of process, and equipment boundaries, manufacturing high-quality thin films through wide regions is challenging. It is important to develop accessible deposition processes.
- **Reproducibility and Control**: It is serious for device dependability to attain reproducible thin-film assets across groups and manufacturing lines. Control and checking of the process are perilous.
- **Cost-Effectiveness**: Emerging low-cost thin-film deposition approaches are essential for commercial feasibility and wide implementation. It is necessary to establish a balance between quality, recital, and cost [29]. To report these issues, frequent research and development in the deposition techniques of thin film, material science, and process control measures are required. Overcoming these difficulties will result in the formation of high-performance, dependable, and climbable thin films for a diversity of applications.

1.5 Material Compatibility and Adhesion Issues

1.5.1 Attributes of Adhesion

Adhesion is the process by which a material attaches to the surface of another. The field adhesion test is one of the highly useful instruments in the workplace where impacts of different cleaning procedures and primers are evaluated and is also a good indicator of the fundamental adhesive qualities of an adhesive on some substrate.

A film's adhesion to a substrate depends heavily on the chemical structure, cleanliness, and microscopic topography of the substrate surface. The higher kinetic energy of incoming species, deposit adsorption energy, and initial nucleation density improve the adhesion of the film. Any contamination that might be present on the surface of the substrate would either add to the strength or weaken the adhesion, depending on whether it raises or lowers the adsorption energy. Another way to enhance adhesion is to provide more nucleation sites on the substrate, which can be done by using a fine-grained substrate or precoating it with the appropriate ingredients. The quality of deposits depends on the conditions under which they form: deposits formed at high supersaturation and low vacuum tend to be loose and porous and therefore less adhesive than compact deposits. Compatibility and adhesiveness of the product are, therefore, part of choosing an appropriate sealant for a given building application in conjunction with the intended substrate, possibly including waterproofing or air barrier [30].

A sealant can be acceptable to a few substrates but does not possess desirable bonding properties to the substrate that work as expected. That is why, before combining any two materials for your project, you must decide if they will bond to and be compatible. Let's see the difference of these two materials and how they influence each other to reduce the chances of incompatibility or adhesion failure and, thus the detrimental effects on a structure.

1.5.2 Characteristics of Compatibility

They are said to be compatible if things that come into contact with each other exhibit no undesirable response or loss of performance attributes. Incompatibility can cause discoloration, streaking, changes in the curing process of either product, damage to the substrate, and so on. Let me state that while compatibility is very essential for performance, it does not guarantee satisfactory adhesion [31].

1.5.3 Compatibility and Adhesion Problems

If you test compatibility and adhesion between any two products you're going to use together or near each other, then you'll never have a bad reaction. We've seen that even though two products are curing within inches of one another but are not touching, they still cause some adhesive and cure problems. Specify testing for adhesion and compatibility, and work closely with the design/build team to review sequencing, consulting with product manufacturers before applying the product to help minimize challenges with sealant compatibility and adhesion to other technologies and substrates. Historical data on previous testing of any two goods are helpful, but nothing replaces field adhesion testing to ensure that there haven't been major substitutions or changes [32]. Single-source solutions from a single manufacturer are recommended wherever possible for complete connectivity confidence.

1.5.4 Film Thickness Uniformity

Another critical area of thin-film uniformity optimization concerns large-area coatings, such as optical coatings that could be meant to tolerate errors in nanometers. Physical vapor deposition is one common technique for depositing thin films. Applications can be quite wide-ranging, including antireflection coatings and photovoltaics. Methods and simulations applied toward making thin films more uniform in the physical vapor deposition process both through evaporation and sputtering techniques including the emission characteristics from material sources, the projection effects on the distribution of film thickness, and geometric and rotational influence from configurations of apparatus have been applied. A new tool involving MathCAD was also developed to model and simulate the homogeneity of thin film in the physical vapor deposition process. The program was applied to create a new tool equipped with models and simulations of thin-film homogeneity in the physical vapor deposition. Its results were compared with theoretically established analytical equations of the thickness distribution and experimental data, displaying good agreement. The program was successfully applied toward the creation of an optimization mask for the thinning film thickness distribution, which led from 4 to 0.56% [33].

1.5.5 Evaporation Source's Thin-Film Thickness Distribution

Three assumptions are made formerly approximating the film thickness:

The pressure at which the process takes place is low enough not to allow collisions between the vapor and other particles, i.e., the mean-free routes are substantially bigger than the dimensions of the deposition system. The strength of the vapor emission from the source can allow for collisions between vapor molecules in a wide area around the source [17].

The sticking coefficient of the molecules of the vapor can be 1, depending on the material which is undergoing evaporation. The sticking coefficient is defined as the ratio of the number of atoms stuck on the surface over the total number of atoms impinging on that surface in the same time period. Multiple beam interferometry was used to investigate the potential uniformity attainable along the source axis in evaporation systems with large source-to-substrate distances and small source areas. Thin films of silver were evaporated onto stationary and rotating substrates 3.86 cm in diameter, symmetrically arranged about the source axis, from a resistance-heated dimple boat operating in standard and ultrahigh vacuum. Measurements of substrate diameters have been conducted to obtain thickness differences along with average film thicknesses. Data were obtained on aluminum, aluminum oxide, and germanium films also. These films all exhibit random variations in thickness that are much larger than the geometrically predicted uniform gradations of thickness. The rotary planetary positions of the substrates can only be partially effective in removing these differences [34].

1.5.6 Temperature-Related Challenges

One of the main obstacles to our society's green transformation is the economical and environmentally responsible production of power. Solid oxide fuel cells have the potential to significantly lower the use of fuel and pollution emissions since they can directly transfer the chemical energy of hydrogen or hydrocarbons into electricity through electrochemical reactions. Fuel cells work by mixing an oxidant gas and a gaseous fuel via porous electrodes and a dense electrolyte that conducts ions. H_2 and CO may be transformed easily. First, hydrocarbons have to dissolve into a synthesis gas, which could occur outside the SOFC or, in the case of methane, partially inside the combustion chamber. SOFCs are classified as either proton-conductor ceramic fuel cells or oxygen-ion conductor ceramic fuel cells, depending on the ionic species that are carried through the electrolyte. To keep the commonly used terminology in the community, they will be referred to as protonic ceramic fuel cells (PCFCs, proton conductors) and oxygen-ion conductors (SOFCs). The first-generation SOFC was built on a thick (> 100 μm) yttria-stabilized zirconia (YSZ) electrolyte that also functioned as the mechanical support. It was designed for stationary high-power (megawatt) applications. To attain reasonable power densities, a temperature above 850 °C is necessary for operation. The integration of a gas turbine with this high-temperature SOFC (HT-SOFC) can optimize the power system's overall efficiency. However, for particular standalone applications, lowering the operating temperature helps to minimize the system's cost and increase life expectancy. Lower temperatures will cause degradation to occur more slowly thereby opening the door to less expensive materials for metal interconnects and balance-of-plant components. The development of the second-generation SOFC was driven by these advantages, which allowed for the production of thinner YSZ electrolytes (≈ 10 μm) and a working temperature in the intermediate temperature range (≈ 700 °C) because the anode provided mechanical support [35]. For instance, a high-performance IT-SOFC with thin YSZ electrolyte and good stability at 700 °C (a two-layer stack that continuously worked for more than 100,000 h at 0.5 A cm^{-2}) was demonstrated by researchers at Forschungszentrum Jülich [36]. In the expanding market for mobile or portable electronics, the unique characteristics of SOFC, such as fuel flexibility, high-energy density, and the ability to deliver continuous power, make them appealing for use as battery replacements or chargers. However, these devices frequently require quick startup times and are subject to increased cost pressure, which calls for a further drop in operating temperature from 700 to 600 °C or even 500 °C. Because the underlying transport and electrode processes are thermally activated, the key problem in this attempt is to maintain sufficiently high performance at such low temperatures. This is because the resistance from the electrolyte and both electrodes grow intensely [36].

1.5.7 *Film Defects and Quality Control*

The performance and functionality of the thin film are greatly influenced by the presence of defects in a thin film. The term quality control can be defined as a significant parameter by ensures that some specific standards and requirements must be fulfilled by a thin film [37]. Some common defects of the thin film and quality control parameters are as follows:

1.6 Issues Related to Adhesion

If the adhesion of the considered surface is poor, it can cause delamination and flake off the thin film from the surface of the substrate. Some factors that can be the cause of peeling off the film from the surface of the substrate are, surface pollutants, not enough preparation of the substrate, or if we do not properly match the coefficients of the thermal expansion [28].

Complete cleaning and preparation of the surface and the right substrate selection can affect the quality control of thin film. The existence of pores in the structure of thin film causes porosity. Some factors that cause porosity are vacuum conditions, the temperature of the substrate, and during the process of deposition coverage of the substrate. In this case, if we properly optimize the parameters of deposition, vacuum levels properly ensure and temperature of the substrate monitors significantly [29].

1.6.1 *Surface Irregularities and Harshness*

The optical electrical and mechanical properties of the thin films are affected by the irregularities and harshness present in the substrate surface. The harshness and irregularities are due to the surface pollutants or incompatible deposition rates [38].

1.6.2 *Fractures and Pinholes*

Small holes and fractures present in thin film affect its integrity. The causes of such fractures and small holes are thermal stresses, thermal expansion's mismatch coefficient, and inappropriate conditions of depositions [39].

1.7 Deposition Techniques

The material's efficiency and behavior during operation can be influenced by its surface characteristics. It is possible to adjust and fine-tune these surface characteristics to specifically satisfy the need for improved performance, and this technique has been widely used in numerous areas of life. This can be accomplished by applying a thin-film coating on the surface. The literature on various deposition methods for coating and surface modification is reviewed in this study [1]. Several procedures known as "deposition" are used to build thin or thick layers of a material on a solid surface, atom by atom or molecule by molecule. Depending on the application, the generated layer deposits as a coating on a surface and modifies the substrate surface's characteristics. The surface that supports a thin-film optical coating is manipulated in a certain way to change its properties; as a result, the coating's deposition originates with this surface. Although a variety of properties, including thermal, chemical, acoustical, and environmental ones, may be expected from the coating, the main goal is usually to achieve a desired set of specular optical properties, which change the quality of light manipulated by the surface without changing its direction [40]. From one atom to the next, all of the forces that maintain thin films together and attach them to their surfaces are short-range. Due to the extreme strength and limited range of these forces, a single contaminant molecule can entirely disrupt the bonds. Therefore, the cleanliness of the surface is much more important than its smoothness [41].

1.7.1 Why Are Deposition Techniques Necessary?

Processing a thin layer of solution on a particular surface is part of the deposition procedure. There are numerous methods for projecting liquids like sol–gel solutions onto surfaces. Spin coating, dip coating, droplet coating, roll-to-roll coating, spray coating, and inkjet coating are some of these methods. The solution surface tension, which attempts to maintain the solution in a drop shape, is resisted by the spreading of the solution on top of a surface but remains intact when the substrate's surface energy is high. To prevent dewetting, a very high chemical contact between the substrate surface and the solution is therefore necessary in every situation. This can be accomplished by activating the substrate surface, particularly for solutions with high surface tension. High-energy surfaces are typically found in oxides with many hydroxyl groups that encourage hydrogen bonding. When roughness is homogenous and formed of sufficiently modest peak/valley features, it is a significant attribute that could contribute to increased affinity. Furthermore, one must remember that no surface is flawless, not even an optical one [42]. Chemical inhomogeneities resulting from inadequate cleaning, topological flaws caused by improper polishing or scratches, or even dust that has been absorbed from prior conditioning and handling might all be present. These represent high-energy zones where the solution can be pinned, resulting in a thickness fluctuation that is confined. Working in a clean room or carefully cleaning must be done to avoid them. The size and morphology of the substrate determine the approach to be used as well as the level of control needed during processing. The control mostly pertains to the homogeneity of the spreading and the amount of delivered solution per unit surface. It could not be simple to achieve the appropriate thickness throughout the surface. As far as we can tell, spreading is in opposition to dewetting, which reduces surface area [43]. Weak chemical interactions, ranging from van der Waals to hydrogen bonding, control this equilibrium. The system can also be seen in terms of energy variation. To maintain the stability of a solution layer on a surface, strong adhesive connections between the liquid and solid phases are required [1].

1.7.2 Significance of Deposition Techniques

Oxide thin films as coating materials have several uses in the fields of biosensors, orthopedics, sensors, electronic and photonics devices, environmental purification, sterilization, and deodorization, as well as self-cleaning surfaces (textiles, windows, etc.). While the original purpose was to distribute metals and oxides as aggregates on an oxidizing substrate, heterogeneous catalysis has found extensive use in the formation of shaped three-dimensional carriers such as monoliths, foams, or reactor walls with structured structures. Since the majority of reactions are highly exothermic, it's critical to eliminate any hot spots that may exist in the catalytic pellets in fixed-bed reactors. Hot areas cause structural deterioration and premature catalyst deactivation. In addition, they typically encourage reactions that result in byproducts, such as carbon oxides in selective oxidation processes [44]. The thin layer of catalyst covering these exposed structures facilitates effective heat and mass transfers between the gaseous reactants, the catalytic active phase, and the reactor wall. Nowadays, monoliths and foams with different substrates and shapes are used more and more. The deposit needs to be thermally, chemically, and mechanically stable, much like in other applications. Its surface must, however, also be highly reactive in catalytic applications and able to endure high reactant/product flow rates without eroding, collapsing, or attrition. In addition to the well-established powder suspension and sol–gel preparation coating techniques, there are several other known thin-film manufacturing techniques including reactive plasma, atomic layer deposition, chemical vapor deposition, and magnetron sputtering.

Our previous results suggest that the application of PECVD is a different approach that merits further investigation. PECVD has not attracted as much attention in thin-film preparation as evaporation approaches, even though a large range of experimental parameters can be adjusted to regulate the microstructure of the films. We are based on a polymerization method utilizing cold plasma as the assistant for tetramethyldisiloxane, which is the precursor of silica. The silica-like layer coated on the stainless steel substrate should be at least in such a position that it prevents Fe diffusion, gives some mechanical stability, and functions as a bonding layer to support the depositions of the active phase, VOx/TiO2. The good results obtained on this catalyst/structured wall reactor or catalyst/3D-carriers system may tempt the use of new reactors with enhanced heat and mass transfers. Its active phase is known well for its applications in the chemical industry, for example, oxidation of o-xylene to phthalic anhydride, and in pollution abatement, deNOx with ammonia, COV, etc. [45].

1.7.3 Types of Deposition Techniques

Physical processes, such as physical vapor deposition, and chemical processes, such as chemical solvent deposition and chemical vapor deposition, make up the majority of the deposition processes. This section presents the methods of chemical vapor deposition (PVD) and physical vapor deposition (CVD).

1.7.4 PVD Procedures

Physical vapor deposition refers mostly to the deposition techniques where material particles enter a vapor phase and subsequently form a condensed thin film on the substrate. These procedures are therefore divided into two broad categories: thermal processes and sputtering techniques. Thermal processes are further divided into the following categories:

- Vacuum evaporation.
- Pulsed laser deposition.
- Cluster ion beam deposition.
- Ion plating.
- Activated reactive evaporation.
- Molecular beam epitaxy.

1.7.5 CVD Procedures

Two main steps define these processes: A volatile substance that is to be deposited must first enter the vapor phase. Second, in the presence of heat sources, the vaporized substance breaks down into atoms and molecules. It then combines with other gases, vapors, and liquids around the substrate to form thin layers [41]. Accordingly, the majority of CVD processes function in the pressure range of a few Torr to a pressure that is higher than the reactants' atmospheric pressure. Moreover, these methods need a high temperature of roughly 1000 °C to operate; however, some modified CVD processes, such as plasma-enhanced chemical vapor deposition (PECVD) and plasma-assisted chemical vapor deposition (PACVD), call for a lower working temperature. In general, there are three categories of CVD processes [45].

- Thermal CVD.
- Plasma CVD.
- Laser CVD, etc.

1.8 Applications of Thin Film

Thin films have a wide range of claims depending on their purpose of usage, and due to the advancement of deposition techniques, it has applications due to the vacuum technology and electrical power. In a wide range of applications of thin film one which is most important is that thin films become the cause of enhancement in the properties of bulk materials because adding the atoms layer by layer causes improvement in the physical and chemical characteristics of materials [46]. A necessary field that has numerous applications of thin film is electronics optoelectronics devices. Various types of transistors, in the field of semiconductor thin films, absorb a great amount of developing factors. This thin film is used due to the reasons of ease of fabrication, resistance to chemicals and high temperatures, and appearance of a large bandgap due to high impedance. Other than this, for the fabrication of optical-electronics, multiple microelectronics devices and optical devices are required. The thin films of Al are commonly used in MOSFETs for operating voltage. Also, thin films made up of Cu are usually used in CMOS due to their higher resistance to electromigration and higher electrical conductivity. Due to the good magnetic properties of thin films, they are also used in data storage which is near to replacing flash memory. The production of electricity by using thin films in the photovoltaic sector is usually taken into account as an effective solution to reduce cost. This effect is making fast progress and rapid penetration in the market due to its higher efficiencies and the advancement of thin-film structures. Another benefit of thin film in the photovoltaic sector is due to the reason of high absorption coefficient of the material that is going to be absorbed will reduce the material's thickness and this thing will automatically reduce the cost of the materials. The use of thin films in the solar spectrum for increment in the conversion of efficiency is very crucial. Initially, amorphous silicon was used to evolve for the state of the art of thin films for PV applications. But later on, we preferred to use materials that are cost-friendly and can be operated at low temperatures, for example, organic and perovskite-based PV cells.

Thin films are used in a vast range of applications where we require enhancement in the mechanical and chemical properties. We are using optical thin films in eyeglasses for the improvement of vision by using optical elements that are based on spectacles coated with polymer-based elements. We can also control the undesired reflections and undesired transmission of ultraviolet light by applying the coatings of thin films that must be able to absorb the wavelength. We can also enhance the energy efficiency of office buildings by drawing architectural glazing on thin films. We can manage the transfer of heat from the outside and inside of the building by the process of spectral areas of light. One of the most unique types of tin film is organic thin film due to the most unique properties of thin film that are low-cost material and flexibility of materials which helps us to expand the use of organic thin films in a vast range of technologies [47].

1.8.1 Optoelectronics and Photonics

There has been a tremendous demand for high-quality optical thin films with well-characterized and controllable physical properties since the advent of widespread laser technology. Since the coatings are often prepared to thicknesses much smaller than wavelengths of visible light, standard microscopy has difficulty characterizing them. Because of their great range of refractive indices, optical absorption, and often near-total reflectivity in the visible spectrum, metal films are particularly a good example of these difficulties. However, because of their very constant crystalline structure, the bulk optical properties of metal thin films are primarily dependent on their thickness. Even in an age of meticulous scientific investigation and technological development, metal thin films were being exploited by ancient metalsmiths to gild less valuable materials in a thin layer of gold and silver [48]. While these artisans may have had very little understanding of the physicochemical processes behind their plating methods, they set in motion centuries of thin-film innovation with uses far beyond their purely aesthetic origins. Building from that basis, metal thin films of today are used for many industrial and scientific applications, none more than in optics. These films offer specific compositions and thicknesses a unique combination of transparency and conductivity. These films also exhibit broadband reflectivity for relatively thicker films. Many different metals have been utilized in thin-film optics, including pure metal films as well as metal-containing composites, such as Indium Tin Oxide (ITO) in applications ranging from optoelectronics to the generation of surface plasmons, depending on the desired film characteristics. Depositing of metal films in nanoscale thicknesses on ceramic substrates is one of the most popular applications for metal films since it generates high-quality broadband mirrored surfaces. Such mirrors have been used in adaptive optics to make deformable mirrors for optical wavefront control, in microscopy for dark-field illumination, and as a replacement for lens arrays in reflecting telescopes to reduce image distortions [49].

1.8.2 Photovoltaic (Solar Cell) Thin Films

Thin-film solar cells are advantageous due to their low material consumption and increasing efficiency. The three main methods for thin-film solar cells are cadmium telluride, copper indium gallium selenoid, and amorphous silicon (α-Si). This section discusses the evolution of each technology in both commercial and experimental environments, providing equal explanations for market share and dependability. Currently, holding a market share of over 55% crystalline solar cells have module efficiencies that almost match those of CdTe and CIGS technologies, and α-Si is nearly non-existent in terrestrial applications due to its low efficiency and light-induced deterioration. The two materials with the most potential for thin-film technology are CdTe and CIGS. Before thin-film solar cells are investigated for use in integrated photovoltaic systems, they must first be proven in terms of longevity, dependability, consumer confidence, and increased investments [48].

To compete with traditional sources, a solar cell a device that converts sunlight into electricity must be dependable and reasonably priced. Numerous solar technologies, such as wafers, thin films, and organic, have been successfully researched to attain high efficiency, dependability, and cost-effectiveness. For example, crystalline silicon accounts for up to 90% of the worldwide PV market and has demonstrated remarkable success in its transition from laboratory to commercial integration. Reducing material use and improving energy conversion efficiency are two examples of cost-effectiveness [50].

1.8.3 Thin-Film Transistor-Liquid Crystal Displays (TFT-LCDs)

Nowadays, thin-film transistor liquid crystal displays, or TFT-LCDs, are the most widely used flat-panel displays due to their many good qualities, such as their small size, low power consumption, and lightweight. Many commercial display modes, such as twisted nematic (TN), in-plane switching (IPS), and multivertical alignment (MVA), have made use of TFT-LCDs. A significant and essential part of TFT-LCDs is polarizer film (POL), which is produced by stretching film with dichroic dye molecules such as iodine. Two polarizer films are sandwiched around liquid crystal (LC). The voltage-on and voltage-off configurations of an LC director for a normally white transmissive TN mode TFT-LCD are demonstrated. The alignment barrier on the substrates enables the LC to rotate 90° in the bulk while it is in the voltage-off state, with its alignment parallel to the transmitted optical axis of the polarizer films [51]. The rear polarizer film generates the incident linearly polarized light, which travels through the front polarizer film and into the white state by way of the twisted LC molecular structure. The LC is vertically reoriented to the substrates in the voltage-on state in response to the electric field [44]. A black condition is formed when the incident linearly polarized light from the rear polarizer film is blocked by the front polarizer film. Gray levels are produced by modulating the output light intensity through the LCD cell using the driving voltage that is applied to the LC. The angle of 90 degrees between the absorption axes of the crossed polarizer films, when observed from the normal direction of an LCD cell, can produce an excellent black state. However, the effective angle is not 90 when observed from an oblique angle. Light leaking consequently happens in the dark state. There is a drop in the contrast ratio (CR). Additionally, there is a decrease in the field of view angle at which the CR is greater than 10:1. Retardation films, sometimes known as compensation films, are inserted into polarizer films to lessen light leakage in the dark state [11].

1.8.4 Displays and Flat-Panel Technologies

A flat-panel display is essentially a published picture with a time-varying dimension added to it. However, there are a lot of technical prerequisites to be met to do this.

Flat-panel displays, or FPDs, first appeared in industrial and military applications some 20 years ago. They have been available in high-volume consumer products for less than five years, and they have been used commercially in the consumer marketplace for ten years. Whereas FPDs have crossed the $1 billion sales mark in one year alone in 1989, they have entered the consumer product stage over the last five years and are entrenched as high-performance electronic components. FPDs with a performance comparable to color CRTs at VGA resolution were made available only in commercial quantities during the last three years. The number of the picture element or pixels varies from 50,000 and 1.5 million and above, in monochrome or full-color versions, from entry-level video games, laptops, and portable TVs to full-color VGA, XGA, and HDTV screens. The markets and display technologies that are available for displays with fewer pixels are quite broad and varied. The origins of this type of display are sometimes associated with Burroughs' initial Nixie Tube manufacture, which occurred around 1955 and whose inventor is unknown.

Billboards, marquees, meters, clocks, and handheld calculators are examples of low-pixel-count displays. Furthermore, the confusing matrix addressing the cross-coupling issue (which will be covered in more detail later) that affects all HIC FPDs and is proportionate to pixel count is not present in low-pixel displays. The initial goal of the market was to switch from CRTs to FTDs, but this goal has not been met. Rather, a whole new line of products has been developed that is unable to use CRTs. Handheld color television displays and personal computers in the form of briefcases and notebooks are two examples of these initial new devices [52].

1.8.5 Light-Emitting Diode LEDs

GaN-based light-emitting diodes, or LEDs, have greatly advanced from their introduction. The improvements observed include all the reorientation of chip architecture, forward voltage, and internal quantum efficiency, thus ensuring impressive performance across a very wide UV-to-green spectral range. With InGaN LEDs that emit green and blue having reached parity with AlInGaP-based LEDs, three-color solutions like LED-backlit liquid crystal displays and direct-view big area advertising can now be sold. Phosphor-converted LEDs are best used now for accent and portable lighting applications since the efficiency of these LEDs for illumination applications is now at par with that of a standard fluorescent light. For such applications as projection, car forward lighting, and general lighting upscale, however, higher power handling and light extraction are required. One of the recent advances to boost the light extraction of GaN-on-sapphire LEDs is the thin-film LED structure. Removal of the substrate with an excimer laser usually results in a thin-film LED structure wherein the n-doped GaN exposed layer subsequently becomes photoelectrochemically roughened. The n-GaN is first roughened and then patterned with N metal, after which a wire bond is used to connect the cathode. A flip-chip device installed on a high thermal conductivity sub-mount has emerged as the second advance in LED technology. The p-GaN has been metalized and attached to another semiconductor substrate that acts as an anode in the aforementioned LEDs; thus, the vertical device architecture has emerged. Removal of wire bonds and growth substrate with n-contact metallization redistribution enhances high-power performance and offers options for optics and surface mounting in flip-chip LEDs. Using a split-wafer experiment, we first compare the performance of standard flip-chip FC, thin-film flip-chip TFFC, and vertical-injection thin-film VTF LEDs. All three types of devices are processed from the same InGaN–GaN multiple-quantum-well epitaxial heterostructures grown on sapphire substrates [53]. Thin-film light-emitting diodes (LEDs) have attracted significant interest in light of the rapidly expanding portable display market because of their inexpensive cost of production and superior efficiency when compared with inorganic LEDs and liquid crystal displays. An emitting layer, a metal cathode, hole and electron transport layers, an Indium Tin Oxide (ITO) anode, and an emitting layer comprise a typical thin-film LED, which has a thickness of less than 200 nm. Thin-film LEDs can be classified as organic LEDs (OLEDs), polymer LEDs, quantum dot LEDs, or perovskite LEDs (PeLEDs) based on the substance that emits light sheets [54].

1.8.6 Medical and Biomedical Applications

Various factors, including mechanical characteristics, biocompatibility, and shape ability, are used to select materials needed for biomedical applications. There aren't many conventional materials available for use in biomedicine. Consequently, there has been an excessive arrangement of attention in emerging new, useful, and biocompatible materials that the body may utilize for an extended period without causing any problems. Numerous medical devices, including heart valves, intraocular lenses, hip and knee joints, and coronary stents, have been effectively surrounded in human bodies. Because of the environment inside the body, the materials used in the implants are expected to interrelate with the tissues and organs of the body, possibly possessing anticorrosive qualities. Protective film coatings may decrease corrosion and wear issues on implants, thereby extending their lifespan for the benefit of patients. Implantable devices can have thin films put on them to help them integrate better with the body. For instance, to prevent device failure and unfavorable biological impacts, Si microchips must be shielded from chemical interactions with bodily fluids by bioinert/biocompatible hermetic encapsulation coatings. Therefore, it is possible to implant Si microchips in human body components through encapsulation or protective coatings. Metallic glasses are one type of material that has been making waves in the biomaterial world. The best qualities of metallic glasses are their exceptional processing ability, high strength, flexibility, and corrosion resistance.

The amorphous nature of metallic glasses gives rise to these special characteristics. They lack grains and their associated boundaries due to this amorphous structure. Consequently, it is possible to achieve high-resolution surface patterns or textures on metallic glasses. Furthermore, the thermoplastic capabilities of these materials at supercooled liquid temperatures, where metallic glasses undergo severe softening, enable the exact production of complicated three-dimensional designs. A novel material for thermoforming-created biomedical microdevices could be metallic glass thin sheets [54].

1.8.7 Biocompatible Thin-Film Coatings

The main purpose of this research is to report on the biocompatibility of hydrophobic coatings and atomic layer-deposited (ALD) alumina (Al2O3). Although these coatings are becoming more widely used in microelectromechanical systems (MEMS) as dielectric layers to stop electrical shorting, this work is the first to ascertain whether ALD coatings are suitable for use in Bio-MEMS. The biocompatibility of ALD alumina-coated glass coverslips with human vascular smooth muscle cells is quantified using optical microscopy in conjunction with an Alamar Blue (AB) assay. The use of Bio-MEMS in biomedical applications is growing rapidly. These include thermal conductivity sensors, microfluidic devices, catheter-based blood pressure monitors, and neurosurgical instruments. The option to permanently implant sensors for in vivo monitoring is appealing. But this poses a serious set of problems for both the dependability of Bio-MEMS and a suitable biological response. To do this, coatings that shield the apparatus from a fluid environment while also reducing the influence of the materials on the biological processes they are monitoring are needed [55].

1.8.8 Drug Delivery Systems

In general, a thin, flexible polymer layer, with or without a plasticizer, is referred to as a thin film. Because of their inherent thinness and flexibility, patients may find them to be less intrusive and more tolerable. The polymeric matrix of the thin film satisfies numerous parameters for effective use as a drug release platform. In general, thin films are great alternatives to tablets or liquid formulations for specific targeting of sensitive locations. Thin films have demonstrated the ability to increase therapeutic efficacy, decrease dose frequency, and accelerate the beginning of action. Likewise, thin films could help reduce extensive metabolism brought by proteolytic enzymes and get rid of a drug's negative effects. It is significantly better than the current conventional dosage forms in terms of increased bioavailability, high patient compliance, and patent extension of active pharmaceutical ingredients (API) [56]. Additionally, thin-film formulations provide several benefits, such as

- Straightforward administration via non-invasive methods.
- Handling convenience during production and shipping.
- Cost-effectiveness in formulation development.

The development of a broad range of thin films has been made possible by the availability of a large variety of appropriate polymers and the change in paradigm in manufacturing technology. As a result, thin films are becoming more and more recognized as cutting-edge drug delivery dosage forms in the pharmaceutical industry. A lot of work has gone into developing polymeric thin films that are often applied topically, oral, sublingual, and ocular. The use of thin films, among other methods, to deliver medication to the buccal or sublingual mucosal has attracted a lot of attention recently. Meanwhile, ophthalmic films are being developed to get beyond ocular barriers and stop medication loss via the lacrimal drainage system. Modifying the composition of polymers with varying grades has made it easier to alter important thin-film traits like mechanical strength, sticky qualities, drug release rate, and other associated attributes [56].

1.8.9 Medical Imaging Devices

In the past few years, the quickly developing fields of thin-film transistors and photodiodes have been leveraged to produce the first two-dimensional, self-scanning amorphous silicon (aSi: H) imaging arrays. These technologies are being developed for active-matrix liquid crystal displays, solar cells, scanners, and copiers. With the use of this new technology, the first high-quality, megavoltage, diagnostic X-ray images with low-contrast anatomical detail have just been shown. At the moment, arrays up to 30 cm across can be created using this flat-panel technology, while realistic limitations limit the size of the greatest imaging surface that is possible. At Xerox PARC, the largest array we have made so far for our collaborative research is roughly 25 by 23 cm. The imaging pixels in the arrays are arranged in a standard two-dimensional matrix [57].

A thin-film transistor (TFT) connected to a-Si: H n-i-p or p-i-n photodiode makes up each pixel. To reach complete depletion, a network of bias wires provides the photodiodes with a reverse bias voltage. The TFT a-Si: H field effect transistor (FET) in the case of the arrays being developed by our collaboration and for similar devices. Other TFT structures are possible, though. A polycrystalline FET is one such alternative; however, unlike a-Si: H FETs, it may have electron mobility that is one to two orders of magnitude higher, accompanied by a correspondingly greater leakage current. Furthermore, it has been suggested that thin-film switching components like diodes can be used in place of TFTs in these kinds of imaging arrays.

An X-ray converter that is positioned on or over the array's surface is used to indirectly detect incident radiation. The converter could be a phosphor screen, a channeled-light scintillator like CsI (TI) or fiber-optic scintillator, or any other appropriate scintillator for diagnostic quality X-rays (20 to 150 kVp). The converter for megavoltage X-rays could be a thick scintillator with light channels or a metal plate/phosphor screen combination. This signal is included in the sensor's capacitance when the FETs are not conducting. The voltage delivered to the FET control lines regulates the conductivity of the FETs [58].

These lines link each transistor's gate along a certain row. The light signal is integrated into the pixel, and the FETs are non-conducting when the bias on these lines is negative. This bias is turned positive for one row of pixels at a time during readout, causing the FETs to conduct. Thus, the integrated charge for every pixel in the row travels to an external preamplification-digitization circuit via the DATA line, the FET, and the pixels themselves. The pixels are also reinitialized during this reading stage [59].

1.9 Biosensors and Diagnostic Tools

Thin-film analytical compact devices are the basis of the next generation of biosensors. The main benefits of these biosensors are their low cost, ease of use, sensitivity, specificity, and quick response time. Each of these characteristics is intimately associated with the film thickness utilized in the sensor's construction. The interaction between the biological analyte and the surface-modified thin film forms the fundamental basis of a biosensor's detecting mechanism [60]. The signal arising from the identification of the biological analyte is transformed into another measurable signal by the thin film, which functions as a physicochemical optical, mechanical, magnetic, and electrical transducer. Biosensors are a broad and diverse topic of study. Since the early reports of the Clark enzyme electrode, biosensor technology has grown into an ever-expanding and diverse discipline. Because biosensors often provide streamlined, reagent-less assessments for a variety of industrial and biomedical applications, this sector has been growing and becoming more interdisciplinary over the past few decades. Reversibility and quick response time are frequently critical for any sensor [6]. Any solid-state sensor necessarily relies on the diffusion of analyte molecules into and reaction with the detecting component in operation, together with any reaction products diffusing out, to work. Of course, this may include increasing the speed and reversibility of sensor response—for instance, the thinner the layer of sensing being utilized, the shorter this process will be. Moreover, expensive molecules can be used inexpensively because the dimensions of molecular films are in the nanoscale; hence, a typical self-assembled monolayer has a mass of around 2×10^{-7} g/cm^2. To immobilize biomolecules, several compounds have been electrodeposited onto electrode surfaces such as polyaniline, polyphenol, and polythiophene. Other relatively thick-film systems that can provide immobilization of biomolecules are, for example, coating slides with cast films of materials like polylysine whose high affinity for biomolecules is largely used in the elaboration of DNA microarrays or by covalent immobilization with linkers such as dicyclohexyl carbodiimide (DCC) [61]. Over many years, much research has been carried out on the manufacture of thin films composed of a wide variety of materials, often only one molecule thick [61, 62].

1.10 Conclusion

In this chapter, basic concepts and historical development of the thin film have been described. Thin film has a wide range of applications in a wide range of applications, including medical and tissue engineering, solar cells for energy requirements, diagnostic tools in various fields, LEDs, and flat display panels. The optical, electrical, and mechanical properties of the thin film have been described, which should affect the growth formation of the thin film. There are two basic methods of thin-film deposition, namely, physical vapor deposition (PVD) and chemical vapor deposition (CVD), and certain types of these two categories. All depositions have their respective properties and limitations that must impact the growth rate of the thin film. Besides, the history of development as well as early discoveries regarding thin film is discussed to study the important aspects of the deposition methods of thin films. It is the uniformity of the deposited layer which, at times, proves challenging to the growth rate of the thin-film coating: It therefore provides us with the ability to overcome some of the deposition-related problems such as adhesion-related issues, thin-film uniformity, and film thickness.

References

1. Abegunde, O.O., et al. 2019. Overview of thin film deposition techniques. *AIMS Materials Science* 6 (2): 174–199.
2. Mao, S.S., and P.E. Burrows. 2015. Combinatorial screening of thin film materials: An overview. *Journal of Materiomics* 1 (2): 85–91.
3. Greene, J.E. 2017. Tracing the recorded history of thin-film sputter deposition: From the 1800s to 2017. *Journal of Vacuum Science & Technology A* 35(5).
4. Shum, K., et al. 2010. Synthesis and characterization of CsSnI3 thin films. *Applied Physics Letters* 96(22).
5. Romankiw, L.T., and S. Krongelb. 2013. Thin film magnetic heads-early inventions and their ongoing impact on magnetic storage and on electrochemistry. *ECS Transactions* 50 (10): 3.
6. Ostroff, R.M., et al. 1999. *Thin Film Biosensor for Rapid Visual Detection of Nucleic Acid Targets*. Oxford: Oxford University Press.
7. Barber, Z., and M. Blamire. 2008. High throughput thin film materials science. *Materials Science and Technology* 24 (7): 757–770.
8. Ahn, C., K. Rabe, and J.-M. Triscone. 2004. Ferroelectricity at the nanoscale: Local polarization in oxide thin films and heterostructures. *Science* 303 (5657): 488–491.
9. Fernandez, A., et al. 2022. Thin-film ferroelectrics. *Advanced Materials* 34 (30): 2108841.
10. Schultz, P.J., and K.G. Lynn. 1988. Interaction of positron beams with surfaces, thin films, and interfaces. *Reviews of Modern Physics* 60 (3): 701.
11. Brody, T.P. 1984. The thin film transistor—A late flowering bloom. *IEEE Transactions on Electron Devices* 31 (11): 1614–1628.
12. Fortunato, E., P. Barquinha, and R. Martins. 2012. Oxide semiconductor thin-film transistors: A review of recent advances. *Advanced materials* 24 (22): 2945–2986.
13. Edelmann, F.T. 2009. Lanthanide amidinates and guanidinates: From laboratory curiosities to efficient homogeneous catalysts and precursors for rare-earth oxide thin films. *Chemical Society Reviews* 38 (8): 2253–2268.
14. Franklin, A.D. 2015. Nanomaterials in transistors: From high-performance to thin-film applications. *Science* 349(6249), aab2750.

15. Gohain, M.B., et al. 2020. Development of thin film nanocomposite membrane incorporated with mesoporous synthetic hectorite and MSH@ UiO-66-NH2 nanoparticles for efficient targeted feeds separation, and antibacterial performance. *Journal of Membrane Science* 609: 118212.
16. Akther, N., et al. 2019. Recent advances in nanomaterial-modified polyamide thin-film composite membranes for forward osmosis processes. *Journal of Membrane Science* 584: 20–45.
17. Yianoulis, P., and M. Giannouli. 2008. Thin solid films and nanomaterials for solar energy conversion and energy saving applications. *Journal of Nano Research* 2: 49–60.
18. Mukhopadhyay, S.M., P. Joshi, and R. Pulikollu. 2005. Thin films for coating nanomaterials. *Tsinghua Science and Technology* 10 (6): 709–717.
19. Rao, M., and M. Shekhawat. 2013. A brief survey on basic properties of thin films for device application. In *International Journal of Modern Physics: Conference Series*. World Scientific.
20. Wei, X., et al. 2022. A critical review on thin-film nanocomposite membranes enabled by nanomaterials incorporated in different positions and with diverse dimensions: Performance comparison and mechanisms. *Journal of Membrane Science* 120952.
21. Hardwick, D. 1987. The mechanical properties of thin films: A review. *Thin Solid Films* 154 (1–2): 109–124.
22. El Shazly, A., et al. 1998. Electrical properties of thin films. *Journal of physics: Condensed matter* 10 (26): 5943.
23. Heavens, O. 1960. Optical properties of thin films. *Reports on Progress in Physics* 23 (1): 1.
24. Aspnes, D.E. 1982. Optical properties of thin films. *Thin Solid Films* 89 (3): 249–262.
25. Macleod, H.A. 1986. Structure-related optical properties of thin films. *Journal of Vacuum Science & Technology A: Vacuum, Surfaces, and Films* 4 (3): 418–422.
26. Wang, H., et al. 2013. Structure and morphology control in thin films of conjugated polymers for an improved charge transport. *Polymers* 5 (4): 1272–1324.
27. Spindt, C., et al. 1976. Physical properties of thin-film field emission cathodes with molybdenum cones. *Journal of Applied Physics* 47 (12): 5248–5263.
28. Litzelman, S., et al. 2008. Opportunities and challenges in materials development for thin film solid oxide fuel cells. *Fuel cells* 8 (5): 294–302.
29. Massines, F., et al. 2012. Atmospheric pressure low temperature direct plasma technology: Status and challenges for thin film deposition. *Plasma Processes and Polymers* 9 (11–12): 1041–1073.
30. Shi, C., et al. 2022. Molecular hinges stabilize formamidinium-based perovskite solar cells with compressive strain. *Advanced Functional Materials* 32 (28): 2201193.
31. Sharma, A. 2021. High entropy alloy coatings and technology. *Coatings* 11 (4): 372.
32. Stachiv, I., E. Alarcon, and M. Lamac. 2021. Shape memory alloys and polymers for MEMS/NEMS applications: Review on recent findings and challenges in design, preparation, and characterization. *Metals* 11 (3): 415.
33. Bhatt, G.R., et al. 2020. Integrated near-field thermo-photovoltaics for heat recycling. *Nature communications* 11 (1): 2545.
34. Schweizer, P.M. 2022. Film thickness and film thickness uniformity. In *Premetered Coating Methods: Attractiveness and Limitations*, 387–452. Springer.
35. Kalmykov, A., et al. 2019. Organ-on-e-chip: Three-dimensional self-rolled biosensor array for electrical interrogations of human electrogenic spheroids. *Science Advances* 5(8): eaax0729.
36. Zhang, J., et al. 2022. Advanced materials for thin-film solid oxide fuel cells: Recent progress and challenges in boosting the device performance at low temperatures. *Advanced Functional Materials* 32 (22): 2111205.
37. Abadias, G., et al. 2018. Stress in thin films and coatings: Current status, challenges, and prospects. *Journal of Vacuum Science & Technology A* 36(2).
38. Gill, J.J., et al. 2001. Manufacturing issues of thin film NiTi microwrapper. *Sensors and Actuators A: Physical* 93 (2): 148–156.
39. Mylvaganam, K., et al. 2015. Hard thin films: Applications and challenges. *Anti-Abrasive Nanocoatings* 543–567.

40. Benjamin, P., and C. Weaver. 1960. Measurement of adhesion of thin films. *Proceedings of the Royal Society of London. Series A. Mathematical and Physical Sciences* 254(1277): 163–176.
41. Macleod, H.A. 2018. Recent developments in deposition techniques for optical thin films and coatings. *Optical Thin Films and Coatings* 3–23.
42. Seshan, K. 2002. *Handbook of Thin Film Deposition Techniques Principles, Methods, Equipment and Applications,* 2nd ed. CRC Press.
43. Mittal, K.L. 2001. Adhesion Aspects of Thin Films: vol. 1. Vsp.
44. Gong, L., et al. 2019. Fundamentals and advances in the adhesion of polymer surfaces and thin films. *Langmuir* 35 (48): 15914–15936.
45. Mah, J.C., et al. 2017. Metallic interconnects for solid oxide fuel cell: A review on protective coating and deposition techniques. *International Journal of Hydrogen Energy* 42 (14): 9219–9229.
46. Waser, R., et al. 2001. Advanced chemical deposition techniques-from research to production. *Integrated Ferroelectrics* 36 (1–4): 3–20.
47. Hass, G. 1952. Preparation, properties and optical applications of thin films of titanium dioxide. *Vacuum* 2 (4): 331–345.
48. Thouless, M.D. 1994. Fracture mechanics for thin-film adhesion. *IBM Journal of Research and Development* 38 (4): 367–377.
49. Yu, X., T.J. Marks, and A. Facchetti. 2016. Metal oxides for optoelectronic applications. *Nature Materials* 15 (4): 383–396.
50. Greenberg, C.B., and C. Steffek. 2005. Bio-adhesion to thin films in relation to cleaning. *Thin Solid Films* 484 (1–2): 324–327.
51. Hu, M.-C. 2012. Technological innovation capabilities in the thin film transistor-liquid crystal display industries of Japan, Korea, and Taiwan. *Research Policy* 41 (3): 541–555.
52. Tannas, L.E. 1994. Evolution of flat-panel displays. *Proceedings of the IEEE* 82 (4): 499–509.
53. Shchekin, O., et al. 2006. High performance thin-film flip-chip InGaN–GaN light-emitting diodes. *Applied Physics Letters* 89(7).
54. Kaushik, N., et al. 2014. Metallic glass thin films for potential biomedical applications. *Journal of Biomedical Materials Research Part B: Applied Biomaterials* 102 (7): 1544–1552.
55. Rajan, S.T., B. Subramanian, and A. Arockiarajan. 2022. A comprehensive review on biocompatible thin films for biomedical application. *Ceramics International* 48 (4): 4377–4400.
56. Karki, S., et al. 2016. Thin films as an emerging platform for drug delivery. *Asian Journal of Pharmaceutical Sciences* 11(5): 559–574.
57. Oliveira, A.M., et al. 2022. Graphene oxide thin films with drug delivery function. *Nanomaterials* 12 (7): 1149.
58. Vannozzi, L., et al. 2018. Nanocomposite thin films for triggerable drug delivery. *Expert Opinion on Drug Delivery* 15 (5): 509–522.
59. Antonuk, L.E., et al. 1994. Thin-film, flat-panel, composite imagers for projection and tomographic imaging. *IEEE Transactions on Medical Imaging* 13 (3): 482–490.
60. Mahadeva, S.K., and J. Kim. 2011. Conductometric glucose biosensor made with cellulose and tin oxide hybrid nanocomposite. *Sensors and Actuators B: Chemical* 157 (1): 177–182.
61. Shkondin, E., et al. 2017. High aspect ratio titanium nitride trench structures as plasmonic biosensor. *Optical Materials Express* 7 (11): 4171–4182.
62. Davis, F., and S.P. Higson. 2005. Structured thin films as functional components within biosensors. *Biosensors and Bioelectronics* 21 (1): 1–20.

Chapter 2
Physical Vapor Deposition Techniques

Abstract In this chapter, different physical vapor deposition techniques have been discussed. Mostly, all the techniques were based on the same principle which is the conversion of the condensed phase to the vapor phase and then back to the condensed phase. This conversion involves different steps that are critical and vary from technique to technique. A vacuum chamber is needed along with a cathode and anode with a source and substrate. The anode and cathode are attached inside the chamber. The vacuum is generated inside the chamber by different pumps. The source is attached through which coating is done on the substrate. Mostly substrate is attached to the anode and the source with the cathode. The source position may vary in different equipment. After creating a vacuum, the process starts, which involves the heating process, vaporization, sputtering, ionization, and laser techniques. These steps vary in different equipment. Then, the thickness of the coating is checked through a quartz crystal. The quartz crystal is a sensitive material through which thickness is measured. Different techniques involve the thermal evaporation process, sputtering, pulsed laser ablation, and ion plating. The sputtering is further divided into DC and RF sputtering. The oldest technique is the evaporation process, the sputtering technique was invented due to high voltage and melting point. But this also generates the secondary electrons due to which the thin film generated is not good in quality. So, for that purpose, DC and RF sputtering techniques were invented, which is also known as magnetron sputtering. As with the technology advancement, the pulsed laser ablation was introduced in which the laser's high power, intensity, and diameter determine the thin-film patterning and coating. Finally, ion platting was invented which is similar to sputtering but is quite better than sputtering. All the processes have their advantages and disadvantages in daily life applications and processes. Every process has limitations also which is discussed in this chapter. Furthermore, the complete process, working principle, construction, advantages, and disadvantages are also discussed.

T. I. Awan et al., *Thin Film Deposition Techniques*,
https://doi.org/10.1007/978-981-96-1364-9_2

2.1 Physical Vapor Deposition

Deposition is required to achieve high performance by increasing their stability, improving the roughness, enhancing their life, and achieving the desired properties such as improving conductive, resistive, or optical properties. In 1857 first deposition was made by Faraday who melted the metal wire in the vacuum to achieve the metal coating [1]. This phenomenon was limited to physics only, but later on, the hype was increased due to its widespread knowledge about the life enhancement and stability of material. PVD technique not only helps in coating but makes the material more efficient, and through PVD, almost any material can be coated easily with different techniques. Because instead of using expensive materials, it is easy to coat the materials and mold them in different ways to achieve the desired properties. This technique is achieved under the vacuum only [2]. For deposition, a substrate is required usually made of metal, steel, ceramics, or glass. It is also known as PVD and is used to produce thin films from the condensed to a vapor phase and then back to the condensed phase in the shape of the thin film. A vacuum chamber is the main part of the deposition; otherwise, the deposition will not be achieved because of non-uniform gasses.

2.1.1 Introduction

A method called physical vapor deposition (PVD) transfers material at the atomic level. It is a vaporization coating method. The procedure can be delineated based on the subsequent succession of steps. Physical vapor deposition is the process of converting a liquid base substance to a gas using physical reactions. Subsequently, the material reverts to its solid form, forming a thin coating on the outermost layer of the substrate. The substance that has to be placed is transformed into a gas using physical methods such as high-temperature vacuum or gaseous plasma. The gas is then moved from its source to the substrate in an area with low pressure. Finally, the gas changes back into a solid state and forms a thin layer on the substrate through a process called condensation [3]. So basically it involves the three-cycle process. Figure 2.1 describes the principle of the physical vapor deposition process.

PVD methods are commonly employed to apply thin films ranging from a few nanometers to thousands of nanometers in thickness. Nonetheless, they are also capable of being utilized to create multilayer coatings, deposits with graded composition, extremely thick deposits, and autonomous structures. For PVD, different components are required, such as the substrate on which the material is deposited. The substrate can be made of any material and material whose coating is desired is also can be made of any material. While, a voltage source, vacuum chamber, quartz crystal, and gas are also major components for the deposition process. The schematic diagram for the general PVD method is shown in Fig. 2.2 [4].

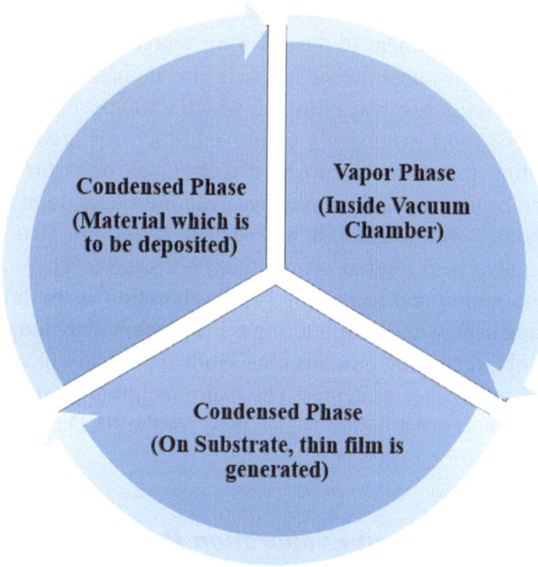

Fig. 2.1 Principle of the physical vapor deposition (PVD) process

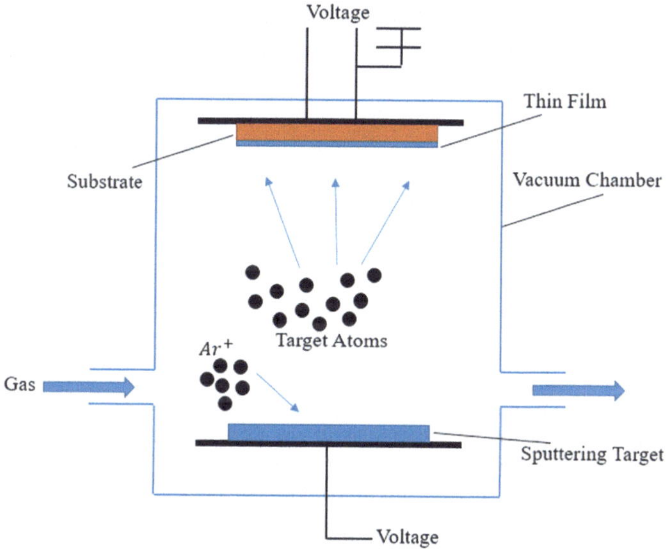

Fig. 2.2 Schematic diagram of general PVD method

PVD method offers numerous properties, including a high vacuum system with low contaminant gas levels and the capacity for input regulated amounts or fractional pressures of one or more operating gases, such as argon, hydrogen, and helium gas. It also has a carefully controlled and often constantly monitored vapor flux known as quartz crystal. It also incorporates a voltage source for regulating the temperature and distance alignment to the covering source, enabling it to precisely and consistently prepare the substrate surfaces. It also demands additional conditions in its further sub-divisions or ways through which PVD occurs [5].

The most extensively employed processes are evaporation, sputtering, and ion-plating processes. Deposition techniques used for PVD thin films include electron beam evaporation, reactive evaporation, and ion plating. Sputtering techniques, whether with plasma or with an ion beam, are also used in PVD. A PVD source can also be a filtered or unfiltered arc source. There are two types of this process: evaporation and sputtering [4]. Evaporation is the process of depositing thin films using heat, whereas sputtering is the process of dislodging atoms or molecules from a solid target using gaseous ions. Both approaches have been expanded upon to create several distinct strategies. The major techniques are evaporation, sputtering, and ion plating. Laser ablation is also used for the PVD process.

2.1.2 Thermal Evaporation Process

Evaporation is the first process that was also used by Faraday in 1857 first time when PVD was discovered. This happened when he melted the metal wire through an evaporation process, where he used a substrate and put the metal on the heating source. Upon applying the voltage the source melted and evaporated, this evaporation generated fumes which then implanted upon the substrate and created the thin film. The thin film ranged from nm to microns. However, the desired property was not achieved. Later on, as technology and advancement increased, the evaporation process was used extensively. The evaporation process is just like the typical physical vapor deposition technique in which substrate, anode, cathode, and voltage are used. It is simply heated and then evaporated which is then implanted on the substrate. This is also done in the three stages as in Fig. 2.3 which describes the evaporation stages. So, usually, it involves heating the substance by providing the voltage, and depending upon the melting point, the voltage is provided.

Through this, the atoms of the material start vibrating through heating and get dislodged (evaporates) and then scatter in the air. Due to uniform gas, the evaporated material goes straight to the substrate and sticks to it. The amount of the coating is then checked or calculated through digital device called quartz crystal. Then thin

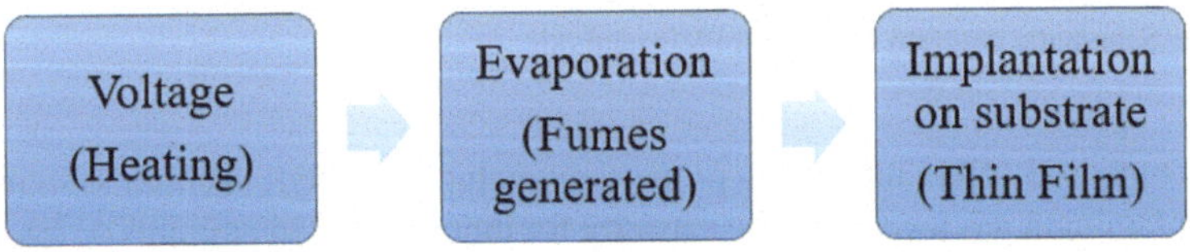

Fig. 2.3 Various evaporation stages

film is further checked for different applications. This process was then used for different commercial process [6]. This method is cost effective, and the quality of film generated through this process is quietly good then all others. The deposition rate in evaporation is very high around 1000 atom/sec. It means that in 1 s 1000 atoms are implanted in the film or the film is generated in just 1 s. Here, source can be easily chosen depending upon its melting temperature [7]. Because the melting point is only the main thing that is noticed during this process, otherwise this process will be failed and we have to move toward other processes. On earlier, this was used extensively but later on this process was less used due to its severe problems regarding the thin-film deposition.

Thermal evaporation occurs when the target material is placed inside a vacuum chamber and is heated by electric current. The evaporation point is reached by heating the target substance inside the evaporation source. The distance between source and substrate should be average 1 m. So, by creating the vacuum the molecules go straight and get deposited. This process is also known as resistive evaporation since heat creation results from the evaporation source's electrical resistance. Following evaporation, the target material's molecules migrate to the substrate and deposit a thin layer on its surface. This process can be used to deposit a wide range of materials, such as steel, gold, copper, iron, and silver [8].

In general, firstly the substrate is attached with the anode and the source is placed on the cathode, near heating element. Then chamber is closed and vacuum is created inside the chamber, and after the desired vacuum, it is closed [9]. Then voltage is applied according to the melting point. If the melting point is higher than the voltage given should also be higher, but on higher voltage dense fumes and bulk fumes are generated which then hit the substrate and create roughness on the surface. So, the voltage can be tuned easily, and upon applying the voltage, the filament starts heating, the material is also heated which then generates fumes, and these fumes are then deposited on the substrate. The amount of deposition required is digitally measured through the electrical device called quartz crystal, which detects the fumes from anywhere inside the chamber. The film is then taken outside the chamber for further tests and further properties. This whole process is known as thermal evaporation process.

2.1.3 Working Principle and Construction

The working principle is based on vacuum, where vacuum is created through pumping system, and then after desired vacuum, the evaporation becomes uniform and flux density of gases becomes same. The equipment involves different components which are required to generate a thin film through thermal evaporation technique. Figure 2.4 illustrates the schematic diagram of thermal evaporation technique. The heated substance is then converted to vapors and vapors got deposited on the substrate. So, it is just like physical vapor deposition three-stage principle, in which condensed phase (the material which is to be deposited, either aluminum, gold, silver) is converted

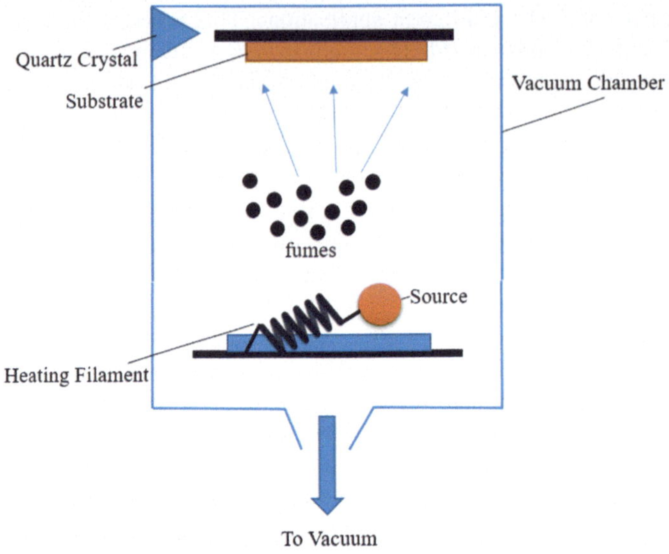

Fig. 2.4 Schematic diagram of thermal evaporation technique

to vapor phase (through heating by applying voltage) and finally again vapor phase converted to condensed phase (on substrate). In this way, the thin film is generated and deposited through evaporation [7].

The construction phases involved different components given as:

- Vacuum chamber.
- Pumping system.
- Substrate.
- Source.
- Filament.
- Voltage supply.
- Quartz crystal.
- **Vacuum Chamber**

A metal or glass chamber can be used for the assembly. This is the major portion for the evaporation because without this the film cannot be generated and cannot be deposited. Because of gas collision and heavy pressure or a different mixture of gases and air, the film is deposited on the substrate. For a more accurate process, the glass bell jar is used as an assembly on which different equipment is connected. The chamber should be such that it can tolerate high voltages and it should be heat resistant. Because of high temperature, it may explode, that's why it should be heat, water, and voltage resistant. Moreover, there should be enough space for all the equipment to be attached inside. It should be in such a way that an external vacuum pump can easily be connected to generate the desired vacuum [10].

ii. **Pumping System**

Different vacuum pumps are used to generate a uniform vacuum inside the chamber. To avoid the further collision of gases, smooth thin film is produced. There are different vacuum pumps for generating vacuum such as: oil-sealed vacuum pumps rotary van pumps and rotary piston pumps. While mostly diffusion, cryogenic, and gathering pumps are used, different pumps have different vacuum speeds and different types of equipment. They are used depending on the desired vacuum and desired range. Typically, a low deposition rate is used for high purity, the temperature should also be low to avoid roughness, the vacuum should be uniform and flux density should also be the same [11].

iii. **Substrate**

Substrate is the material on which deposition is made to generate a thin film. It can be also made of any desired material. The substrate is usually placed on the anode or the electrode, and it sticks until deposition occurs; after that it is taken out of the chamber. The substrate should be such that it can withstand the pressure of the deposition and ions. If it breaks with the maximum pressure, then film is not generated and desired properties are not obtained.

iv. **Source**

A source is a material that is deposited on the substrate. It can be of any type such as gold, silver, chromium, magnesium, and nickel. However, the choice of the source should depend upon the melting point. If the melting point is low, it can be easily used and the film generated is smooth. But if the melting point of the material is high, then high voltage is applied, and upon high voltage, the filament heats faster and generates bulk fumes which are deposited on the substrate and create roughness. So, melting point is one of the main reasons on which roughness or smoothness of film depends.

v. **Filament**

The filament or heater is usually placed on the cathode and one point is attached to the voltage supply, while the other point is with the source. Upon applying the voltage, it starts heating and glowing up. When it grows faster, it means that it also transfers heat fast to the source and the source then melts and converts to vapor form. It should be simple. The low-voltage and high-power transformers are used to provide the voltage to the filament. Usually, graphite filaments or heaters are used for high temperatures [7].

vi. **Voltage Supply**

A voltage supply is necessary to start the reaction. Here, low-voltage high-power transformers are used to provide the necessary voltage to the filament. Usually, a high-voltage supply is used for the process.

vii. **Quartz Crystal**

A quartz crystal is digital equipment attached inside the chamber and shows the digits outside. It detects the fumes deposited on the substrate and tells the thickness of the deposition [12]. With the help of quartz crystal, we can easily control the thickness and set the required thickness of the thin film.

2.1.4 Advantages and Disadvantages

- One of the major limitations or drawbacks of evaporation is that it is not feasible for sources that have high melting points. Because at a high melting point, high voltage causes faster heating and more bulk fumes deposited, which generates rough thin films [5].
- It also shows lower throughput because of low vacuum, as in the evaporation technique high vacuum or ultra-high vacuum can't be generated, which is a major drawback of the evaporation technique.
- It also shows poor step coverage, which means that through this technique, the source is not fully deposited because of low vacuum and voltage supply.
- It is difficult to form alloys through this deposition technique.
- The material is wasted through this technique as more heating is required than deposition of material.
- Maintaining the purity is also difficult because the material can also react with other substances leading to roughness and not inaccurate deposition [13].
- As for high vacuum, the pumping speed is less and it consumes more time and electricity.
- Deposition of multiple materials on a single substrate is also difficult through this technique, as materials may stick to the filament and the cleaning process is also difficult.
- Mostly this technique is not suitable for many materials; it is limited to only a few materials that are also expensive.
- Adhesion strength to the substrate is also limited, which decreases life time of the thin film.
- For large-scale production, this technique is not used.
- Achieving uniform deposition across the substrate is also difficult, as some portions get more deposited and some portions remained same.
- The major advantage of the evaporation technique is that its deposition rate is higher which is nearly $\sim 1000 atom/s$. This means that during 1 s 1000 atoms get deposited on the substrate [7].
- Here, choice of source is very easy as we can easily choose any type of source for the deposition, but depending upon melting point.
- Here, replacement of source is also easy.
- It is cost-effective because of the low vacuum process.
- The quality of the film is quite good as we can tune the voltage according to our own will.
- Low-level impurity is generated [14].
- It is good for materials with low melting points.
- It gives uniformity if the substrate is flat.

- It gives precise measurement as voltage can easily be tuned and controlled.
- This technique is very simple and easy to use.
- It is suitable for temperature-sensitive materials, because of the low-temperature substrates and the sources.
- It converts the source material into vapor phase without any further steps and saves a lot of time.
- It can be easily integrated with other techniques including photolithography.
- This is well-suited for the organic semiconductors.

2.1.5 Applications

There are different applications, regarding the evaporation technique. Some of them are given below.

2.1.5.1 Evaporation of Droplet

In order to determine the wettability of the solid material, evaporation of droplet is used. A droplet is made of any liquid substance. While droplets combine to form a drop. This drop can be of any shape symmetric, asymmetric. If it is on the straight substrate, it is symmetric, but if the substrate is rotated, it can be changed to different shapes leading to change in symmetry. This application is related to the evaporation technique, here liquid droplet containing substance such as solid material is evaporated, and then, this sticks to the substrate leaving behind the solid material, and in this way, we can also generate the thin film and also tells the wettability or liquidity present in the solid material by comparing its mass before and after [15]. This also involves rapid or slow drying process which is much related to the voltage tuning technique.

2.1.5.2 Optical and Protective Coatings

The evaporation deposition is also used in deposition of optical films onto lenses, mirrors, filters and other optical components by using a vacuum-coating chamber with a lattice of quartz substrates. TFT displays, OLEDs, and others also use evaporation deposition technique. Through this, it makes it possible to accumulate layer organic and inorganic materials with thickness control at a precise accuracy through voltage and heater and it helps lengthen the lifespan and performance of display devices. The evaporation technique is very useful in OLED and LED lights and also in transistors. Such coatings function as diffusers, reflectors, light transmitters, absorbers, and other optics, thereby becoming integral parts for optics, photonics, and lasers systems [16].

Thin-film deposition by evaporation is the method utilized to coat the surfaces with the protective materials in order to secure the best resistance against abrasion, corrosion, and general wear. This is basically done to prevent from the rust and corrosion and to make it resist toward the heat, temperature, weather, and water. The materials used for the coating have been applied to the automobile, aerospace, packaging, and other industries which aim to enhance the performance outputs and make the constituent materials long-lasting [17].

2.1.5.3 Photovoltaic Cells

There is also a use in photovoltaics (solar cells) of the evaporation deposition, which is used to allow the thin films of semiconductor materials such as cadmium telluride (CdTe) or copper indium gallium selenide (CIGS) to deposit onto substrates. These solar cells required deposition through evaporation of the material. As evaporating the substance, the substrate gets enough coating and thin film generated may have all the desired properties along with greater light capturing and electricity transferring properties. With this type of deposition, the activity may enhance. Such solar cells required rapid thermal evaporation deposition [18]. The different films are employed as active layers in thin-film solar cells. There is an improvement in flexibility, light-weightiness', and cost-effectiveness compared to the traditional bulky silicon-based solar cells. The cadmium selenide is now used for the deposition through evaporation. As newly launched solar cells can easily stick to the walls and have greater light capturing ability and also cost effective.

2.1.5.4 Thin-Film Capacitors

Evaporation deposition technique is very useful for the production of a thin-film resistor and capacitor having exactly required electrical properties. The constituents boast to be significant in the case of miniaturization and performance tuning of digital devices which involve ICs, sensors, and memory devices. In order to enhance their storage capacity and life span, deposition is required. The evaporation deposition is used to fabricate thin films for MEMS devices as one of the widely utilized methods, which are the latter including micro sensors, micro actuators, and microfluidic systems. Here, Mems are the devices known as microelectromechanical systems. Such thin films offer mechanical reinforcement, electrical insulation, and also many useful coatings to MEMS, and they are consequently used to be integrated into many different areas of the use such as automotive, biomedical, and consumer electronics [19].

2.1.5.5 Application in Electronic Devices

Thermal evaporation is the also used in industrial processing to produce ultra-thin films made of active TFT layer materials such as a-Si, p-Si, and IGZO metal oxides. Flat panels constitute the most popular type of display, such as LCDs, OLEDs, and flexible electronics, since the most common semi-conductive thin-film layers make up that too. Cadmium sulfide is usually used for this process through thermal evaporation deposition, in order to enhance the life span, ability, resistivity, conductivity, and other optical properties [20].

For a stronger bonding of dielectric materials in the form of depositing thin films, e.g., oxides of silicon (e.g., silicon dioxide, silicon nitride) are used by evaporation. Such thin-film dielectric layers are responsible for insulation, shielding, and confinement aspects of semiconductor elements, yielding the device robustness, permanence, and enhanced performance including life time, optical and absorbance properties.

2.2 Sputtering

Sputtering is a popular method for depositing thin coatings on surfaces. The method involves subjecting the target, a source material, to an ion bombardment. The target material sputters under ion bombardment, which is a purely chemical reaction that produces a vapor. As such, this method falls within the category of physical vapor deposition methods. Here, in this method the gas is used which ionizes on applying voltage, and after that, the ions oblates the source surface and then further it ionizes the gas and produces the secondary electrons; these secondary electrons go further on substrate and damages it surface [21].

So for that purpose, the RF sputtering is used by using a magnetic source to bombard the target with positive ions from the charged particles of a magnetically enhanced discharge is the most popular method for producing thin films via sputter deposition called RF. Several methods, such as applying current and radio frequencies (RF) for electrically inert targets or direct current (DC) for metallic targets, can be used to power the target. Reactive sputtering, which adds chemistry to the plasma spontaneously by introducing a reactive gas, is necessary since the process of sputtering is a solely physical process [22].

As this is the gas that ionizes and starts the process of oblation the source and starts deposition. A prime instance of this is the development of a compound layer. The reactive gas's unfavorable reactivity with the target material causes the deposition parameters to behave nonlinearly as an effect of the reacting gas flow. The fluxes of the different species toward the target need to be identified to mimic this behavior. The species fluxes occurring at the substrate, however, are just as significant as they regulate the formation of the intended film in addition to impacting the reactivity of

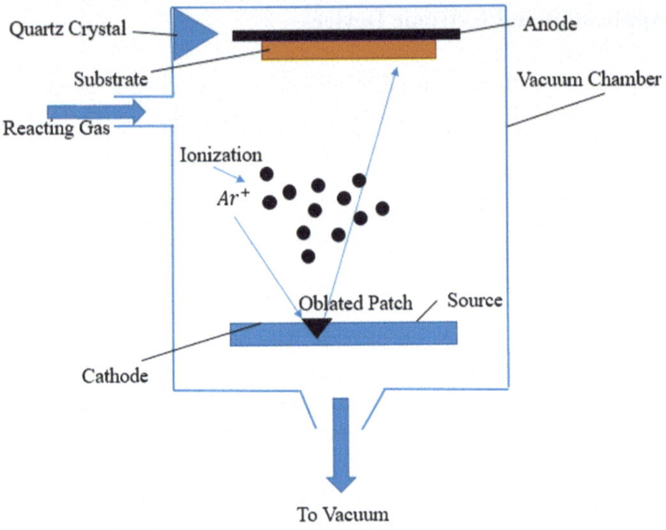

Fig. 2.5 Schematic diagram of the sputtering process

the sputter deposition process. The type of particles that arrive defines the microstructure of films produced by magnetron sputtering. The schematic diagram of sputtering process is given below in Fig. 2.5.

Further, the sputtering also generates secondary electrons due to oblation which damages the substrate and it is the biggest limitation of sputtering, that's why it is further divided into two types.

1. DC sputtering.
2. RF sputtering.

2.2.1 DC Sputtering

DC sputtering is the major type of sputtering in which direct current is applied, to deposit material layer on the substrate. For large-area vaporization and for the uniform plasma, the dc sputtering is used, mostly for the metals. In this process, the cathode which is the source should be a good electrical conductor, so, that ion bombardment occurs very easily and it deposits the layer on the substrate. The method is easy to use but time-consuming and very expensive because we have to make another electrode for the source. The generation of secondary electrons through the cathode also occurs here, but here we use the magnets around the cathode to deflect these electrons and they glow up around the cathode, moving around the cathode. This is also known as magnetron sputtering. Here, plasma is confined to the cathode only and does not go to the reactive gas for the sputtering which is a drawback of DC sputtering. For that purpose, the arrangement of the magnets is changed, but it shows less efficiency and

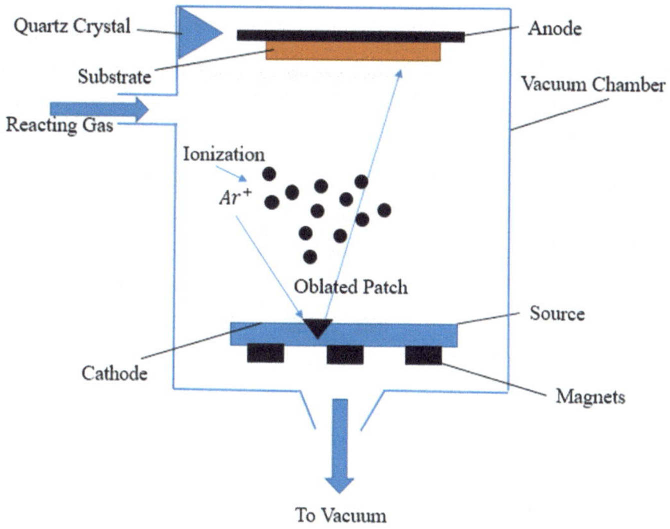

Fig. 2.6 DC sputtering mechanism

plasma is also not uniform through this process [23]. Figure 2.6 illustrates the DC sputtering mechanism.

So, this method involves using magnets around the cathode to stop the further generation of secondary electrons, and the secondary electrons due to the magnetic field move around the cathode and do not go to the substrate for further deposition and this helps in smooth thin-film formation.

2.2.2 *RF Sputtering*

RF sputtering also known as radio frequency sputtering or AC sputtering is the process that also involves the magnets for the sputtering process, but in RF sputtering a specific radio frequency is applied to deflect the secondary electrons. Here, a specific frequency below 50 kHz is applied, which reverses the applied potential periodically and the electrons in this way move alternately around the cathode, so that the plasma is not only confined to the cathode for reactive gas sputtering. The plasma will be uniform through this process. Above 50 kHz, the electrons do not discharge from the cathode and the potential that is applied flows between the two electrodes and the ionization occurs sufficiently. Here in alternating voltage, the cycles work. During the half cycle, the electrons oblate the surface and go to the substrate for sputtering, but during the next half cycle, these electrons just prevent any build of the charge on the surface. This method is useful for insulating materials that are brittle, not hard. So this is the biggest disadvantage of RF sputtering that insulating materials have less thermal conductivity, and if high voltages are used in this process, it will break

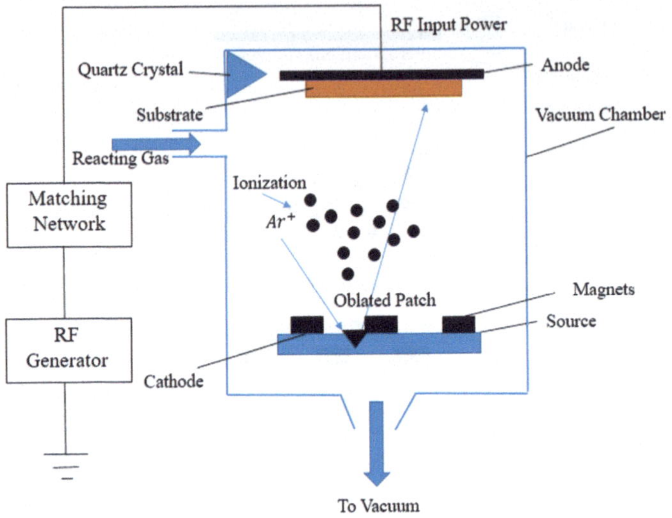

Fig. 2.7 Schematic diagram of RF sputtering

the target material and further deposition will not occur [21]. The schematic diagram of the RF sputtering is given in Fig. 2.7 [24].

So, in this method, the radio frequencies below 50 kHz are used producing alternating voltages and preventing further deposition of secondary electrons, this also generates glow around the cathode, and the thin film produced is quite good than reactive gas sputtering [25].

2.2.3 *Working Principle and Construction*

The principle is based on the process of physical deposition because this is also the type of physical deposition. This also involves conversion of condensed matter. But here, vapors are not generated, but instead, a patch of the source is oblate through ion bombardment. The ions are generated through the ionization process when gas is added and voltage is provided. On applying the voltage, the gas ionizes and oblates the patch from a source which is then deposited on the substrate and generates a thin film. The RF and DC sputtering is also based on this principle but involves a magnet that deflects the secondary electrons. The film produced through DC and RF sputtering is better than normal sputtering, but also has some limitations and drawbacks.

The construction involves different equipment such as:

- Vacuum chamber.
- Substrate and cathode.
- Reactive gas.

- Magnets.
- RF generator.
- **Vacuum Chamber**

A vacuum chamber is made up of glass, steel, or metal. It should be such that it can bear the temperature, resistance, water, light, heat, and all other harsh conditions. It should be unbreakable and has great strength. It should have enough space for all equipment. It should be leak-proof, heat, and water-resistant. It should have pressure regulation and high pumping speed with an ultimate vacuum level. The seals, valves, and feed-through should be properly checked and environment friendly [26]. On these bases, a vacuum chamber is chosen and made for such type of deposition. It is the same as that for the physical deposition.

II. **Substrate and Cathode**

The substrate for thin-film deposition can be made of any material, but depending upon the temperature range, it is chosen. For metals, conductors, and insulators, reactive gas sputtering is used. While for conducting medium, the DC sputtering equipment is used, and for insulating mediums, the RF sputtering equipment is used. Because these are made for specific materials as they have specific properties. Similarly, for the source, a complete electrode is generated which is called the cathode. The cathode is made of the source that is deposited on the substrate. So, it is also nature-specific for all types of sputtering.

III. **Reactive Gas**

The gas that is used for the sputtering process is argon or helium gas. Because of their high ionization ability. Before entering the gas, the desired vacuum is generated through pumping systems that are mentioned above including oil-sealed vacuum pumps, gattering pumps, cryogenic, diffusion, or turbo molecular pumps depending upon the desired vacuum required with the desired pumping and pressure speed. After a specific vacuum range, the reactive gas is added for the ionization process. The voltage is provided, gas ionizes, and the ions generated oblate the surface. That oblate patch then goes on the substrate and is deposited there.

IV. **Magnets**

Magnets are also used in the sputtering technique. In normal reactive gas sputtering, the magnets are not used, but later on, the magnets are used in DC and RF sputtering. These magnets help in stopping the production of secondary electrons that damage the substrate. The magnets are of different shapes and also include balanced arrangement and unbalanced arrangement [27]. In the balanced arrangement, the magnets are arranged in a specific order through which the electrons deflect and move around the cathode and glow up, but in DC sputtering through this process, the plasma is only confined to the cathode and not helps in the reaction with the gas, that's why the unbalanced arrangement is used through which some of the electrons go to the substrate and some of them revolves around the cathode [22]. In this way, magnets are used.

V. **RF Generator**

The RF generator is used in the RF sputtering; here, specific range below 50 kHz is used, because above 50 kHz, the electrons do not discharge from the cathode and the voltage only remains between two electrodes which will cause roughness in the film. The range is measured through the matching network, which is a digital circuit that detects the radio frequency only and then illustrates it after calculation [28]. This is attached through the anode and RF input power is given to the anode, while the whole chamber is grounded.

2.2.4 Advantages and Disadvantages

- The sputtering technique is costly.
- Deposition rate is minimum then thermal evaporation process, i.e. 1 atom.
- This is time-consuming technique.
- The choice of source is not easy as we have to make a completely new electrode, say cathode for the deposition as the cathode is used for the deposition on the substrate.
- The replacement of source is also not easy, as in the evaporation, the source can easily be replaced and used for deposition, but in the sputtering, the new cathode is produced which consumes lot of time and is very expensive.
- The major drawback of sputtering is that it produces secondary electrons during the oblation process, further ionization occurs and electrons are generated which are called secondary electrons and these electrons directly hit the substrate and damages the substrate, causing roughness of the thin film.
- The quality of the film is not good because of high temperature, energy, and generation of secondary electrons.
- One of the major drawbacks of DC sputtering is that, during this process, the plasma is only confined to the cathode and not interacts with the gas which causes less deposition and not covers the complete surface area.
- There is also a limitation in RF sputtering because if an RF frequency of above 50 kHz is applied then further electrons will not generate from the cathode and further no deposition will occur as voltage then felt between two electrodes.
- The major drawback in magnetron sputtering is use of arrangement. There are two arrangements balanced and unbalanced. But depending upon requirements, it can be changed, which is also time consuming and changes the whole scenario of deposition if arrangement is different.

Following are some specific advantages of sputtering:

- This technique is not temperature sensitive. As here we can deposit any type of material, including insulator, conductor, or metal.
- The deposition process is easy.

- Once the arrangements are made and equipment is set up, then it is easy to handle and use the whole apparatus for the deposition process.
- It also covers up the large surface area easily as plasma is uniform throughout with the same flux density.
- For achieving higher deposition rates, RF sputtering is used.
- The DC and AC sputtering helps to prevent further charge buildup and helps in the deposition process, making the thin film smoother.
- The thickness of the film can easily be controlled through the matching network by controlling the voltage and radio frequency with current.
- It can be used for materials with high-temperature ranges because here evaporation is not involved, but sputtering is involved through which ions bombard and oblate the patch easily.
- Through this technique, the three-dimensional structures can easily be covered with the coating material, and also it allows the reactive sputtering to enable stoichiometric and other properties.
- This technique enables directional deposition as magnets are involved which can easily change the direction by controlling the electrons and deflecting in a specific manner.
- This does not involve toxic or hazardous chemical materials and it is environment friendly.

2.2.5 *Applications*

The applications for the sputtering technique are given below.

2.2.5.1 Surface Modification

It is important to modify the surface of any material according to the desired requirements which is a difficult step. Surface modification also has greater importance in industrial processes. Through sputtering the surface can easily be modified in any direction or any dimension by coating with the material and in the desired direction as magnetron sputtering deflects the electrons in a specific manner, and by changing the arrangement the material can be further modified. The balanced and unbalanced magnetron sputtering helps in a deposition by preventing the charge deposition on the surface and the substrate gets easily deposited also the thickness can be controlled to detect the range and the pressure [25]. Through this technique, the surface can easily be modified by coating with metal, conductor, insulator, or any other material. The temperature, friction, hardness, voltage, frequency, current, and pressure can easily be controlled.

2.2.5.2 Immune to Corrosion

This technique helps in preventing the film from corrosion and rust by increasing its lifetime through the deposition of nitrides as it enables the stoichiometric and further properties helping in preventing the corrosion and making them resistant. Through the PVD technique, the corrosion comes quickly as the voltage is not uniform between the coating and the steel, but here voltage is uniform, and plasma is also not confined but uniform because of the flux density is the same. Further, the addition of a magnetic field through magnetron sputtering helps in increasing lifetime by immunizing them from corrosion and stops the generation of secondary electrons that damage the substrate. Therefore, mostly nitrides are used for the coating on the substrate as they are corrosion-resistant and have a better lifetime [25].

2.2.5.3 Flat Panel Coatings and Hard Coatings

Sputtering plays a crucial role in the manufacture of electronic displays like liquid crystal displays (LCDs), organic light-emitting diode displays (OLEDs), and thin-film transistors (TFTs). In general, this technology is used to deposit transparent conductive oxides (TCOs), metal electrodes, and metal-oxide films onto glass or flexible substrates to increase display resolution and luminance figures. Besides, the sputtering process is used to coat hard materials and wear-resistant films onto cutting tools, mechanical components, and sliding surfaces. The involvement of magnets enhanced its features by deploying further properties and enabling more abilities efficient in the construction of electronic devices. These thin films better withstand the environment, high temperature, humidity, and chemical (acid) attacks which help the applied materials live up to the life and performance expectations [29]. In this way, it also helps in the prevention of corrosion and rust. Further, it helps in increasing the life span and making more stable design of photovoltaic cells by coating them with specific materials including conductors and insulators and changing the silicon with other efficient materials that control the reflections and polarization of light.

2.3 Pulsed Laser Ablation

A straightforward yet flexible experimental technique, pulsed laser ablation is applied in several fields such as thin-film deposition and multilayer study to shape a broad variety of materials. The method is theoretically straightforward as well, at least on the surface, but this seeming simplicity conceals an abundance of intriguing, and as of yet, little-understood, molecular physics. The development of material from target destruction to the deposited film, describing the processes involved in the first laser-target interactions that allow solid material to enter the gaseous phase, the processing and spread of material in the material cloud that is ejected, and the final accommodations of gas phase species onto the structure that is to be coated [30].

The first accounts of the application of laser radiation bursts to "ablate," or remove, particles from a solid (or liquid) target, appeared soon after the first ruby lasers became accessible in the early 1960s. It wasn't long before the pulsed laser method was used as a pathway to thin-film deposition, considering the material ablation process's evident efficiency. In the meantime, pulsed laser ablation (PLA) has taken many different forms and is currently used in many areas of modern engineering, research, and medicine.

In this process, a high-energy laser beam is used which is directed toward the target material, and when it hits the target material due to high energy, it heats up and converts to the vapors, which are then deposited on the substrate and thin film is generated [31].

Additionally, laser ablation techniques are widely employed in the removal of biological tissue (such as corneal tissue), in the micro-structuring and sequencing of monomers and wide bandgap semiconductors like silicon and fusion silica, and are even utilized to clean the sensitive surfaces of artwork. The discovery that whole molecules, even big macromolecules, could be ejected using laser ablation techniques into the gas phase gave rise to innovative analytical approaches like matrix-aided laser desorption/ionization. It is currently regarded as crucial in many life science fields.

In this experiment, a pulsed laser concentrates on a target material in a vacuum (usually better than 10^{-6} Torr) or with minimal background gas pressure. Typically, the target is turned to prevent ablation from occurring again from the same location on the target. When it comes to making the best use of the produced target surface, rotation is the preferable method. The output wavelengths of the lasers utilized in PLD research span from the mid-infrared to the ultraviolet (UV), through the near-IR and visible spectrum. Excimer laser beams, which function at several UV wavelengths, are used in a large portion of current PLD research [32].

With a largely uniform distribution around the target surfaces, the discharged cloud of neutral material and plasma extends out from the contact volume. Film deposition happens in PLD when the discharged flux is positioned to impact the target substrate. Numerous essential factors, such as substrate selection, substrate temperatures, and both the absolute and comparative kinetic energies and onset rates for the different components within the cloud, will often affect the development and quality of the final film. The selection of the excitation wavelength, time frame, energy, and level of the emitted laser pulse, the existence of any background gas, and any additional plasma excitation in the target-substrate gap can all have an impact on the latter.

2.3.1 Working Principle and Construction

The principle is based on the physical deposition process in which condensed matter is converted to vapor phase and then again condensed phase. The illustration of the three-stage principle is given in Fig. 2.8.

In this method, a high-energy laser beam of a specific wavelength falls on the target material or source which is continuously rotated to prevent the ablation from

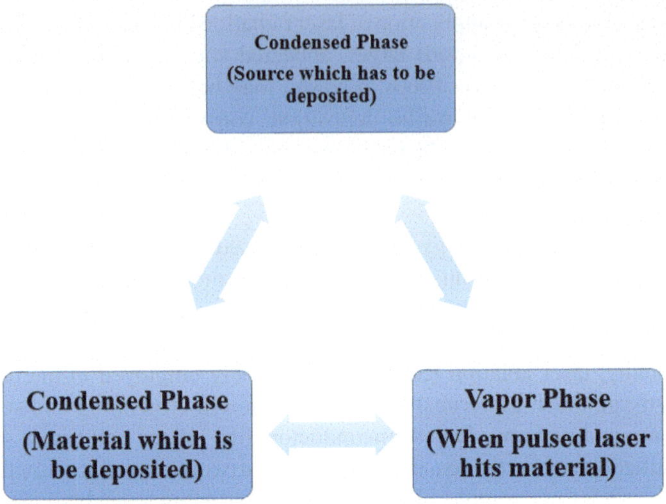

Fig. 2.8 Principle of pulsed laser ablation

the same face. After the beam hits the target, it ablates the surface and converts it into vapors due to high temperature and then these vapors attach with the substrate and thin film is generated.

The construction is very different from the other processes, as this technique involves the rotator along with a pulsed laser. The schematic diagram of pulsed laser ablation is given in Fig. 2.9 [30].

The construction involves different components and all have different properties.

- Laser beam.
- Focusing lens.

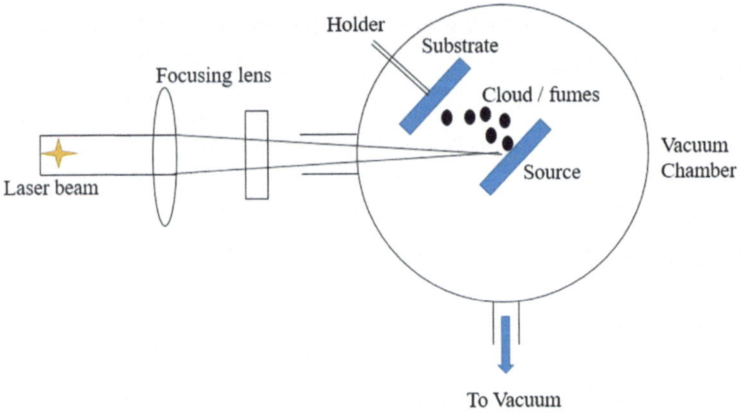

Fig. 2.9 Schematic diagram of pulsed laser ablation

- Rotor.
- Source and substrate.
- Vacuum pump.
- **Laser Beam**

The laser beam is used with specific wavelength according to desired properties of thin film. It ranges from UV region to IR region and can be of any wavelength. It was first discovered in 1960 as mentioned above. The laser beam has different ranges from nm to micro meters with different temperatures. Some of beams have high energy and high temperature relative to other lasers; for laser ablation method, high energy beam is utilized for fast process. So, when it hits the source it easily ablates the surface, upon hitting the surface melts and converts to vapors which are then deposited on the substrate. The laser beam is selected depending upon different parameters such as wavelength, pulse duration, divergence, and its availability. If wavelength is shorter, then energy would be high which means high temperature and high pulse duration. For better response, short pulse duration is chosen because it does not burn the source, as if source is heated extra, it will burn and also damages the source and substrate [33].

Similarly, the laser beam should be stable and show no divergence if it diverges upon hitting other surfaces, it will also damage other parts of the equipment. Mostly laser beams are not available for specific reasons and demands, that's why they are also costly. So, this method proves to be very costly.

(2) **Focusing Lens**

The focusing lens that is used here is the one that focuses the beam right toward the source. Mostly, convex lenses are used that converge the light to one point and go straight in direction. If the diverging lens is used, then it will hit the other parts and also the laser light resulting in damage to the setup. It also depends upon different parameters including, focal length and diameter, coating properties, and damage threshold.

The distance from the lens to focal point determines the focal length, and it is very critical because if the lens is near the source, then the light will continuously damage the source and deposition will not occur; similarly if it is far away from the source, then only some part of the source is ablated, while other portion remains same. So, focal length and its diameter are two main points. The coating properties of the focusing lens should be such that it may not damage the source; it should have good transmission and less reflection and have high optical characteristics. Similarly, the damage threshold should be such that the source or target on which it falls can sustain its intensity; if it is higher intensity, then it will damage and burn the source material, and it should also be environment friendly because watching the beam with naked eye can cause eye burn and damage to eye tissues and biological tissues [34].

It also depends upon the desired spot size and other contamination reasons, and further, it is chosen on a different basis according to the desired requirements.

(3) **Rotor**

The rotor or vacuum chamber is used in this process. The rotor is used because upon hitting the beam on the same surface of the target may damage the vacuum chamber also, that's why the source is attached to the rotor which continuously rotates and the beam falls on different faces of the source and ablation occurs perfectly. Similarly, the substrate is also rotated because deposition will occur on the same spot and will not cover the large surface area, so, to cover the whole surface area, the substrate is also connected to the rotor. The rotational speed and the temperature management should be controlled in a specific way because if rotational speed is low or maximum, it may also damage the source and the substrate and deposition will not occur fast. Similarly, if it revolves fast, then the temperature will increase, and the beam temperature is also high, which causes the setup.

(4) **Source and Substrate**

The source and the substrate are dependent upon different conditions mainly temperature because of the high temperature of the laser beam that falls and hits the surface of the source. It should be such that they can bear the high temperatures and high-intensity lights. It also depends upon its optical properties including absorption and transmission. It should absorb the intensity of light that falls and also the melting temperature because on absorbing it melts and gets converted to vapors which are then deposited. Chemically, it should be stable because of reactivity with the highly intense light and the substrate size should be according to the source, it should not be larger or very small. To cover a large surface area, the size should be per the source or target material. The substrate also depends upon temperature, because it also revolves continuously and gets deposited by high-temperature vapors. These are selected on the basis of electrical conductivity, resistivity, optical characteristics, and thermal stability.

(5) **Pumping System**

For pumping, different vacuum pumps are used which are mentioned above. But for this selective pumping is used with high pumping speed for achieving ultra-high vacuum level. The vacuum pump should be efficient and reliable. Having good energy and size with greater pumping speeds and ultra-high vacuum, high quality thin film can be deposited. It should be cost effective, and maintenance cost should be less. It should be reliable. It should also include all the safety features [35].

2.3.2 *Advantages and Disadvantages*

- One of the major disadvantages is that the whole apparatus is temperature sensitive; if the temperature of any component is not achieved according to the requirement, then it may cause serious damage to all the other components resulting in roughness of the film.

- When laser light falls on the source the vapors are ejected from the surface, but instead of depositing on the substrate, these particles also deposit on the nearby surfaces because of rotation, as these vapors get deflected by rotating.
- Another bigger disadvantage of the apparatus is that it has serious safety issues; if it is not used by a professional, then it may cause damage to the eyes, nerves, biological tissues and even it may explode if not properly used.
- The ablation speed is high, because of the high-intensity beam, and it may ablate the complete depth of the source, so achieving the depth of ablation is a very difficult process. It may break the source if not properly controlled, and once the source is broken, then further deposition will not occur according to desired requirements.
- The process is very costly, as every component is imported from other countries and even laser light is not available which is also expensive. So, if any damage occurs to the setup, then the maintenance cost is also very high. The whole method is very costly.
- The equipment is used in the proper room having good ventilation and safety requirements, because it is not environment friendly, as laser light may generate hazardous fumes that are not good for the health and environment.
- The process completion is also very difficult, because of the complexity it can only be used by a professional physicist or engineer, which is very difficult to find, and most of them don't know how to use this and handle all the setup.

Some advantages are as follows:

- This process is highly controllable, which is biggest advantage because we can control every component according to our requirement and can get a very precise coating on the whole surface area, by controlling the temperature, rotation, intensity, voltage, and material we can get a fine thin film coating.
- This process doesn't involve any use of lubricants or other fluids, so this is also a clean and dry process. In this way, it is also called environment friendly because it doesn't pollute the environment.
- It is versatile because it can be applied to a range of materials according to the wavelength and intensity of light being used. Laser light is available in all wavelengths ranging from nm to microns. It also has high and low intensities with specific energies and temperature, so it is versatile.
- The ablation spot can also be controlled by selective ablation. So, it does not damage the other areas or not ablates the other portions, and in this way, it also gives us a selective ablation.
- Laser ablation can be done through a computer, meaning that it is a remote-based system, through which if material and substrate are placed, then it can be controlled through the computer-based systems and helps in remote processing.
- We can fabricate different materials from macro to micro scales through this process, as the laser beam can easily be controlled with a specific resolution.
- It provides high purity and the quality of film generated is high pure and fine.

2.3.3 Applications

It is used in different ways from medical to engineering fields.

2.3.3.1 Biological Tissues and Treatment

The laser ablation technique is used to treat the cancer cells. The cancer cells are first identified through other systems, and then a high pulsed laser beam is fallen on these tissues which breaks them into different particles and these small particles are then collected or discarded away. Similarly, it is also used to break the kidney stones. The stones break into small particles and are discarded away through the urination process. The tumors can also be removed through this process, and even the tattoos are removed in this way by using laser beams [32].

It damages less to the skin and is also used to break the dark spots on the skin, and this technique is majorly used to remove tattoos. Also, it is used to reshape the tissues or remove any unwanted tissues. It has fewer issues than the traditional way of surgeries.

2.3.3.2 Laser Welding and Machines

This technique is also used in the welding process. For welding the metals and large surface areas with thick and dense volumes, the laser with high-intensity light and temperature is used. Laser welding is also used in welding larger tanks and ships. Laser machines are also used, such as laser wood cutting, and this technique is called CNC machines. In which laser light is used to cut the wood for specific designs. This is also computer-based. Through which the wood piece is attached under the laser light. Upon giving the design through a computer it makes the design by cutting with the help of a laser and gives the pattern according to the requirements [36].

Currently, water laser technology is used, in which high-speed water is used in a specific direction with high intensity that can even cut the metal.

2.3.3.3 Geological Studies

Laser ablation is also used for measuring specific directions, with geological studies. It is employed for mineral exploration as a highly intense beam can ablate the surface and helps in finding the specific minerals, or metals. For inner analysis of rocks, and sediments this technique is used. While for environmental monitoring the laser ablation technique is employed [37].

2.4 Ion Platting

The early 1960s saw the first description of ion plating, which was then applied to increase surface coverage and film adherence. Subsequent research demonstrated that film characteristics like density, shape, index of refractive index, and leftover film stress may be altered by carefully calibrated bombardment. In more recent times, responsive and deposition methods have employed bombardment to intensify chemical reactions. Currently, "directed deposition" is utilized to increase surface detail fulfillment in semiconductor fabrication by ionizing and accelerating the process of deposition film atomic particles, or "film ions."

Here, ions are used which are continuously bombarded on the substrate for achieving the adhesion and coating properties and increasing its life span. Ion plating is a type of atomic vacuum deposition in which energetic, atomic-sized, neutral, or reactive particles are constantly or intermittently blasted into the depositing film, potentially influencing the film's development and characteristics. The ions are generated through different techniques as studied earlier. Atoms can be deposited via arc vaporization, sputtering, thermal evaporation, or chemical vapor precursors. The two most common types of bombarding species are atoms from an "ion source" or ions from a plasma generated in the coating chamber (also known as "plasma-based" or "vacuum-based" ion plating). Ion plating may be divided into three stages: film growth, nucleation interface generation, and surface preparation [38].

Sputter cleaning and keeping the surface "clean" until the film starts to develop are the key adhesion-related advantages of ion plating. Additionally, it enhances diffusion, chemical reactions, and other processes by providing a high energy flow to the substrate surface, which raises the surface temperature without requiring bulk heating. By adding high defect levels, physically combining the film and substrate material, and affecting the nucleation and development of the depositing film, it may also change the surface and interfacial structure. In certain systems where traditional deposition techniques yield subpar results, ion plating proves to be especially successful in achieving good adherence. The extremely high "throwing power" of the deposited film material is another result of the ion-plating process [39].

In this process the ions from any source or previous deposition techniques are used, which are continuously thrown on the substrate or the thin film in order to increase its efficiency and life span and making it corrosion-resistant along with greater adhesion properties that previous techniques not provided.

The process involves three stages of surface preparation, nucleation, and film growth. Among the preparatory processes in ion plating are surface cleaning and degreasing, where all mpuritics that may obstruct the layers from connecting are eliminated. In order to improve upon this, surface roughening techniques such as chemical etching or abrasive blasting may be applied to strengthen the surface attachment and create mechanical interlocking between the coating layer and the substrate. Controllable pre-treatment techniques, such as plasma cleaning or ion bombardment, can be used in addition to one another to improve coating adherence on highly active substrate surfaces. That being stated, the next phase denotes nucleation, which is the moment at which atoms or molecules separate from the vapor phase and adhere to the substrate surface [40].

Secondly, nucleation occurs where determining the adhesion and structure morphology of the formed coating is part of nucleation. Controlling the nucleation conditions—such as substrate temperature, ion energy, and deposition rate—allows for the manipulation of nucleation density and orientation, leading to improved performance in tailored film attributes [38].

The film growth is most likely the last step. According to nanoscience processes, this is the location of film deposition, where atoms or molecules display and build layer after layer until the film is continuous. Fast-moving ions improve film consolidation, accelerate in-plane reordering, and increase atom mobility, all of which contribute to changing the dynamics of film growth. The thickness, structure, and other characteristics of the film may be accurately controlled since the controller can alter variables including secondary ion energy, substrate temperature, ion fluency, and deposition rate. The illustration of ion gauge mechanism is given in Fig. 2.10 [41].

It is further divided into vacuum-based ion gauges and plasma-based ion gauges.

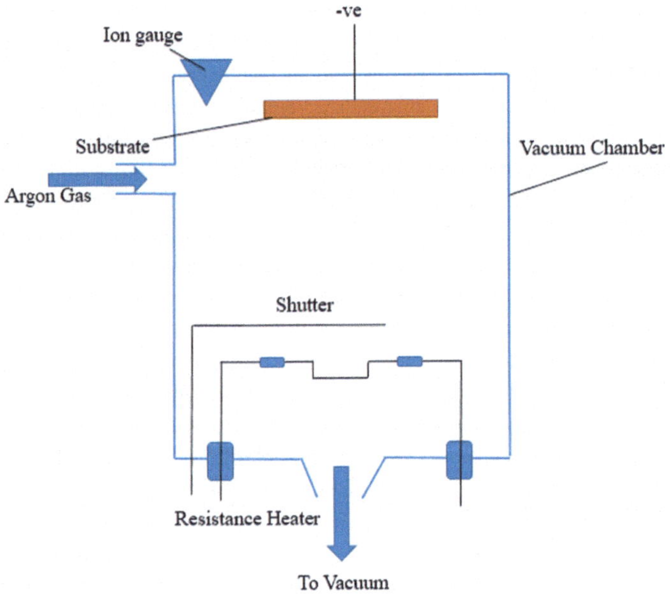

Fig. 2.10 Ion platting mechanism

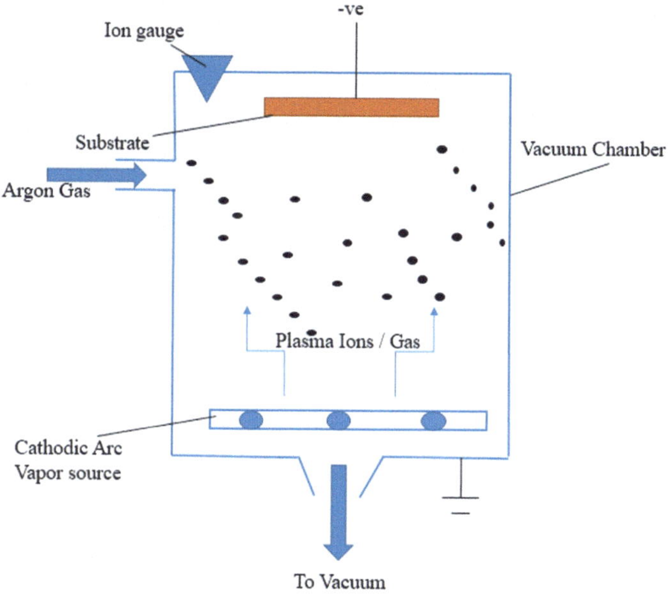

Fig. 2.11 Schematic diagram of plasma-based ion gauge

2.4.1 Plasma-Based Ion Gauge

During a plasma-based ion gauge, a plasma beam is used that causes the plasma gas discharge. A plasma ion gauge is based on the mechanism of creating a plasma discharge in a vacuum chamber, ionizing gas molecules. The produced ions are attracted to the electrode that has been positively charged; the corresponding ion current enables calculation of the gas pressure via an application of Ohm's law. These ions are then bombarded on the substrate directly to enhance its properties. The schematic diagram of a plasma-based ion gauge is given in Fig. 2.11 [38].

2.4.2 Vacuum-Based Ion Platting

The ionized particles are produced and accelerated from gas sources against the target material into a vacuum chamber in the vacuum-based ion-plating process using a cannon. To create a thin film, the embedded atoms are ejected from the target surface and deposited onto the substrate. Based on the electron beam source, the target sub-capacity, the acceleration, and finally the film deposition, beam line ion vacuum evaporation is performed. Here, the thermal evaporation process is used for creating vapors. Figure 2.12 describes the vacuum-based ion platting [38].

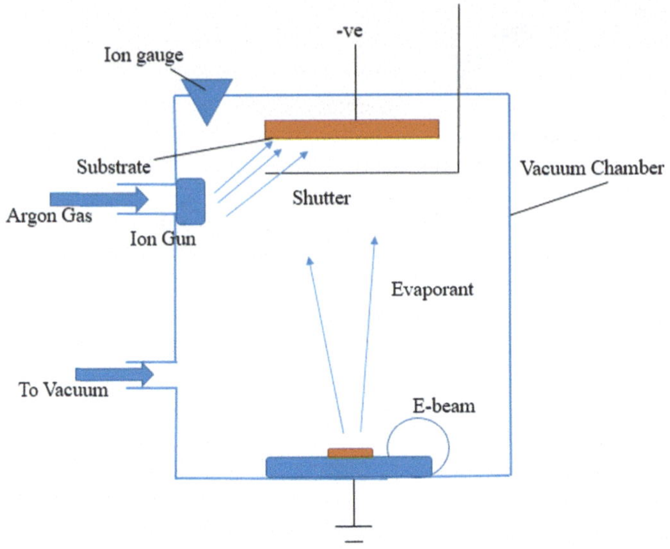

Fig. 2.12 Vacuum-based ion platting

2.4.3 Working Principle and Construction

The working principle is dependent on all the techniques mentioned above, because of the vapor process. Here, in this process, the vapors can easily be generated through evaporation, sputtering, electronic beam, laser beam, ionizing gas, or any other arc deposition method. Where the condensed matter is converted to vapors, directly vapors are entered in the chamber. or gas is entered which then ionizes and converts to vapors, and these vapors are then deposited on the substrate. The construction involves different components such as:

- Shutter.
- Vapor source.
- Plasma gas.
- Ion gun.
- **Shutter**

The shutter's function in ion plating is to reject material removal during the process's first deposition stage, allowing for the proper surface preparation and polishing. By simply opening and closing the controlled apertures, it also offers a very precise control over the deposition process by facilitating or stopping the film growth through the competitive transition of the ion beam. By obstructing deposition flux when it is not needed, the Gate also helps to maintain homogenous layer thickness and minimize particle contamination. Lastly, it highlights the regular production of new layers by the application of diverse coatings with particular properties. Ion shutters also provide a significant contribution to rate management and quality enhancement since they help to achieve this.

To achieve consistent, precise, and variable film thickness and characteristics, the shutter is utilized to regulate the length of time the substrate is exposed to the vapor flow.

(2) **Vapor Source**

A single metal, an alloy, or a mix of elements can be the target material, depending on the strength and corrosion resistance that the deposited coating will need. The source that powers the evaporation or sublimation process can be heated using a variety of techniques, including as resistive heating, electron beam bombardment, or radiofrequency induction [42].

However, one may take use of the sputtering process, which involves the impact of energetic ions on a target surface, to cause the atoms to be dislocated and generate a cloud of vapor. Following its entry into the porous carrier media, the vaporizer condenses to produce a thin film. When taking into account the ion-plating process, the source of vapor and the deposition technique are dictated by several criteria such as the film composition, the rate of deposition, and special process requirements. Coatings with expected qualities will be more consistent and repeatable if a vapor source is stabilized and characterized.

(3) **Plasma Gas**

There are several important components to the plasma gas-based ion-plating process. It makes it easier for ions to be produced, which are subsequently used in the procedure. Usually, this is accomplished by creating positive or negative ions by the disintegration of atomic or molecular neutral gases. In addition, the plasma gas is essential to maintaining the required pressure inside the chamber, which is what allows ionization, film deposition, and other processes to proceed as best they can. In addition to the denser gradients, gas plasma concentrations can also indirectly alter the energy state and trajectory of the ions, which eventually produces a different film in terms of density, adhesion, and structures [43].

Particularly, argon, nitrogen, and hydrogen are examples of plasma gases that are typically chosen for ion plating due to their ionization potential and biocompatibility with the target material and deposition process. Correct ionization and consistent ion flux are achieved by careful selection and management of the plasma gas, resulting in high-quality coating.

(4) **Ion Gun**

For ion acceleration and emission, an ion gun found in ion plating is a crucial component of the procedure. This kind of study is often carried out with an ions source that ionizes a portion of the gas atoms, and an extraction electrode that gives the ions the kinetic energy they need to move toward the substrate. The gas flow rate, ion energy, and beam divergence curve customizable characteristics of the ion cannon allow for fine control over ion flux and deposition.

2.4.4 Advantages and Disadvantages

- One of the major drawbacks of ion plating is that it is very difficult to control the equipment and understand the whole apparatus, as it involves all the techniques which are mentioned above in detail, it also involves the thermal evaporation process, the sputtering process, and the plasma process and the new ion gun all of them can be used here to create the vapors for the deposition but the understanding of the apparatus is a difficult task.
- It gives a low deposition rate; as compared with all others, this technique gives a very low deposition rate as gas is not uniform and vapors can go anywhere and can even stick with the walls of the vacuum chamber.
- The optimization is very difficult, as of complex setup skilled engineers are required who can use such equipment and these are more expensive than all other setups because this is used for increasing properties.
- Using ion platting only a few materials can be deposited or can be used for deposition because in this process most materials cannot be used and this also does not involve using multiple materials simultaneously, that's why it is the biggest limitation.
- As due to high energy some of the ions have high energy which may damage the substrate and produce thin film is not quite good. These ions can only damage the substrate if it is sensitive.
- The temperature can also be varied during this process, as the continuous bombardment of ions causes' temperature disturbance and the substrate may heat up leading to the distortion and roughness of the film.
- It is also not environment friendly, because during vapor formation it generates toxic gases that can be harmful to the environment and also generates a lot of waste because of vacuum pumps that are continuously used in ion platting.
- The initial cost is very high, while the maintenance rate is also very expensive.
- Achieving uniformity in ion platting is very difficult, it is not available in normal ion platting, while in plasma-based, the plasma gas also reacts generating more ions, and uniformity is still not achieved as uniformity is checked through an ion gauge, but ion gauge shows continuous disturbance in the chamber due to formation of vapors; also during vacuum-based ion platting, the ion gun further disturbs the chamber, as it reacts with the vapors produced through the thermal evaporation process, so uniformity not achieves leading to uncovered surface area, the larger surface area of the substrate is not covered through this way, which is the major drawback.

Following are some important advantages:

- One of the major advantages of ion plating is that it is used to enhance the properties, including adhesion property, protective coating, prevention from corrosion, resistance from heat and water, increasing electrical conductivity, and enhancing life span.

- The pores generated in the thin film can easily be covered through this process, and it reduces the porosity making the thin film smoother and finer in quality.
- After initial cost, it can be used for a longer time and it does not get damaged easily; it has a larger life span which makes it more cost-effective.
- Through ionized vapors, the deposition rate is also very higher than other techniques and consumes less time, making it time effective and management.
- Here, thermal distortion or damage is very low, because this process occurs at very low thermal temperature for achieving high-quality films with enhanced properties.
- To achieve desired properties, the ions are bombarded at a specific speed, so the ion's speed and their ejection can be controlled easily.
- The surface can also be modified along with dimensions by coating with other materials and bombarding in a specific manner may change its dimensions leading to alteration in the surface of the material.
- Here, any type of substrate can be used, including metals, non-metals, semiconductors, supermaterials, Nanomaterials, conductors, insulators, polymers, ceramics, or superconductors. So, it is also versatile leading to great properties and enhanced abilities.
- It generates less waste as compared to previous techniques and makes it environmentally friendly.
- The gas composition and ion energy can easily be controlled leading to precise coating on the surface and desired properties are achieved.
- The ion gauge helps in detecting the ion bombardment and specific composition of gas in the chamber, which is important for fine and precise thin film.

2.4.5 Applications

This technique is used in a lot of applications, some of which are given below.

2.4.5.1 Nanomaterials Modification

This technique is used to modify the nanomaterials by enhancing their properties. Any type of nanomaterial can be used in this process to enhance its characteristics. Mostly nanomaterials are used for photocatalysis or electrolysis. In photocatalysis achieving a higher degradation rate is difficult. So, this can be achieved by enhancing the bandgap of the material, increasing its optical absorbance and emission characteristics, and decreasing the charge recombination rate, all can be achieved through this technique by coating it with specific ions [44]. We can also generate enhanced antimicrobial activity of the materials through this process. The advanced Tin can also be used to enhance its strength through ion bombardment. For ion platting, specific gas atmosphere is required such as nitrogen or argon depending upon the temperature and stoichiometric properties. Higher deposition may be achieved through this process as the temperature of most gases is low and it mostly works at low temperatures leading to less distortion and fine film production [45].

Similarly, it can also be used for enhancing the qualities of electrodes in electrolysis. The surface can be modified by increasing the defects that help in trapping more oxygen species and the phase can also be changed through this method. The morphology and crystallinity can also be changed with the help of the ion-plating technique.

2.4.5.2 Decorative Coatings

Ion plating is used to coat the decoration pieces and artwork. The gold water is also platted through ion platting making it corrosion-resistant. The durability and look also changed through ion platting. Gold platting or any other platting can be achieved through this process. The ions are bombarded to enhance the life span and look of the decorative pieces.

The ion plating can also be used for mirrors and lenses. To enhance their focusing ability, reduce reflection and enhance transmission rate. This can also be used to dye tools and other devices such as machines, stamps, and molding to reduce friction and enhance durability [46].

The heat and water resistance, improved electrical conductivity, corrosion resistance, enhanced optical properties, high hardness, higher strength, and improved life span all can be achieved through the ion-plating technique.

2.5 Conclusion

The physical vapor deposition process is used to deposit layers on the different samples to produce a thin film. This process is further divided into different categories depending upon their advantages and disadvantages. Mostly, the principle is the same as that of the evaporation process, in which the condensed phase changes to the vapor phase and then back to the condensed phase. The evaporation process is not efficient for metals with high melting points, that's why sputtering was introduced in which it is further categorized as RF and AC sputtering. Here, magnets are used to eradicate the limitations of the evaporation process. To get enhanced film deposition further pulsed laser ablation is used. This process is efficient in generating thin films with different patterns due to its high-intensity laser beam which is monochromatic. The ion platting also uses the same principle to deposit the layers, and it is further divided into different types to enhance the deposition process. All the types of PVD have their limitations and advantages. They are used to modify the surface, and also in future, they can be integrated with AI technology to enhance the film growth and deposition process to produce a fine thin film.

References

1. Colligon, J.S. 1999. Physical vapor deposition. In *Pergamon Materials Series*, 225–253. Elsevier.
2. Makhlouf, A. 2011. Current and advanced coating technologies for industrial applications. In *Nanocoatings and Ultra-Thin Films*, 3–23. Elsevier.
3. Moshfegh, A. 2003. PVD growth method: Physics and technology. *Physics and Technology of Thin Films: IWTF* 2004: 28–53.
4. Rigsbee, J. 1989. Physical vapor deposition. In *Surface Modification Engineering*, 231-255. Boca Raton, CRC Press.
5. Rossnagel, S. 2003. Thin film deposition with physical vapor deposition and related technologies. *Journal of Vacuum Science & Technology A: Vacuum, Surfaces, and Films* 21 (5): S74–S87.
6. Popov, Y.O. 2005. Evaporative deposition patterns: Spatial dimensions of the deposit. *Physical review E* 71 (3): 036313.
7. Deshpandey, C., and R. Bunshah. 1991. *Evaporation Processes*. Boston: Academic.
8. Shah, S.I., et al. 2010. Evaporation: Processes, bulk microstructures, and mechanical properties. In Handbook of Deposition Technologies for Films and Coatings, 135–252.
9. Westwood, W. 1989. Physical vapor deposition. In *Microelectronic Materials and Processes*, 133–201. Springer.
10. Safarian, J., and T.A. Engh. 2013. Vacuum evaporation of pure metals. *Metallurgical and Materials Transactions A* 44 (2): 747–753.
11. Weston, G. 1985. Pumping systems. *Vacuum* 35 (10–11): 493–497.
12. Couturier, G., et al. 2009. Quartz crystal microbalance and evaporation of sessile droplets. *Journal of Applied Physics* 106 (5).
13. Jamkhande, P.G., et al. 2019. Metal nanoparticles synthesis: An overview on methods of preparation, advantages and disadvantages, and applications. *Journal of drug delivery science and technology* 53: 101174.
14. Jilani, A., M.S. Abdel-Wahab, and A.H. Hammad. 2017. Advance deposition techniques for thin film and coating. *Modern Technologies for Creating the Thin-Film Systems and Coatings* 2 (3): 137–149.
15. Zang, D., et al. 2019. Evaporation of a droplet: From physics to applications. *Physics Reports* 804: 1–56.
16. Du, P., et al. 2022. Thermal evaporation for halide perovskite optoelectronics: Fundamentals, progress, and outlook. *Advanced Optical Materials* 10 (4): 2101770.
17. Paton, B., and B. Movchan. 1978. Production of protective coatings by electron beam evaporation. *Thin Solid Films* 54 (1): 1–8.
18. Li, K., et al. 2021. Rapid thermal evaporation for cadmium selenide thin-film solar cells. *Frontiers of Optoelectronics* 14 (4): 482–490.
19. Maddocks, F., and R. Thun. 1962. Properties of evaporated film capacitors. *Journal of The Electrochemical Society* 109 (2): 99.
20. Dey, M., et al. 2019. *Deposition of CdS thin film by thermal evaporation*. In *2019 International Conference on Electrical, Computer and Communication Engineering (ECCE*. IEEE.
21. Depla, D., S. Mahieu, and J. Greene. 2010. Sputter deposition processes. In *Handbook of Deposition Technologies for Films and Coatings*, 253–296. Elsevier.
22. Depla, D., and S. Mahieu. 2008. *Reactve Sputter Deposition*, vol. 109. Berlin: Springer.
23. Safi, I. 2000. Recent aspects concerning DC reactive magnetron sputtering of thin films: A review. *Surface and Coatings Technology* 127 (2–3): 203–218.
24. Kiani, A. 2014. *Analysis of metal oxide thin film transistors with high-k dielectrics and source/drain contact metals*. University of Cambridge.
25. Alfonso, E., J. Olaya, and G. Cubillos. 2012. Thin film growth through sputtering technique and its applications. *Crystallization-Science and technology* 23: 11–12.
26. Stuart, R.V. 2012. *Vacuum Technology, Thin Films, and Sputtering: An Introduction*. New York: Academic.

27. Sigmund, P. 1993. Introduction to sputtering. *Mat. Fys. Med. Dan. Vid. Selsk* 43: 7–26.
28. Rossnagel, S. 2001. Sputtering and sputter deposition. In *Handbook of Thin Film Deposition Processes and Techniques*, 319–348. Elsevier.
29. Bouazza, A. 2022. Sputtering of semiconductors, conductors, and dielectrics for the realization of electronics components thin-films. *International Journal of Thin Film Science and Technology* 11 (2): 225–232.
30. Ashfold, M.N., et al. 2004. Pulsed laser ablation and deposition of thin films. *Chemical Society Reviews* 33 (1): 23–31.
31. Shirk, M., and P. Molian. 1998. A review of ultrashort pulsed laser ablation of materials. *Journal of Laser applications* 10 (1): 18–28.
32. Vogel, A., and V. Venugopalan. 2003. Mechanisms of pulsed laser ablation of biological tissues. *Chemical Reviews* 103 (2): 577–644.
33. Lunney, J.G., and R. Jordan. 1998. Pulsed laser ablation of metals. *Applied Surface Science* 127: 941–946.
34. Stafe, M., A. Marcu, and N.N. Puscas. 2014. *Pulsed Laser Ablation of Solids*, vol. 10, 978-3. Berlin: Springer.
35. Hablanian, M. 1984. Comments on the history of vacuum pumps. *Journal of Vacuum Science & Technology A: Vacuum, Surfaces, and Films* 2 (2): 118–125.
36. Panzner, M., et al. 1998. Experimental investigation of the laser ablation process on wood surfaces. *Applied Surface Science* 127: 787–792.
37. Woodhead, J.D., et al. 2007. Isotopic and elemental imaging of geological materials by laser ablation inductively coupled plasma-mass spectrometry. *Geostandards and Geoanalytical Research* 31 (4): 331–343.
38. Mattox, D.M. 2000. Ion plating—past, present and future. *Surface and Coatings Technology* 133: 517–521.
39. Mattox, D.M. 1973. Fundamentals of ion plating. *Journal of Vacuum Science and Technology* 10 (1): 47–52.
40. Aisenberg, S., and R. Chabot. 1973. Physics of ion plating and ion beam deposition. *Journal of Vacuum Science and Technology* 10 (1): 104–107.
41. Kajdas, C., E. Wilusz, and S. Harvey. 1990. *Encyclopedia of Tribology*, vol. 15. Amsterdam: Elsevier.
42. Spalvins, T. 1980. Survey of ion plating sources. *Journal of Vacuum Science and Technology* 17 (1): 315–321.
43. Streiff, R. 1993. Protection of materials by advanced high temperature coatings. *Le Journal de Physique* **IV 3** (C9), C9-17–C9-41.
44. Barhoum, A., and A.S.H. Makhlouf. 2018. *Emerging Applications of Nanoparticles and Architecture Nanostructures*. Amsterdam: Elsevier.
45. Zega, B., M. Kornmann, and J. Amiguet. 1977. Hard decorative TiN coatings by ion plating. *Thin Solid Films* 45 (3): 577–582.
46. Chen, N.-H., et al. 2013. Antimicrobial and decorative ion-plated copper-containing ceramic coatings. *Surface and Coatings Technology* 236: 29–35.

Chapter 3
Chemical Vapor Deposition Technique

Abstract In this chapter, the most efficient and widely used deposition technique Chemical Vapor Deposition (CVD) has been explained, and also its some types such as Atmospheric Pressure Chemical Vapor Deposition (APCVD), Low-Pressure Chemical Vapor Deposition (LPCVD), and Ultra-high Pressure Chemical Vapor Deposition. Each technique has been explained according to its basics, introduction, chemical process, Figures, Explanation of the figures, advantages and disadvantages as well as their advanced applications in a wide range of fields. Each CVD method is different from one another as atmospheric pressure CVD requires a moderate range of pressure to perform the deposition process, while the Low-Pressure CVD process cannot operate to deposit film on atmospheric pressure, it must require a low-pressure range and high temperature to perform deposition process. Ultra-high pressure requires a high range of pressure and temperature to deposit an identical layer of material on the substrate. Every technique has different properties and variation in the thickness of the thin film deposition layer. Overall, CVD is widely used due to the diverse range of properties and applications according to the properties.

3.1 Chemical Vapor Deposition

One of the most effective bottom-up methods for generating thin films and nanoparticles is chemical vapor deposition (CVD). This technique generates constant solid yields by initiating one or more vaporous adsorption classes to respond to or disrupt an impassioned surface. However, additional complex than physical vapor deposition, chemical vapor deposition has benefits such as generating uncontaminated thin films or nanoparticles and the simplicity of ascending. A specific category of procedures known as chemical vapor deposition (CVD) is the method by which a firm material is deposited from a vapor via a chemical reaction that happens on or near a substrate surface that is typically heated. One single crystal, thin layer, or powder is the subsequent solid substance [1]. Materials with extensive physical, tribological, and chemical features may be technologically advanced by modifying the investigational circumstances, such as substrate material, substrate temperature,

T. I. Awan et al., *Thin Film Deposition Techniques*,
https://doi.org/10.1007/978-981-96-1364-9_3

reaction gas mixture composition, total pressure gas flows, etc. [2]. The remarkable throwing power of the CVD process is one of its distinguishing features; it allows for the fabrication of coatings with consistent thickness and low porosity on surfaces with complex shapes. Another crucial aspect is the capacity to perform selective or localized deposition on patterned surfaces [3].

A wide range of thin film applications, such as dielectrics, conductors, passivation layers, oxidation barriers, conductive oxides, heat-resistant coatings, tribological and corrosion-resistant coatings, and epitaxial layers for microelectronics, use CVD and related techniques [4]. High-temperature materials (tungsten, ceramics, etc.) may be prepared using CVD, and they can also be used to make solar cells, high-temperature fiber composites, and particles with precise dimensions. This method has correspondingly in recent times been used to generate carbon nanotubes and high Tc superconductors [5]. Superconductivity can be achieved without annealing in oxygen because the vapor's oxygen activity can be accurately regulated throughout the deposition process [5].

3.1.1 Introduction

Numerous thin film applications using CVD and related technologies include dielectrics, electrodes, passivation coatings, corrosion fences, conductive oxides, heat-resistant coatings and corrosion-resistant layers, and epitaxial layers for the application of microelectronics [6]. The manufacture of high-temperature tools (tungsten, ceramics, etc.) and the generation of solar cells, high-temperature fiber composites, and particles with precise sizes are further used for CVD. This method has also recently been used to generate carbon nanotubes and high-temperature-based superconductors [7]. Superconductivity may be achieved without annealing in oxygen as the vapor's oxygen activity can be accurately regulated throughout the deposition process.

A vacuum-based deposition technique known as chemical vapor deposition (CVD) is used for fabricating solid materials with exceptional performance and quality features [8]. Thin film creation is a common use of this method in electronics manufacturing. In the conventional CVD process, one or more impulsive precursors are familiarized with the substrate, and following that they respond and/or break down on the substrate exterior to procedure the intended deposit. Impulsive by-products are also often generated and eradicated by the chamber's gas flow [9].

CVD is often used in microfabrication techniques to deposit materials in various morphologies, such as unstructured, polycrystalline, monocrystalline, and epitaxial growth of materials deposition. These constituents include graphene, carbon (fiber, nanofibers, nanotubes, diamond, and nitride), silicon (dioxide, carbide, nitride, and oxynitride), fluorocarbons, strings, tungsten, titanium nitride, and diverse high-κ dielectrics [10].

John M. Blocher, Jr. originally used the word "chemical vapor deposition" (CVD) in 1960 to distinguish it from "physical vapor deposition" PVD. CVD is used in numerous altered scenarios. The procedures used in these measures to proceed with chemical reactions often vary [11].

Classified according to the operational environment:

1. Pressure in atmosphere CVD under atmospheric pressure is called APCVD.
2. Below-atmospheric pressure CVD is known as low-pressure CVD (LPCVD). Lower pressures often lessen undesirable gas-phase responses and enhance the homogeneity of the coating throughout the substrate.
3. CVD at extremely low pressure, usually less than 10^{-6} Pa (about 10^{-8} torr), is known as ultrahigh vacuum CVD (UHVCVD). In other domains, the boundary between high and ultra-high vacuum is sometimes lowered to 10^{-7} Pa [12].
4. CVD at sub-atmospheric pressures is known as sub-atmospheric CVD (SACVD). uses silicon dioxide (SiO_2) to fill high aspect ratio Si structures using tetraethyl orthosilicate (TEOS) and ozone [13].

3.1.2 Working of CVD Reactor

CVD generation of graphene is a chemical synthesis method that produces SLG (Single Layer Graphene) or FLG (Few-Layer Graphene) on any type of substrate under-regulated reaction conditions by putting the substrate to gas-phase precursors at atmospheric pressure. Because CVD is so flexible, it involves complex mixtures of both heterogeneous and homogeneous gas-phase surface processes. In general, homogeneous gas-phase reactions and subsequent identical nucleation grew fine as the fractional pressure or temperature in the reaction material rose. We are required to reduce this homogenous nucleation to develop a high-quality graphene layer [14].

The following eight steps comprise the general process for CVD-based graphene formation on catalytic metallic substrates, which results in the formation of an even and extremely crystalline graphene coating on the external surface of the materials:

(1) Mass transmission of the reactant,
(2) Film precursor response,
(3) Gas molecule dispersion,
(4) Precursor adsorption,
(5) Precursor flow into substrate,
(6) Material exterior reaction,
(7) Desorption of product, and
(8) Removal of by-product.

Figure 3.1 illustrates the working mechanism of chemical vapor deposition.

Thin films are widely used in the creation of solid-state-based devices and materials. A lot of gadgets are "built-up" from layers of sequential film, each with an exact layout carved into it. Films with thicknesses varying from a few nanometers to several microns can operate as a protective layer or modify a device's properties, or they might be an active component of a working device [13, 15]. Microelectronics (transistors, integrated circuits (ICs), superconducting devices), magnetic tools (recording

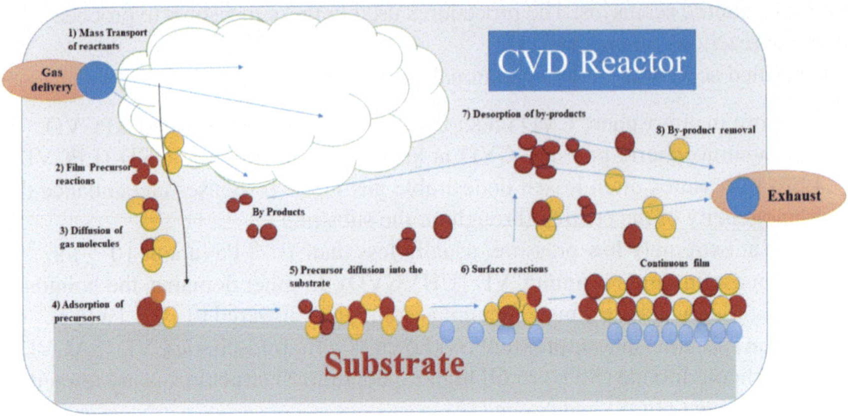

Fig. 3.1 In-side working mechanism of CVD reactor

adhesive tape), optical devices (solid state-based devices and semiconductor lasers, photosensitive fibers, xerography, glass, or window coverings), solar energy alteration (photovoltaics, photoelectrochemical resources, antireflection coatings), and attractive, passivation, and wear-resistant layers are just a few of the many uses for protecting, semiconducting, or conducting coatings [16]. Figure 3.2 describes the various types of CVD technique of thin film deposition.

The aforementioned technical capacities require film installation or formation methodologies that can consistently produce exact electrical, optical, mechanical, or chemical characteristics in a wide range of constituents, frequently at extraordinary efficiency [17]. Various techniques have been devised to achieve these goals. For instance, the most popular methods include evaporation, chemical vapor deposition, and sputtering.

Chemical vapor deposition is a versatile and highly efficient technology that has been used to pledge a wide variety of thin film constituents. Chemical vapor deposition provides consistent step coverage, appropriate deposition speeds, and superior control over film composition and structure in comparison to other film-forming techniques [2]. Chemically reacting vapors are used in this technique, which goes by the term chemical vapor deposition, to fabricate or deposit a layer of the desired material. The energy needed to break a bond is often provided by heat, however, photons or releases can furthermore be used. Chemical responses can happen either way, but homogenous nucleation should be avoided in both scenarios to reduce particle production [18].

A chemical reactor is all that a CVD system entails. Therefore, to achieve consistent film layers, reactant vapor flow rates and patterns as well as substrate temperature must be properly regulated. In turn, given the right reactions and reactor designs, a forecast of deposition rates and consistency necessitates an extensive knowledge of thermodynamics, kinetics, fluid flow, and mass transportation occurrences. Lastly, identifying nucleation processes as well as the inclusion of defects and impurities

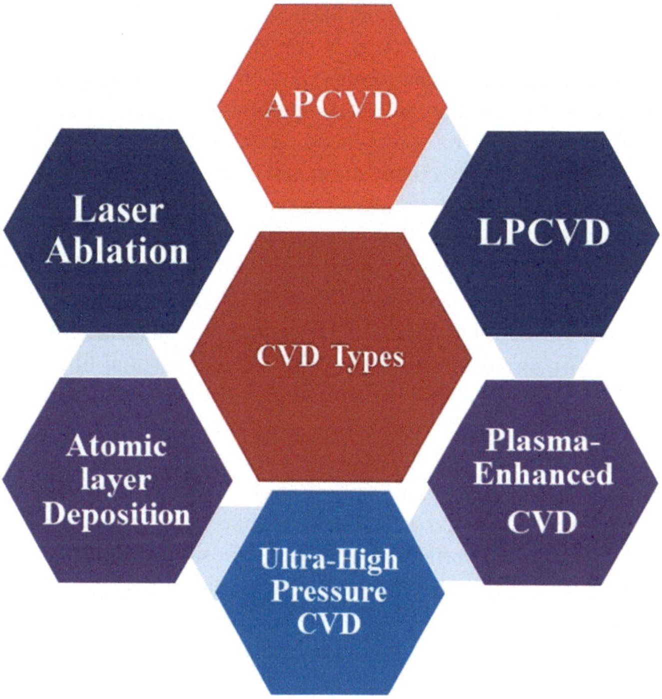

Fig. 3.2 Describe the various types of CVD

during development is necessary for controlling the structure and characteristics of films [19].

3.1.3 *Significance of CVD*

The goals of CVD are to yield thin films with qualities that are repeatable and controlled, at affordable growth rates and expenses, and with minimal influence on the outcomes of earlier processes. Film width, arrangement, cleanliness, crystalline excellence, and superficial shape are among the crucial characteristics [20]. The amount of variation allowed for each of these attributes varies depending on the application; nevertheless, the tolerances for processing electronic materials are far tighter than those usually found in the chemical process industry [21]. Molar stream rates, energy contribution, structure pressure, and substrate situation are the process factors that may be controlled for a particular deposition chemistry and reactor structural design. For various chemicals and reactor types, it would be ideal to be talented in estimating process outcomes as a function of procedure factors [22].

With the aforementioned information, the best CVD procedure for a certain application might be achieved. inevitably, it is sometimes difficult to understand how process variables and results relate to one another. Thus, the majority of the work has been focused on "letting nature" determine how film qualities relate to process variables, with minimal focus on comprehending the chemistry and reactor design. In addition, variations in impurity composition and reactor design make it challenging to compare the findings of existing investigations [23].

The optimum values for the film's properties depend on the application. The material that is deposited may be a dielectric (like SiO_2), conductor (like W), semiconductor (like Si), or amorphous (like SiO_2), polycrystalline (like poly-Si), or single crystal-like (like Si); the microstructure may be amorphous (like SiO_2), polycrystalline (like poly-Si), or single crystalline (like Si); the essential film width may be dense (like > 10 μιη for dense film regulators), thin (like epitaxial coatings ~ 1 μιη), or atomically sized (like superlattices); the sought after superficial morphology may be atomically even (like the first epitaxial film) or macroscopically irregular (metallization), the quantified transparency smooth may be negligible (like ppb (parts per billion) for microwave devices) or substantial (like numerous atomic percentage Cu in Al metallization). Nevertheless, epitaxial deposition of semiconductor films frequently gives rise to severe specifications for the characteristics of CVD electrical constituents [24].

In general, uniformity of thickness is necessary to ensure that parameters that depend on thickness (such as capacitance) are satisfied and that pattern transfer operations may be appropriately regulated. In multi-substrate procedures, the required uniformity is usually $\pm$ 5% throughout the substrate and $\pm$ 5% from substrate to substrate. However, in particular applications, specific regions of the substrate require preferred material growth [25]. This is achieved by working in an area of restriction space governed by surface procedures and supplying a spatially specified seed crystal. Growing a thin coating of imported material on a substrate of the material to be placed is one way to generate the seed (e.g., produce a film of SiO on a Si substrate). Subsequently, "windows" of exposed substrate are formed by lithographically removing the foreign film selectively [26]. The film can be selectively placed in the uncovered portions because heteroepitaxial development requires greater amounts of supersaturation than homoepitaxial growth to achieve nucleation. An overgrowth structure may arise if growth is allowed to continue up until the film width goes beyond the gap elevation and the exposed crystallographic plane normal to the substrate is a slow-growth plane in comparison to an adjacent plane [27]. Designated as graphoepitaxy, this technique may be applied, for instance, to the growth of epitaxial Si on SiO_2. Controlling the apparent temperature of the substrate surface or coordination spatially and subsequently, the response rate is more approach to attaining selective development [28].

For instance, lithographic techniques may be used to generate patterns in single-crystal semiconductors, which can then be selectively filled eliminating the requirement for a mask. The reactants need to be moved to the hard surface to be integrated into the crystal lattice during crystal formation. Moreover, the amount of transference or the level of a surface procedure limits the thickness or its time-derivative,

Table 3.1 Difference between the chemical vapor deposition (CVD) and physical vapor deposition (PVD) processes

Chemical vapor deposition (CVD)	Physical vapor deposition (PVD)
Require high temperature (> 1000 °C)	Require low temperature (< 500 °C)
Film thickness range (1–10 μ m)	Film thickness range (0.1–2 μ m)
Excellent properties	Good properties
Hard, corrosion and wear resistant	Hard, wear-resistant, and low friction
Applications such as cutting instruments, wear tools, and medical implants	Applications such as Cutting instruments wear parts, and jewelery optics

growth rate. Thus, the composition of the gas stage, pressure of the system, temperature range, and substrate orientation may all affect the growth rate. The commercially employed chloride technique for GaAs deposition, which is utilized to produce microwave devices, is a hypothetical example [29].

Table 3.1 illustrates the general difference between the Chemical Vapor Deposition (CVD) and Physical Vapor Deposition (PVD) processes.

3.1.4 Atmospheric Pressure CVD (APCVD)

One type of CVD process is atmospheric pressure chemical vapor deposition. Three main parts typically contribute to the basic CVD system:

(1) Precursor supply
(2) The CVD reactor
(3) Combustion regulation.

One CVD technique used at ambient pressure is atmospheric pressure chemical gas confession. It is employed in the confession of numerous oxides. The reaction chamber of an atmospheric pressure chemical vapor deposition (APCVD) system is pressurized to atmospheric pressure, or 1 atm [30]. Because APCVD may be used with vacuum-free, continuous in-line processes, it is a particularly appealing option for high-volume, cost-effective industrial applications such as the production of photovoltaic cells. Hard metallurgical coatings such as epoxy and compound films are put in cold barrier apparatuses using the high-temperature APCVD method [31].

Reactor Types

1. **Cold Barrier Reactor**

Only the substrata of a "cold wall reactor" get heat from a graphite susceptor. The source of heat is RF induction. Since just the substrates and susceptor are heated, this reduces the amount of deposition on the ramparts of the apparatus. High temperatures within the substratum might affect gas flow and result in non-uniform depositions (microstructure and thickness). Gas flow dynamics in these systems hence require greater caution and attention [32].

2. **Hot Wall Reactor**

The substrate and reactor wall are at similar temperatures in "hot wall reactors." Additionally, there is deposition on the walls, which may result in the breakage of grains and fragments onto the substrate. This particular type of reactor is also frequently used for homogeneous vapor phase processes. On the other hand, these reactor types are easy to set up and operate well for batch processing. SiO_2 and BPSG glasses are two examples of the various insulating film layers that may be deposited using the low-temperature APCVD method. Because of the comparatively low temperature, the deposited oxide has a low density and only moderate coverage [33].

3.1.5 Working Principle and Construction

Traditionally, APCVD, LPCVD, or PECVD are used to generate graphene on metallic stops (copper, Ni, Pt, etc.). The development and quality of graphene are significantly influenced by the temperatures, compressions, and attentions of forerunner airs within the heater. To achieve the intended growth effects, these variables must be adjusted. Generally, optical microscope (OM) then scanning electron microscopy (SEM) are used to study external morphology; Raman spectra, UV–visible spectroscopy, then transmission electron microscopy (TEM) are used to examine the quality of the graphene; and the four-point investigate technique is used to measure the page resistance (Rs) of manufactured graphene [34].

In comparison to LPCVD or PECVD, nearly all graphene films generated with APCVD are monolayers of better quality under atmospheric conditions. APCVD synthesis research would play a major part in the desired growth of graphene and similar materials.

Most experimental investigations have generally observed that non-uniform thick graphene layers are produced using LPCVD. APCVD would be an improved choice for high-quality single-layer graphene development, nevertheless, because of the catalyst's decreased reactive species' mean free path and diffusion coefficient. When comparing APCVD to LPCVD, the diffusivity coefficient is substantially less with $Dg \sim 1/(\text{total pressure})$. As a significance, likened to LPCVD, the APCVD growth method outcomes in fewer multilayer graphene islands developing on an entire significant monolayer graphene island. Figure 3.2 illustrates the transport and reaction processes involved in the CVD growth mechanisms (APCVD and LPCVD) of graphene [35].

In another situation of the development mechanism of graphene, hydrocarbon molecules are absorbed and dispersed on Cu, which mains to the creation of active carbon classes by a dehydrogenation response. To generate graphene nucleation seeds, these species spread on both edges of copper foil then agglomerate on its lively positions. Most CVD techniques need the introduction of H_2 gas as an essential part of graphene formation [36]. The general mechanisms of Cu-graphene growth are explained. The three primary determined phases are.

(i) Nucleate-grow,
(ii) Diffuse-desorb,
(iii) Adsorb-decompose.

Typically, thermodynamically stable species quickly agglomerate with active carbon species to generate graphene nucleation germs by the following responses:

$$(\mathrm{CH}x)\mathrm{s} + \text{graphene} \rightarrow (\text{graphene-C}) + (\mathrm{CH}x)\mathrm{s}, \quad \text{where } x = 1, 2, 3$$

until those seeds formed [37]. The working mechanism of the APCVD reactor is shown is Fig. 3.3. The choice of the precursors that are deposited APCVD is volatile which means only liquid or gaseous precursors that can be deposited must be vaporized before entering the mixing chamber. By controlling the rate of flow of precursors and their related concentrations, the required precursors must be mixed with some carrier gas, and then the reaction mixture is transported to the substrate surface, where some chemical reaction happens and bubbles are formed from the mixing chamber which is sometimes beneficial and as a last step of deposition, by-products are removed out from the exhaust.

In general, catalytic stimulation of chemical responses of forerunners at the development substratum external/boundary in an appropriately optimized atmosphere is required for APCVD formation of 2D resources (e.g., graphene). systematically, the primary elements influencing the ultimate quality of the formed 2D materials are precursors, circumstances (such as rapid growth rates, huge area sizes, or extremely high crystal-like excellence), environment, substrata, and substances. Although a lot of investigation has been conducted thus far to prepare high crystalline 2D resources

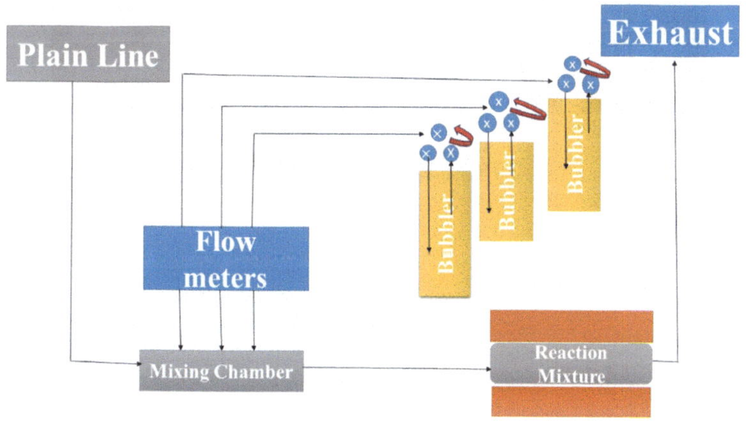

Fig. 3.3 Working mechanism of APCVD reactor

(like graphene), there are still a lot of obstacles to be resolved. For instance, it is very challenging to develop homogeneous, tall-excellence graphene since of the rough surface of the catalytic metal. Original forerunners (e.g., hard precursors such as gas precursor, or solid precursor combined through) are also of attention to the 2D solid study public [38].

The impact of growing degree on the catalytic metallic external on graphene excellence is an additional question. At this point, it is hard to identify a precise response because research is moving at a very exponential speed [39].

Neither theoretically nor experimentally, the general process of the APCVD development of graphene has recently been clarified to a sufficient degree. To produce large-scale, high-quality graphene, having a wide range of properties and applications in various fields of interest, the 2D material research community will thus be very interested in comprehending the graphene development mechanism and the impact of diverse growth circumstances [40].

3.1.6 Advantages and Disadvantages

Following are some advantages of the APCVD technique, Figs. 3.4 and 3.5 generally describe the advantages and disadvantages of the Atmospherics Pressure Chemical Vapor Deposition Technique, respectively.

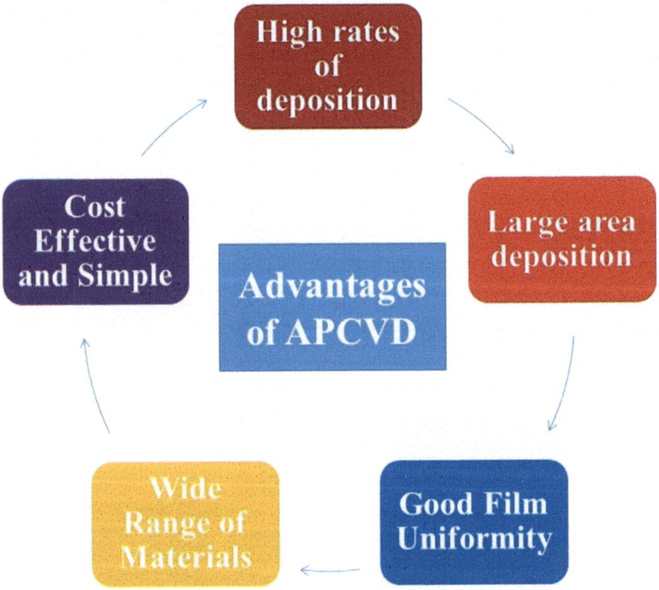

Fig. 3.4 A few advantages of APCVD

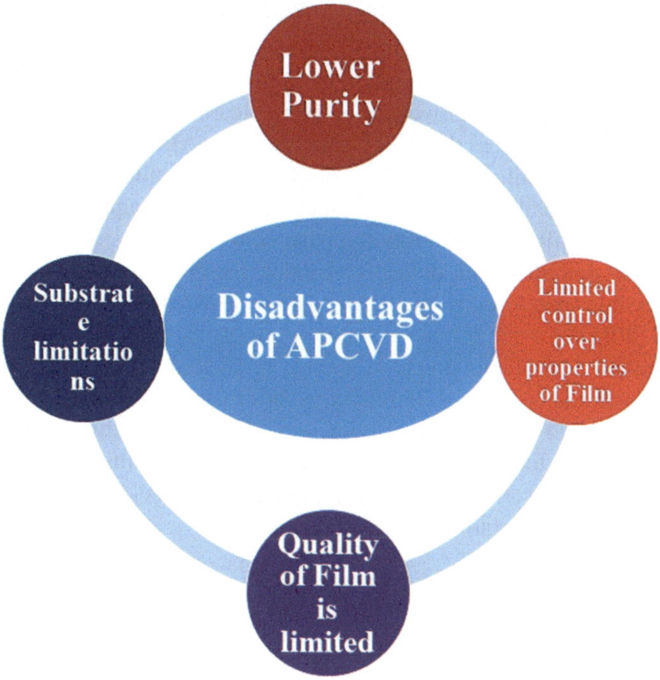

Fig. 3.5 Certain limitations of APCVD

3.1.7 Applications

One common CVD method is atmospheric pressure chemical vapor deposition (APCVD), which is used to deposit coatings fabricated from gaseous precursors onto an appropriate substrate. Additionally, the reaction chamber pressure in the APCVD process is near atmospheric heaviness, preferably about single unit of atmospheric heaviness. Forerunners must be instable fluids or low-melting objects that can be transported to the response site by a transporter vapor. Two different kinds of reactions might manifest inside the reaction chamber: mixed responses that take apartment at the vapor-hard external contact and homogeneous reactions that only happen in the vapor phase [41].

Precursors having single or dual sources of origin can be employed in a range of CVD methods. Precursors from a single source include every group or component needed to produce subsequent coatings. Every precursor that is utilized needs to be compatible with the APCVD technique, meaning that it needs to be volatile and react appropriately to produce the desired material, preferably producing only gaseous byproducts. A solo basis forerunner must generally be humble, instable, then comprise every atom needed for the synthesis of the substance [42].

By reducing wounded in replication and engagement at the see-through electrode/lively coating boundary and increasing preoccupation competence in the lively coating achievable through the usage of TCOs through unique visual and electric goods the competence of photovoltaic change in carbon-based solar cells can be improved. SnC12 was used as the initial substrate in the APCVD process to generate container oxide SnO_2 slim films. Using scanning electron microscopy, the external geomorphology of the films placed on crystal substrata was examined. The Four-Fact investigation was utilized to quantity the page confrontation, and the ellipsometry was cast-off to determine the deflective catalog for the films put at 480 °C. By using UV–visible spectroscopy, the transmission of SnO_2 films formed on ITO was determined [40].

SnO_2 films with an 11-min preparation time, 19.57 Ωcm^{-2} sheet resistance, more than 80% transmittance, and a 1.75 refractive index can be utilized as an interfacial layer in organic solar cells to reduce reflection. Using these films results in SnO_2/P3HT: PCBM with a total reflectance of less than 3%. SnO_2 films can also be interfacial coatings in applications involving upturned solar cells [43]. APCVD also has several applications in certain fields of interest such as,

- In the fields of Glass Coatings
- Microelectronics Industry
- Coatings for Protection
- Solar Cells Fielda.

3.2 Low Pressure CVD (LPCVD)

Below-atmospheric pressure CVD is known as low-pressure CVD (LPCVD). Lower pressures often eliminate undesirable vapor-stage responses and enhance the homogeneity of the coating throughout the cracker. The most common depositing substrates are polysilicon and dielectric materials. Compared to other atmospheric CVD methods, the technology has simpler control over film composition, fewer particle contamination, better film conformation, and higher wafer processing capacity. Particularly effective as a disinfection and insulating material is silicon nitride. The majority of research on the LPCVD nitride process consequently has been experimental. These investigations have described the effects of reactor operating settings on the electrical characteristics, stress, covering-up ability, and nitride film stoichiometry. There are currently no supporting experiments for theoretical methods to understand the significance of chemistry and dynamics in the nitride process [44].

A method to generate polysilicon slim films is little-heaviness chemical vapor deposition, which uses silane (SiH%) thermal decomposition. The microstructure of polysilicon, which is based on the deposition circumstances, has a important influence on its powered characteristics. At the lowest growth temperatures (below C570 °C), the films are formless. At intermediate growing temperatures (between C570 °C and C610 °C), they exhibit fine (C0.1 μm diameter) ellipsoid-formed ounces. At higher growing temperatures, the films cover columniform 110′-coated grains through a slim, fine-grained nucleation coating at the substratum boundary. The uniform nucleation then development of Si crystallites from an amorphous silicon film as it is deposited forms the fine-grained microstructure that is randomly oriented at intermediate deposition temperatures. Stated differently, the statement frequency is somewhat greater than the crystallization frequency under this growth structure [45].

3.2.1 Working Principle and Construction

Due to the simplicity of growing large-area films, the chemical vapor deposition (CVD) generation of graphene has attracted a lot of attention in recent times. However, large-domain or large-grain-size single-crystal growth is yet to be investigated. Nevertheless, the domain size was only a few tens of micrometers, and the surface nucleation and a two-dimensional growth process are the major mechanisms for graphene development on Cu by CVD. It has been shown that domain borders negatively affect transport characteristics. Although the exact mechanism of degradation is yet unknown, structural flaws are known to encourage surface interactions with deposited dielectrics or adsorbates from the surrounding environment [46].

Reactors with quartz tubes with cold or hot walls can be employed for the LPCVD process. High efficiency is possible with sequential processing in hot-walled furnaces. Additionally, they produce homogenous films by ensuring strong thermal homogeneity. Hot wall systems have the drawback of depositing material on the furnace walls as well, prompting more powerful maintenance in the form of cleaning or tube replacement in the long run to prevent cracking of the put solid and succeeding element pollution. Reactors with unemotional walls require less maintenance since their walls are not formed by film deposition [47].

The pipe is expatriate to little compressions during LPCVD, which can be anywhere between 10 mTorr and 1 Torr. The pipe is heated to deposition temperature the temperature at which the precursor gas splits down after it has been placed under vacuum. The method and the reactive gases being employed can determine the temperature range between 425 and 900 °C. The solid phase material is generated by the gas that is pumped into the pipe, anywhere it disperses then responds through the substrate's external. With that, any extra gas flows out of the pipe then passes complete a reduction scheme. When comparing films produced via PECVD and PVD processes, LPCVD films usually show better step coverage, are more uniform, and have fewer flaws [48].

Materials

The materials deposited by LPCVD procedures are:

1. Oxide at low temperature (LTO)
2. Oxide at high temperature (HTO)

3. Nitride
4. Oxynitride
5. Minimal tension nanowire
6. Polysilicon
7. Annealed silicon polymer
8. Silicon amorphous
9. N-category doped Si polymer
10. P-category doped Si polymer
11. Crystal made of phosphorus (PSG)
12. TEOS.

3.2.2 *Advantages and Disadvantages*

- **Figures of Merit**

One can adjust the following characteristics of LPCVD films. These characteristics and process temperature frequently have to be negotiated off. greater temperatures utilized in thermal oxidation, or HTO, typically result in films with better quality—that is, films with greater breakdown voltages, denser, and more uniformity. In addition, films with high process temperatures typically have lower pinhole density, more standard indices of refraction, and more predictable stoichiometry. Deposition rates will typically rise for lower temperature operations; stoichiometry and directory of bending will be additional flexible; and pinhole density may rise [49].

- **Index of Refractiveness**

An visual feature of the film that too provides info on the compactness, insulator continuous, and stoichiometry of the film is its refractive index. Ellipsometry can be utilized to quantify energy. For instance, the value of n for silicon nitride ranges from 1.8 to 2.2, where 2.0 denotes high-grade, stoichiometric silicon nitride. A Si-rich film is indicated by $n > 2.0$, while an profusion of oxygen is often indicated by $n < 2.0$. Similar variations exist for silicon oxide's refractive index, which ranges from about 1.44 to 1.47 based on the method of deposition, density, and film quality [49].

- **Rate of Wet Etch**

Etch rate measurement provides details on the quality of the film. The quantity and density of SiO_2 in the film have an impact on the etch rate [1]. The concentration and quantity of SiO_2 increase with decreasing etch rate. An elevated dielectric constant is also correlated with a slower HF etch rate. The etch rate of thermal oxide, which is the highest grade oxide, is 230 Å/min in a 10:1 HF: H_2O solution [50].

- **Rate of Deposition**

There are several ways to estimate the deposition rate at the substrate, which is commonly represented in Å/sec. It is determined using the sputter tools' deposition time and measured in real-time in the evaporators.

- **Consistency**

The thickness variation across a substrate is measured by uniformity, which is typically given as a percentage.

(Depth Max − Depth Min)/Depth Average is the usual definition of homogeneity.

- **Spatial Stress**

The force that the film applies to both the substrate and itself is measured as stress. Stress is commonly restrained in Megapascals (MPa), where confident stress is denoted as "tensile" then undesirable stress as "compressive." Pressure in slim films can negatively impact adhesion and other properties, in addition to devices then substrates.

- **Analysis of Steps**

The quantity of coating on a feature's sidewall or bottom compared to its top or field portions is measured as step coverage. The geometry of the topographies then the category of statement being cast-off are major determinants. HTO and TEOS offer the best sidewall coverage, respectively. LTO won't provide much to the sidewall coverage [51].

3.2.3 Applications

LPCVD can be used to deposit silicon dioxide, silicon nitride, silicon oxynitride, and polysilicon. Most LPCVD coatings can provide sidewall protection for structures that need electrical isolation since they are relatively convex. The temperature, form of LPCVD statement, and, in the circumstance of a tunnel eye, the feature's exact feature relation all affect how much border coverage is applied. In essence, improved conformity results from higher process temperatures are obtained. Even though the films are fairly linear, the keyhole's tool emergence during ditch filling can still be a problem since top and sidewall deposition rates can differ. It is possible to deposit polysilicon undoped, P or N-fixed, and P+ or N-fixed after statement, allowing you to adjust the resistance to your particular requirements. Polysilicon is frequently used in wire connections for MEM and IC devices, such as neurological (brain) sensors [52]. Parasitic layers are found in MEM devices such as pressure sensors. It is also frequently utilized in surface micromachine devices as a structural layer. Additionally, radiation detectors can take advantage of it. To adjust the stress, the polysilicon can be annealed after it has been deposited. Depending on the required material qualities, Si nitrogen compound can be put in either stoichiometric (Si_3N_4) or little-pressure (Si-rich) procedure. It is possible to generate skins that are resilient to HF engraving using low-stress nitride. In MEMS devices, brain probes, and integrated circuits, stoichiometric Si nitrogen compound is employed as an soundproofing, insulator, and chemical and/or water wall. Si oxide can be put in three dissimilar customs, and individually technique has dissimilar goods. High-temperature, is put at about 900 °C and is rather conformal, creation it appropriate for sidewall covering and approximately ditch refill requests providing the feature relation is not too plain. HTO is the maximum excellence LPCVD oxide creation it appropriate for requests anywhere a high quality insulator is obligatory. The finest quality oxide is a thermally full-grown film somewhat than LPCVD [53].

There are three distinct methods for depositing silicon oxide, and each has unique characteristics. Providing the feature relation is not excessive, high-temperature oxide (HTO), which is put at about 900 °C and is relatively linear can be used for dig replacement and sidewall coating. Since HTO is the best LPCVD oxide available, it can be used in applications where a premium dielectric is needed. Thermally generated films are superior to LPCVD films in terms of quality.

Tetraethyl orthosilicate, or TEOS, can hydrolyze into silicon dioxide to deposit oxide. The most conformal LPCVD technique produces silicon dioxide, which is ideal for tunnel replenishment, coating features with a higher aspect ratio, and passing through-wafer channels that need to be electrically isolated. The characteristics of TEOS material are comparable to those of LTO, although they can be changed to more closely resemble thermal oxide after annealing in steam. Little-temperature oxide (LTO) is the lowermost grade LPCVD oxide, with fewer conformality and fewer gap-fill capabilities, and is located at the little finish of the temperature spectrum at 400 °C [54].

- Transistors with thin films.
- Photovoltaic slim-film solar cells.
- Controllers.
- Insulators for capacitors.
- Passivation of MEMS.
- Coatings of anti-reflection.
- Replenishing trenches.

3.3 Plasma-Enhanced CVD (PECVD)

In the recent historical, there has been a lot of attention in slim-film PECVD (plasma-enhanced chemical vapor deposition). The capacity of high energy electrons in radiofrequency radiance releases, or plasms, to dissolve chemical bonds then hence enhance chemical reactions at or near room temperature, is a major source of this attraction. These kinds of things matter especially when film is being deposited onto substrates that are not meant to be exposed to high temperatures. However, the very reactive plasma environment also has the benefit of allowing materials with distinct chemical, physical, and electrical characteristics to emerge [55].

3.3.1 An Overview of PECVD's History

Although the earliest observations currently forward to the 18th period, it can be said that CVD was invented in the late 1800s to enhance bright tuber fibers by the application of carbon or Ni coatings. With uses throughout the instrument and subsequently the semiconductor industries, CVD has impacted the industry after these pioneering achievements. Parallel to this, plasma-assisted CVD methods, also known as "plasma deposition" or "glow discharge deposition," have been industrialized to lower the procedure temperature by activating the precursor through the plasma phase. These days, PECVD, or plasm-aided chemical gas confession, is the term used to describe these deposition techniques [56]. The PECVD of diamond gem from acetylene in the attendance of Hg gases was first stated in 1911. Other notable reports include the primary confessions of diamond-like carbon (DLC) in 1953 and the discovery of formless carbon-founded films on the ramparts of release apparatuses covering hydrocarbon molecules. One may now better comprehend the development apparatuses of technically significant mixtures like Si and Si oxide from silane releases thanks to the decades-long study of PECVD techniques. Significant technological advancements in the technique were made during the latter part of the 1960s and the early 1970s. On the one hand, additional process temperature lowering was made possible by the creation of an RF (radiofrequency) ambitious similar platter PECVD apparatus, and by utilizing it is quite likely to produce visual coverings [57]. On the additional pointer, Mattox designed the "ion plating" method, which is used to produce metallic and carbide coatings, by significantly reducing the manufacturing pressure and concurrently introducing a bad prejudice to the substratum. Throbbed plasm bases (DC and bipolar) were introduced into PECVD systems in the 1990s, offering even more process control and flexibility. Specifically speaking, the initial reports of hard sums from carbon-based compounds date back to the end of the 18th period and pertain to the PECVD of PPFs, or carbon-based-similar coverings. Before it was demonstrated that a 1 μm profuse coating of plasm-polymerized cinnamene could serve as an adequate dielectric for a nuclear battery, they were thought to be an inconvenience. PPFs were investigated first then for periods later. PPFs have been researched as pinhole-allowed bodily walls for decades now. This is because PPFs are deposited dry and by vacuum, which prevents the absorption of wetness, airborne, solvents, or additional liquids, which is a common occurrence in dyes and rainy-chemical coverings. These applications include corrosion protection and food packaging. Nonetheless, meanwhile the 1980s, PP has been additional advanced in the exploration for PPFs with regulated then customized interaction for high-tech applications, especially in the biomedical sector, where it is used to create platforms for the immobilization of biomolecules, antibacterial coatings, or controlled drug release [58].

3.3.2 Working Principle and Construction

Fundamental Principle

In the case of the PECVD method, vaporizing a forerunner (or monomer) molecule in a statement apparatus with an ignited plasm is the initial step in the formation of also carbon-based or inorganic slim film solid. Resources dispensation using "cold" (thermodynamically non-equilibrium) plasm can be categorized into three steps founded on the discharge conditions, employed vapors or vapors identified, and the part of the preliminary vapor-stage procedures and plasm-external connections (connecting particles and molecular remains radicals, ions, and energetic photons). The PECVD of carbon-based slim films (for example PPFs, easy resources, and others) or mineral slim films (for example DLC films, oxides, nitrides, and carbides of alloys or semiconductor device). It is helpful to clarify what we mean when we talk about "organic" and "inorganic" thin-film materials in this context [56].

(1) The PECVD community generally agrees that organic molecules consist of a primary chain made up of carbon atoms that may contain other elements like nitrogen, oxygen, or hydrogen. DLC thin films lost their organic qualities during film development due to densification and sp3 hybridization, and are now often regarded as inorganic materials.
(2) The unpackaged plasm limits typically regulate the frequency of generating energetic species (electrons, ions, and photons) and chemically lively forerunner type (molecular remains, or allowed radicals) during a deposition for PECVD synthesis of both inorganic and organic materials. The chemical arrangement of the put films is mostly determined by vapor-stage chemical reactions in combination through the chosen vapors. Additionally, substrate surface conditions and plasma-surface interactions influence the microstructure and external morphology of the films.
(3) Dry material removal by plasma etching or chemical etching produces volatile compounds from the chemical interactions between surface atoms and free radicals produced by the plasma, which are often ablated with further support from ion bombardment.
(4) Surface modification allows single to "engineer" the exterior and boundary goods (advance bond, controller irregularity, wettability, etc.) without adding or removing material in significant amounts. Instead, the surface and/or the nearsurface region's composition and structure are altered by plasma exposure [57].

- **Plasma Inductively Coupled**

Without the need for paired electrodes, the ionized gas is created in the inductively coupled plasma (ICP) deposition process by join the electromagnetic field generated by a loop inside the response compartment. Thin film deposition may be accepted out in a extensive variety of conditions using the radiofrequency (RF)-ICP current plasm arrangement, including oxidizing, dropping, slow, then numerous additional sensitive

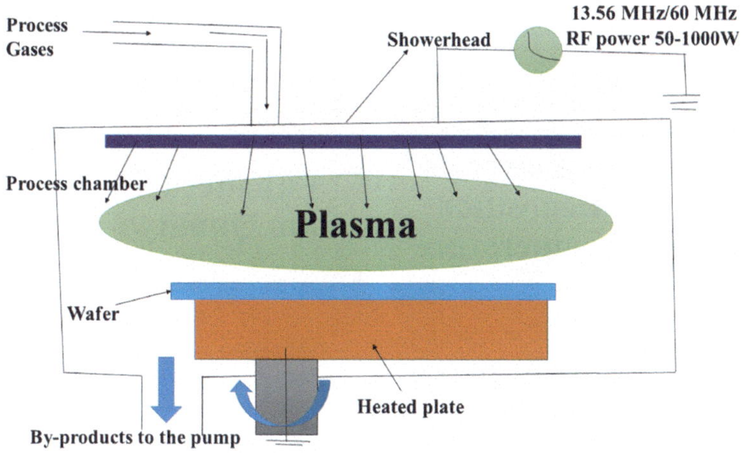

Fig. 3.6 Deposition mechanism of low pressure CVD

atmospheres. It is important to note that the ICP-PECVD setup provides a very pure environment suitable for the synthesis of nanomaterials that need precise morphological and chemical control. Because it can accommodate a larger energy density than its equivalent, the ICP arrangement is more favorable [59]. Figure 3.6 explains the internal working mechanism of the Little-Heaviness Chemical Air confession technique.

- **Electrostatically Coupled Plasma**

The majority of non-thermal radio frequency plasm in the plasma confession arrangement manufacturing is produced by capacitively join dual metallic conductors that are located in the response compartment at a short distance from each other. One of the electrodes is linked to a microwave RF influence basis with a solo rate of 13.56 MHz, while the other electrode is grounded. Atoms are ionized when an electric field appears between the electrodes, releasing electrons that are then faster by the RF electronic field to form secondary electrons, which cause an electron slide as a result of the exponential field rise. Because of the abundance of bind-free electrons, an electron-avalanche breakdown will subsequently allow perfect surface coating and render the gas electrically conductive. The CCP's schematic principle is comparable to that of a traditional capacitor [60].

3.3.3 Advantages and Disadvantages

As like other techniques of CVD, Low Pressure CVD has also several advantages as well as disadvantages. Figures 3.7 and 3.8 describes the advantages and disadvantages of the Low Pressure CVD.

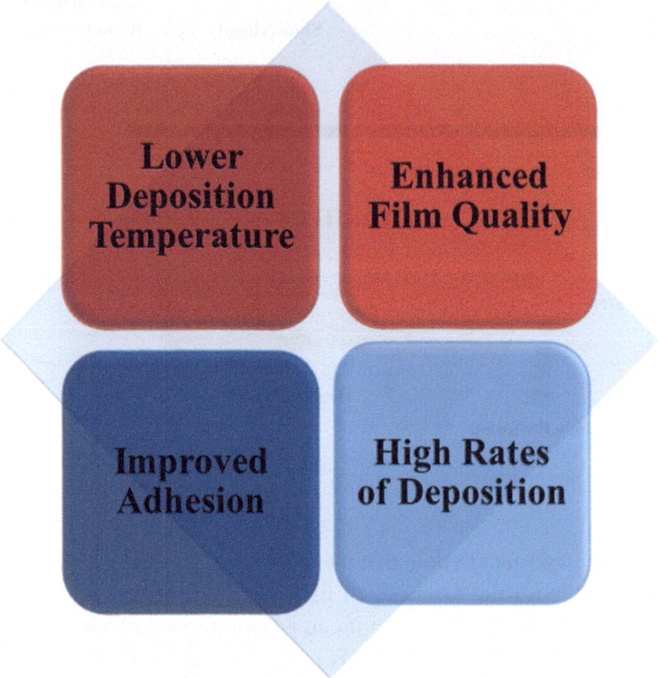

Fig. 3.7 Some advantages of low-pressure CVD

3.3.4 *Applications*

- **Application of Technology**

When it comes to the complete nanofabrication of semiconductor-based devices, PECVD seems promising, but there is still an opportunity for development when it comes to actual device commercialization. It is significant to message that important advancements in solar cells, light-emitting diodes (LEDs), sensors, photocatalysis, triboelectric nanogenerators (TNGs), thin-film transistors (TFTs), integrated circuits, neutron detection, diamond growth, photonic waveguide, energy storage, and UV photodetectors are necessary for the fruitful growth of the upcoming age group of electrical plans that will be commercially viable [61].

- **Using 2D Materials Produced by PECVD**
- **Field-Effect Transistor**

A category of stratagem composed of channel materials, dielectrics, then source, drain, and gate electrodes is called a field-effect transistor (FET). An electric field regulates the channel materials' conductivity. Performance is influenced by various

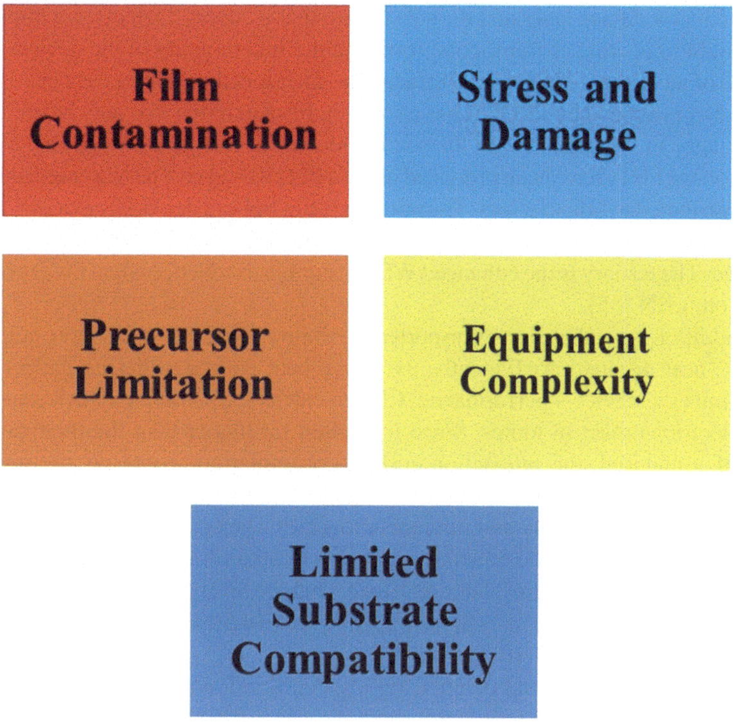

Fig. 3.8 Some disadvantages of low-pressure CVD

parameters, including interfacial contact, absorbents, flaws, and so on. PECVD-grown 2D materials do not have these issues, which are often associated with post-growth transferred materials in FETs. It was using c-E PECVD, graphene crystals doped with nitrogen and pristine formed on dielectrics and then effortlessly integrated into FETs, eliminating the need for post-growth transfer. As a result, it eliminated the chance of adding contaminants or splits and guaranteed atomically pristine surfaces and interfaces. The mobility of pristine graphene (PG) was found to be similar to that of devices manufactured under similar circumstances employing peel-off graphene (850–2200 $cm^2 V^{-1} s^{-1}$) and CVD graphene (600–1500 cm2 $V^{-1} s^{-1}$) [62].

The mobility for NG was restrained to be 100–400 $cm^2 V^{-1} s^{-1}$, which is equivalent to that of NG (200–450 $cm^2 V^{-1} s^{-1}$) and greater than that of formless carbon–nitrogen films (10 $cm^2 V^{-1} s^{-1}$) made by metallic-allowed PECVD. Under the estimate, the Dirac point moved to the bad part of Vg. The mobility dropped to about 32 $cm^2 V^{-1} s^{-1}$ if we transported the NG by means of poly(methyl methacrylate), demonstrating the benefits of PECVD in the manufacturing of high-performance FETs [61].

- **Interface of Device Dielectrics**

The dielectric interface has a significant impact on the operation of electric plans because control transporter carriage often takes place at the boundary among the semiconductor device layer and the substratum that it covers. Defects and impurities are thought to operate as traps, compromising the functionality of the devices in the absence of an optimal dielectric interface. The mentioned problems may be resolved since h-BN produced by PECVD has a smooth, fresh external that may be utilized as the insulator in interaction with the semiconductor device. SiO_2/Si, WSe_2, $MoSe_2$, and graphene FETs have been produced using a PECVD layer. The increased mobility can be attributed to other factors. The initial is the h-BN van der Waals external free of charge traps and hanging promises. The another is PECVD h-BN's pristine dielectric interface. The tertiary is the enhanced WSe_2 example excellence when CVD growing occurs on h-BN [55].

In addition to enhancing transporter flexibility, PECVD h-BN too helps the devices' heat dissipation. Typically, excessive heat generated during gadget operation limits the device's performance. Charge carriers go through a device and heat the dielectric contact in joules. Since it is often far higher than the thermal resistance of a material, the interaction current confrontation crossways the insulator boundary is regarded as the expedient's current debauchery bottleneck. Following PECVD, h-BN was cleanly and intimately touched when conformally coated on the dielectrics [55]. It overcame substrate roughness by straddling the vibrational range at the semiconductor device/insulator interface. On bare SiO_2/Si, the interfacial current confrontation of graphene or WSe_2 dropped by almost 50% or 77%, respectively, to about 4.3×10^{-9} or 4.2×10^{-8} m^2 K W^{1-}.

The h-BN interface must be pure, tightly contacted, and free of interstices and contaminants. When post-growth transported CVD h-BN was employed at the insulator boundary, the interfacial current confrontation of WSe_2/h-BN/SiO_2 arose by 1×10^{-7} m^2 K W^{1-}. Following PECVD, Raman analysis revealed that, at a influence thickness of (9.70–9.84) × 103 W cm^{-2}, the station temperature of graphene FETs on h-BN/SiO_2/Si was approximately 93 °C lesser than that on bare SiO_2/Si. Compared to CPUs, which presently have a power density of about 100 W cm^{-2}, the saturation influence thickness of graphene or WSe_2 FETs rose by two to three times, to 3.45 × 105 or 4.23 × 103 W cm^{-2}, respectively [58].

- **Applications Prototypes**

Because PECVD is compatible with mass production, it has been utilized for decades in the microelectronics field. We developed many prototypes founded on a basic unit with pyramidic microstructures and integrated 2D materials generated by PECVD. Massive production by PECVD opens the door to large-scale dispersed device fabrication. The form of things, such as a cup or a person's hand, was measured using distributed pressure sensors with a 16 × 16 array size and was visible at low weight. In the meantime, a heavyweight person's sitting motion could also be continually observed, demonstrating its wide compatibility. To simulate a human hand touching braille, heaviness devices founded on PECVD GNWs can be placed into a glove's fingers [55, 57].

By using the finger to slide over the written form, the mechanical support was immediately connected to a processer, which allowed for the recording of sliding signals and the recognition of the character "ACR." Conversely, GNW/PDMS

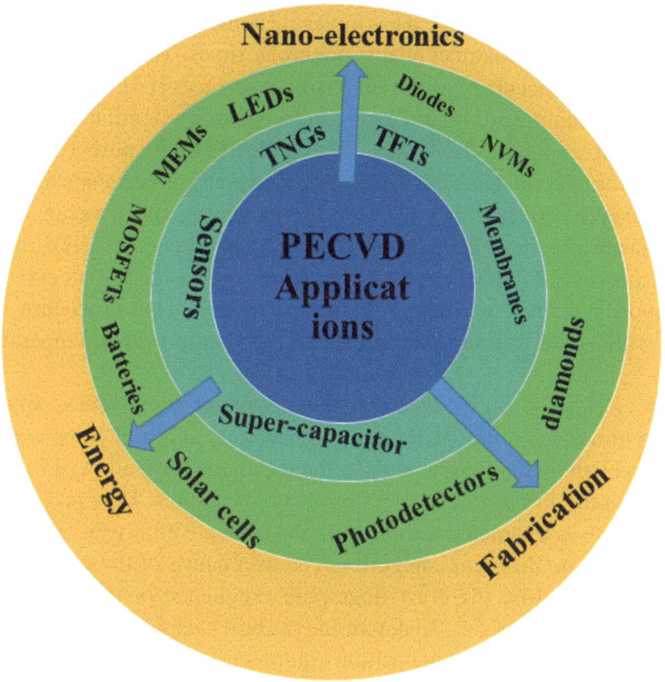

Fig. 3.9 Applications of PECVD in various fields

composites benefit from stretchability and flexibility, then their confrontation is correlated with the finger winding angle. As a result, we developed clever handbags with GNW/PDMS sensor units that might capture hand gestures. It is anticipated that this smart glove would facilitate sign language communication for the deaf. The relevant health monitoring equipment was also used to accomplish other functionalities, such as measuring the pulse. As technology advances, additional 2D resources produced using PECVD will be used to meet consumer demand for a range of commercial goods, including wearable electronics [63].

Figure 3.9 illustrates the applications of Loe Pressure CVD in various fields of interest.

3.4 Ultra-high Pressure CVD

The vacuum region known as ultra-high vacuum (UHV) is defined as compressions lesser than approximately $1 \times 10 - 6$ pascals (1.0×10^{-8} mbar; 7.5×10^{-9} Torr). In American English, ultrahigh is frequently spelled as UHV. The process of urging gas out of a UHV compartment generates UHV conditions. The air is in allowed molecular movement at these low pressures because a gas molecule's mean free path is longer than about 40 km. Before a gas molecule collides with another gas molecule, it will collide with the chamber walls numerous times. As a result, nearly every molecule interaction occurs on different chamber surfaces [64]. UHV environments are essential to scientific study. It's common for surface scientific studies to need a sample surface free of undesirable substances and chemically pure. UHV conditions are necessary for the broadcast of electron or ion rays in surface analysis instruments like little-energy ion sprinkling and X-ray photoelectron spectroscopy. Ray piping in element gases like the Big Hadron Collider are maintained at UHV for the same reason. The process of depositing homoepitaxial Si films for technical requests has remained mostly unchanged during the last ten years. This procedure is usually performed with a cold wall/hot susceptor statement equipment at temperatures over 1000 [59].

Future devices requiring sudden changes in dopant attentiveness among head-to-head solo mineral coatings will require a new generation of Si epitaxial growth techniques, even though advancements in the technique have allowed for the reduction of auto doping through lowering processing pressures. Low process temperatures are a common characteristic of these novel deposition processes, which are being developed into different categories. The excellence of the resulting epitaxial coating is determined by the crystallographic excellence of the early Si external upon which epitaxy is to occur. To determine the best cleaning practices for a silicon surface before it is placed into deposition equipment, systematic studies have been conducted. But often overlooked is the state of the ecosystem into which this clean Si is being introduced [65].

Fundamental Surface Investigations

Gations of the Si/OJH ~ O/SiO ~ symmetry scheme by Ghiddini and Smith have been used in this instance to develop quantitative standards for the processing environment, wherein the system's background in terms of water vapor and oxygen are both such that those species can effectively etch silicon, promoting the preservation of a silicon surface free of oxide. Based on the data they provided, it can be inferred that to attain an oxide-allowed external at 800 °C, a incomplete heaviness of fewer than 10^{-8} torr H_2O must be maintained. The oxygen data indicated slightly fewer requirements need of $-\ 10^{-7}$ torr at 800 °C . Process pressures typically range from 10 to 100 torr when using traditional RPC'WI) (condensed heaviness CVD) epitaxial Si dispensation [66]. When the hydrogen carrier gas employed in this process enters the process environment, it at most stays pure to 1 ppm H_2O. Processing is therefore carried out in an environment with an incomplete heaviness of $-\ 1 > -\ 10^{-4}$ torr of oxygen and water vapor, which needs to be considered as a lower limit as contributing have been disregarded by the system's surfaces releasing leftover water vapor [56].

3.4.1 *Working Principle and Construction*

Figure 3.2 depicts the UHV/CVD setup that is being employed here. Upon system startup, the apparatus's UHV section is pumped and baked until it reaches a base heaviness within the variety of 10 s Torr. The hydrogen plasma afterward generated by the RF loops at the furnace's upriver finish is subsequently used to clean the system. Following a restoration to baseline pressure, the system remains in a vacuum. To introduce the sample, a magnetically associated load lock is used. There are up to 35 wafers arranged coaxially inside the furnace tube of this hot wall isothermal system. Fast in-presence system diagnostics were made possible by a mass spectrometer installed inside the apparatus's UHV portion, which is necessary to ensure that the system is operating inside.

Rapid in real-time diagnostics of the system were made possible by a mass spectrometer installed inside the apparatus's UHV section. This is necessary to ensure that the system is operating within the strict vacuum standards that we use. The first use of the system was to investigate HC1's suitability as a preclear in this low-temperature range [67].

To find out if using in real-time HC1 etching of the substrates before deposition is feasible, several etching studies were carried out. To enable etch rate measurements, examples decorative through SiQ (SiO^ is used as an indicator as it is non-etched) were laden into the transporter, baked, and propelled inside the weight lock cavity to a pressure under 10-′~torr. The transmission into the UHV arrangement was carried out through together the load compartment then the chief structure turbopumps running at a pressure of more than 200 m torr and a 600 sccm hydrogen movement. As a result, the air movement is laminal into the load chamber and the main system, splitting at the air entrance to avoid cross-adulteration [68].

Si etch charges were examined at $T = 825$ ~ HC1/H, the upper limit of the variety of suggested statement temperatures. At an entire current of 175 sccm, the solvent species ratio of 1.33/1 was used. Only the origins/rotating pump structure was used throughout the engraving procedure. It operated at a set heaviness ($P = 215$ m torr) by connecting the pumps in series with a feedback-controlled mechanical valve. Samples were taken out and examined for engrave frequency and exterior geomorphology afterward the etching investigations. Several UHV/CVD growth experiments were carried out after this investigation into sample pretreatment. Doped 10′~ B/cm ~ wafers were added to the UHV/CVD structure in the above-labeled way. Film deposition was started after allowing samples to acclimate to a 600 sccm hydrogen flow for five rains. Using a 2 sccm Sill J20 sccm H_2 statement basis, development circumstances were 750 · < T < 850 ~ for two hours ot deposition at an entire heaviness of 1–2 m torr [69]. Figure 3.10 describes the working of Ultra-high Pressure CVD. The vapors that need to be deposited are vaporized first and introduced in the reactor, as the ultra-high pressure is required to maintain so the high pressure environment is provided to the vapors, and then those vapors is put on the external of the substratum.

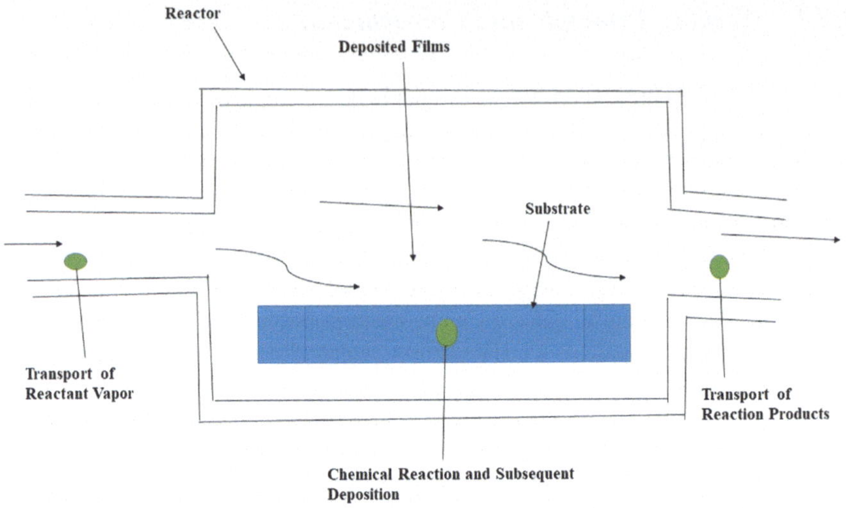

Fig. 3.10 Graphical mechanism of ultra-high pressure CVD

3.4.2 Advantages and Disadvantages

As with other techniques of CVD, Ultra-high Pressure CVD has also several advantages as well as disadvantages. Figures 3.11 and 3.12 describe the advantages and disadvantages of the Ultra-high Pressure CVD.

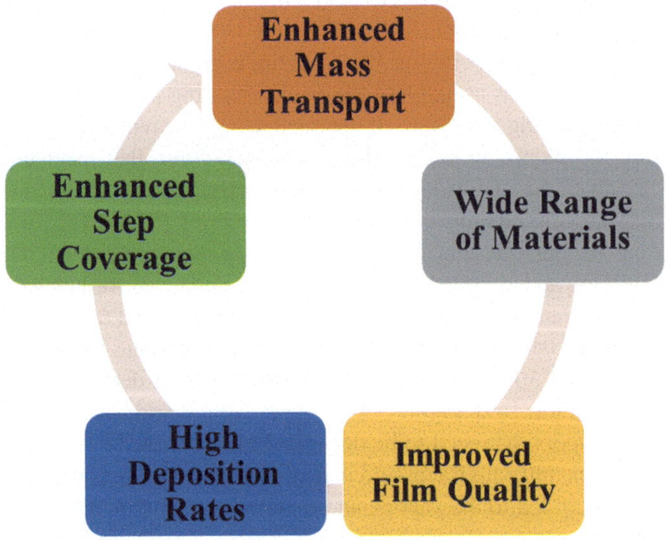

Fig. 3.11 Some basic advantages of Ultra-high pressure CVD

Fig. 3.12 Some disadvantages of Ultra-high pressure CVD

3.4.3 Applications

Ultra-high pressure CVD method has several applications in the following fields,

- External Examination
- High-energy physical science
- Molecular Beam Epitaxy (MBE)
- Thin-Film Deposition
- Nuclear Physics
- Space Simulations
- Analytical Instruments.

Using extreme-slim a-Si pre-statement on a SiO_2 film, the sophisticated UHV-CVD process has effectively produced very solid and two dimensionally filled Si nanocrystals. Grain growth in this process is constrained by the source of Si particles from the pre-deposited a-Si coating, and nucleation takes place on an a-Si surface. Controlling the a-Si depth is crucial to achieving a tall amount of thickness, a nm-gage extent, and a restricted extent distribution since the grains grow considerably faster than the rate of nucleation. By means of a sub-nm thick a-Si coating, we were able to generate Si nanocrystals through a amount compactness of 1:10 1012 cm^2

and a grain extent of 9:1 ~ 3:0 nm. A second a-Si deposition on SiO_2/Si nanocrystals/SiO_2/Si stacked structures was successfully created [70].

- Many external logical methods, with X-ray photoelectron spectroscopy (XPS), require an ultra-high vacuum.
- Electron spectroscopy (AES) using Auger.
- Mass spectrometry of secondary ions (SIMS).
- Spectroscopy of thermal desorption (TPD).
- Techniques for growing and preparing thin films that have strict purity criteria include UHV pulsed laser deposition (PLD), atomic layer deposition (ALD), UHV chemical vapor deposition (CVD), and molecular beam epitaxy (MBE).
- Photoemission spectroscopy with angle resolution (ARPES).
- Both field ion and field emission microscopy.
- Tomography using Atom Probes (APT).

To minimize surface contamination in various applications, UHV must lower the quantity of molecules that arrive in the sample within a specific time frame. Long-term experiments require much lower pressures because it only takes 1 second to cover a surface with a pollutant at 0.1 millipascals (7.5×10^{-7} Torr) [60].

Also, UHV is necessary for:

Accelerators of particles

There are three vacuum systems for UH at the Large Hadron Collider (LHC). The proton beam rushes through pipes close to the interaction (collision) locations, where the lowest pressure is found. In this case, cryopumps are also helium-cooling pipes. The highest pressure that may be used is 1×10^{-6} pascals (1.0×10^{-8} mbar). LIGO, VIRGO, GEO 600, and TAMA 300 are examples of gravitational wave detectors. The 10,000 cubic meters (350,000 cu ft) emptiness space housing the LIGO experimental apparatus is operated at 1×10^{-7} pascals (1.0×10^{-9} bar) to prevent temperature changes and then comprehensive waves that would shake the glasses too greatly to detect gravitational waves [71].

3.5 Conclusion

In this chapter, the chemical vapor deposition process has been explained and described in detail. Chemical Vapor Deposition is a method that operates to deposit a thin film at a high-temperature range and deposit a uniform layer of thin film on the substrate. Some specific types such as Atmospheric Pressure CVD, Low-Pressure CVD, and Ultra-high Pressure CVD have also been explained. These deposition processes are widely used in a wide range of applications according to the extensive range of properties. Every deposition process has different properties and applications depending on the employed method and conditions of substrate deposition.

These methods of CVD are also affected by various parameters such as environmental conditions, temperature range, required pressure range, substrate range, thin film thickness, uniformity, etc.

This chapter also explained the introduction, working principle, working diagrams, and advantages and disadvantages as well as the properties of the CVD methods.

References

1. Carlsson, J.-O., and P.M. Martin. 2010. Chemical vapor deposition. In *Handbook of Deposition Technologies for Films and Coatings*, 314–363. Elsevier.
2. Dobkin, D., and M.K. Zuraw. 2003. *Principles of Chemical Vapor Deposition*. Berlin: Springer Science & Business Media.
3. Zhang, Y., L. Zhang, and C. Zhou. 2013. Review of chemical vapor deposition of graphene and related applications. *Accounts of chemical research* 46 (10): 2329–2339.
4. Dapkus, P.D. 1982. Metalorganic chemical vapor deposition. *Annual Review of Materials Science* 12 (1): 243–269.
5. Pierson, H.O. 1999. *Handbook of Chemical Vapor Deposition: Principles, Technology and Applications*. William Andrew.
6. Jasinski, J., B. Meyerson, and B. Scott. 1987. Mechanistic studies of chemical vapor deposition. *Annual Review of Physical Chemistry* 38 (1): 109–140.
7. Park, J.-H., and T. Sudarshan. 2001. *Chemical Vapor Deposition*, vol. 2. ASM international.
8. Crowell, J.E. 2003. Chemical methods of thin film deposition: Chemical vapor deposition, atomic layer deposition, and related technologies. *Journal of Vacuum Science & Technology A: Vacuum, Surfaces, and Films* 21 (5): S88–S95.
9. Gates, S.M. 1996. Surface chemistry in the chemical vapor deposition of electronic materials. *Chemical Reviews* 96 (4): 1519–1532.
10. Dinata, A.A., et al. 2018. *A Review Chemical Vapor Deposition: Process and Application*.
11. Creighton, J., and P. Ho. 2001. Introduction to chemical vapor deposition (CVD). *Chemical Vapor Deposition* 2: 1–22.
12. Rees Jr, W.S. 2008. *CVD of Nonmetals*. London: Wiley
13. Pattanaik, A., and V. Sarin. 2001. Basic principles of CVD thermodynamics and kinetics. In *Chemical Vapor Deposition*, 23–43. ASM International Materials Park.
14. Beshkov, G., et al. 2002. IR and Raman absorption spectroscopic studies of APCVD, LPCVD and PECVD thin SiN films. *Vacuum* 69 (1–3): 301–305.
15. Arnold, H. 1984. Basic steps and layer homogeneity in CVD (I). *Crystal Research and Technology* 19 (9): 1189–1194.
16. Vahlas, C., et al. 2006. Principles and applications of CVD powder technology. *Materials Science and Engineering: R: Reports* 53 (1–2): 1–72.
17. Sussmann, R.S. 2009. *CVD Diamond for Electronic Devices and Sensors*. London: Wiley
18. Liu, D., et al. 2019. Prevalence and prognosis significance of cardiovascular disease in cancer patients: A population-based study. *Aging (Albany NY)* 11 (18): 7948.
19. Jenkins, D.J., et al. 2018. Supplemental vitamins and minerals for CVD prevention and treatment. *Journal of the American College of Cardiology* 71 (22): 2570–2584.
20. Morgan, A.D., R. Zakeri, and J.K. Quint. 2018. Defining the relationship between COPD and CVD: What are the implications for clinical practice? *Therapeutic Advances in Respiratory Disease* 12: 1753465817750524.
21. Liu, K., et al. 1982. Relationship of education to major risk factors and death from coronary heart disease, cardiovascular diseases and all causes, Findings of three Chicago epidemiologic studies. *Circulation* 66 (6): 1308–1314.

22. Tresch, D.D., et al. 1985. Prevalence and significance of cardiovascular disease and hypertension in elderly patients with dementia and depression. *Journal of the American Geriatrics Society* 33 (8): 530–537.
23. Matsui, K., et al. 2016. Clinical significance of urinary liver-type fatty acid-binding protein as a predictor of ESRD and CVD in patients with CKD. *Clinical and Experimental Nephrology* 20: 195–203.
24. Eisenmann, J.C., et al. 2005. Aerobic fitness, body mass index, and CVD risk factors among adolescents: The Quebec family study. *International journal of Obesity* 29 (9): 1077–1083.
25. Massy, Z., et al. 1994. Hyperhomocysteinaemia: A significant risk factor for cardiovascular disease in renal transplant recipients. *Nephrology Dialysis Transplantation* 9 (8): 1103–1108.
26. Oguoma, V.M., et al. 2015. Prevalence of cardiovascular disease risk factors among a Nigerian adult population: Relationship with income level and accessibility to CVD risks screening. *BMC Public Health* 15: 1–16.
27. Stoffel, A., et al. 1996. LPCVD against PECVD for micromechanical applications. *Journal of Micromechanics and Microengineering* 6 (1): 1.
28. Ficcadenti, M, and R. Murri, Basics of thin film solar cells. *Silicon Based Thin Film Solar Cells* 3
29. Matsuura, M., et al. 1991. Film characteristics of APCVD oxide using organic silicon and ozone. *Japanese journal of applied physics* 30 (7R): 1530.
30. Van Deelen, J., et al. 2013. APCVD of ZnO: Al, insight and control by modeling. *Surface and Coatings Technology* 230: 239–244.
31. Krumdieck, S. 2010. *Biomaterials by CVD Methods, Introduction to the CVD Special Section*, 9–11. WILEY-VCH Verlag, Weinheim.
32. Al-Sarraf, A.H.R., et al. 2018. Preparation and characterization of ZnO nanotripods and nanoflowers by atmospheric pressure chemical vapor deposition (APCVD) technique. In *AIP Conference Proceedings*. AIP Publishing
33. Saikia, P., A. Borthakur, and P. Saikia. 2011. Structural, optical and electrical properties of tin oxide thin film deposited by APCVD method. *Indian Journal of Physics* 85: 551–558.
34. Ivanova, T., et al. 2001. Preparation and optical study of APCVD mixed metal oxide films. *Le Journal de Physique IV*, **11**(PR3): p. Pr3–385-Pr3–390.
35. Han, T., et al. 2019. Probing the growth improvement of large-size high quality monolayer MoS2 by APCVD. *Nanomaterials* 9 (3): 433.
36. Ullah, Z., et al. 2018. A comparative study of graphene growth by APCVD, LPCVD and PECVD. *Materials Research Express* 5 (3): 035606.
37. Vasilyev, V.Y., and Y.N. Popov. 2008. *Equipment and technologies for thin film deposition: analysis of development, domestic opportunities and future trends.*
38. Barron, A.R. 1996. CVD of SiO2 and related materials: An overview. *Advanced Materials for Optics and Electronics* 6 (2): 101–114.
39. Guo, Y., X.-W. Zhang, and G.-R. Han. 2006. Investigation of structure and properties of N-doped TiO_2 thin films grown by APCVD. *Materials Science and Engineering: B* 135 (2): 83–87.
40. Drosos, C., and D. Vernardou. 2015. Perspectives of energy materials grown by APCVD. *Solar Energy Materials and Solar Cells* 140: 1–8.
41. Hatem, D., and M.S. Belkaid. 2013. SnO2 thin films prepared by APCVD for organic solar cells application. *Advanced Materials Research* 685: 166–173.
42. Konar, R., and G.D. Nessim. 2022. A mini-review focusing on ambient-pressure chemical vapor deposition (AP-CVD) based synthesis of layered transition metal selenides for energy storage applications. *Materials Advances* 3 (11): 4471–4488.
43. Fraga, M., and R. Pessoa. 2020. Progresses in synthesis and application of SiC films: from CVD to ALD and from MEMS to NEMS. *Micromachines* 11 (9): 799.
44. Roenigk, K.F., and K.F. Jensen. 1987. Low pressure CVD of silicon nitride. *Journal of the Electrochemical society* 134 (7): 1777.
45. Li, X., et al. 2011. Large-area graphene single crystals grown by low-pressure chemical vapor deposition of methane on copper. *Journal of the American Chemical Society* 133 (9): 2816–2819.

46. Bartl, A., et al. 1996. A comparison of low-pressure CVD synthesis of diamond and c-BN. *International Journal of Refractory Metals and Hard Materials* 14 (1–3): 145–157.
47. Broadbent, E., and C. Ramiller. 1984. Selective low pressure chemical vapor deposition of tungsten. *Journal of The Electrochemical Society* 131 (6): 1427.
48. Gohier, A., et al. 2009. Early stages of the carbon nanotube growth by low pressure CVD and PE-CVD. *Diamond and Related Materials* 18 (1): 61–65.
49. Endler, I., et al. 2010. Aluminum-rich TiAlCN coatings by low pressure CVD. *Surface and Coatings Technology* 205 (5): 1307–1312.
50. Fay, S., et al. 2005. Low pressure chemical vapour deposition of ZnO layers for thin-film solar cells: Temperature-induced morphological changes. *Solar Energy Materials and Solar Cells* 86 (3): 385–397.
51. Vernardou, D., et al. 2015. Low pressure CVD of electrochromic WO3 at 400° C. *Journal of the Electrochemical Society* 162 (9): H579.
52. Kern, W., and G.L. Schnable. 1979. Low-pressure chemical vapor deposition for very large-scale integration processing—A review. *IEEE Transactions on Electron Devices* 26 (4): 647–657.
53. Dischler, B., and C. Wild. 2013. *Low-Pressure Synthetic Diamond: Manufacturing and Applications*. Berlin: Springer Science & Business Media.
54. Hegde, R.I., et al. 1993. Thin film properties of low-pressure chemical vapor deposition TiN barrier for ultra-large-scale integration applications. *Journal of Vacuum Science & Technology B: Microelectronics and Nanometer Structures Processing, Measurement, and Phenomena* 11 (4): 1287–1296.
55. Yi, K., et al. 2021. Plasma-enhanced chemical vapor deposition of two-dimensional materials for applications. *Accounts of Chemical Research* 54 (4): 1011–1022.
56. Batey, J., and E. Tierney. 1986. Low-temperature deposition of high-quality silicon dioxide by plasma-enhanced chemical vapor deposition. *Journal of Applied Physics* 60 (9): 3136–3145.
57. Martinu, L., O. Zabeida, and J. Klemberg-Sapieha. 2010. Plasma-enhanced chemical vapor deposition of functional coatings. In *Handbook of Deposition Technologies for Films and Coatings*, 392–465.
58. Hamedani, Y., et al. 2016. Plasma-enhanced chemical vapor deposition: Where we are and the outlook for the future. *Chemical Vapor Deposition-Recent Advances and Applications in Optical, Solar Cells and Solid State Devices* 4: 243–280.
59. Bo, Z., et al. 2013. Plasma-enhanced chemical vapor deposition synthesis of vertically oriented graphene nanosheets. *Nanoscale* 5 (12): 5180–5204.
60. Kushner, M.J. 1988. A model for the discharge kinetics and plasma chemistry during plasma enhanced chemical vapor deposition of amorphous silicon. *Journal of Applied Physics* 63 (8): 2532–2551.
61. Ceiler, M., P. Kohl, and S. Bidstrup. 1995. Plasma-enhanced chemical vapor deposition of silicon dioxide deposited at low temperatures. *Journal of the Electrochemical Society* 142 (6): 2067.
62. Hofmann, S., et al. 2003. Low-temperature growth of carbon nanotubes by plasma-enhanced chemical vapor deposition. *Applied Physics Letters* 83 (1): 135–137.
63. Vallée, C., et al. 2020a. Plasma deposition—impact of ions in plasma enhanced chemical vapor deposition, plasma enhanced atomic layer deposition, and applications to area selective deposition. Journal of Vacuum Science & Technology 38 (3).
64. Perrin, J., et al. 2000. The physics of plasma-enhanced chemical vapour deposition for large-area coating: Industrial application to flat panel displays and solar cells. *Plasma Physics and Controlled Fusion* 42 (12B): B353.
65. Li, M., et al. 2016. Controllable synthesis of graphene by plasma-enhanced chemical vapor deposition and its related applications. *Advanced Science* 3 (11): 1600003.
66. Yang, Y.-T., et al. 2007. The latest plasma-enhanced chemical-vapor deposition technology for large-size processing. *Journal of Display Technology* 3 (4): 386–391.
67. Zhou, F., et al. 2022. Direct plasma-enhanced-chemical-vapor-deposition syntheses of vertically oriented graphene films on functional insulating substrates for wide-range applications. *Advanced Functional Materials* 32 (42): 2202026.

68. Qin, L., et al. 1998. Growing carbon nanotubes by microwave plasma-enhanced chemical vapor deposition. *Applied Physics Letters* 72 (26): 3437–3439.
69. Teo, K., et al. 2003. Plasma enhanced chemical vapour deposition carbon nanotubes/nanofibres—How uniform do they grow? *Nanotechnology* 14 (2): 204.
70. Yeh, N.-C., et al. 2019. Single-step growth of graphene and graphene-based nanostructures by plasma-enhanced chemical vapor deposition. *Nanotechnology* 30 (16): 162001.
71. Bertran-Serra, E., et al. 2023. Advancements in plasma-enhanced chemical vapor deposition for producing vertical graphene nanowalls. *Nanomaterials* 13 (18): 2533.

Chapter 4
Atomic Layer Deposition

Abstract Atomic layer deposition (ALD) is a procedure for depositing diverse thin-layer substances from the vapor stage. ALD holds significant potential to explore electronics and conversion of energy techniques. Atomic layer implantation (ALD) employs detrimental connections among reactive gases and an accessible solidifying layer to build exceptionally conforming coatings with sub-monolayer. They involve the consistent fabrication of uniform coatings with customized thickness, on top of convoluted three-dimensional substrates, which may enhance the functionality of electrical equipment. This method of production has drawn vital curiosity, for the sake of an elementary comprehension of how ALD can produce new functional materials and for a wide range of real-world uses, specifically sophisticated nanopatterning for nanotechnology, power storage, desalting, catalytic activity, and medicine. Atomic layer deposition (ALD), which used to be called atomic layer epitaxy, is a vapor-phase mounting strategy that produces ultrathin layers enabling fine evolution regulation. ALD has recently undergone significant development, owing primarily to the ongoing tendency of electronics miniaturization. ALD film development is accomplished by executing an assortment of multiple self-destructive surface chemical reactions in which the relevant reacting agents are constantly supplied upon the backing material with intermittent furnace evacuation. The reversible behavior of ALD processes allows for exquisite sub-nanometer supervision of formation thicknesses. ALD is an innovation that permits atomic and near-atomic scale manufacturing (ACSM) of substances, structures, gadgets, and systems for a wide range of activities. The adoption of theoretical computation, multiscale modeling as well as innovative methodologies would push ALD forward, allowing it to meet the demands of ACSM.

4.1 Introduction

A method that permits the formation of ongoing and fragile films of substrate dielectric coatings is crucial for transistor efficiency enhancement and miniaturization. In the installation of dielectric material layers, the approach known as the atomic

T. I. Awan et al., *Thin Film Deposition Techniques*,
https://doi.org/10.1007/978-981-96-1364-9_4

layer deposition (ALD) method fosters consistent deposition of the thin film at the nanometer level with outstanding dependability [4]. Due to its simplicity of thickness modulation and different thin film qualities depending on the type of precursors, ALD has garnered interest owing to its exceptional dependability [7]. Atomic layer deposition (ALD), a chemical vapor coating process that employs successive self-terminating gas–solid interactions, has been used for almost 40 years to create conventional inanimate substrate layers with thicknesses in the nanoscale range. Although multiple lucrative uses of ALD-based material development, many of the physical and chemical reactions that regulate ALD growth remain not well understood [8]. ALD is a vapor coating process that uses consecutive switching bursts to distinguish between the ALD antecedents and co-reactants, showing restricting interactions [9]. As an outcome, Atomic layer deposition (ALD) has developed to become an advantageous technique for generating extremely -thin coatings for several industries. Tuomo Suntola popularized ALD as an "atomic layer epitaxy" in the 1970s by depositing ZnS layers for flat-screen monitors. Few investigations have employed vertical layers such as silicon and GaAs, hence the general term ALD has been utilized to refer to this method until now [10]. ALD, or atomic layer deposition, has drawn interest for growing scientific consideration Over the last couple of generations, due to its unrivaled features for regulating the expansion of materials with atomic accurateness, deceiving intriguing nanostructures, phoning the chemical makeup of materials, providing transparency options, resulting in symmetrical and consistent films inspection, and eligibility for temperature-sensitive materials [11]. Atomic layer deposition (ALD) is a beneficial methodology for developing micron-sized membranes with atomic scale clarity, great homogeneity, exceptional conformality, and outstanding permeability on a variety of surfaces. Lately, various novel approaches for preparing nanostructures that are unable to be produced easily by traditional ALD have emerged. For example, ALD launch is an effective approach for creating independent oxide nanotechnology with accurate and predictable diameters [12]. ALD demonstrates uniform coating capacity on complicated materials exhibiting substantial aspect ratios and/or significant surface regions, where competing thin-film coating methods, such as physiological vapor testimony, vapor deposition with chemicals, and sustained laser testimony, frequently underperform. Furthermore, altering the total amount of ALD cycles allows for fine-tuning of the depth of layers at the level of particles [13]. ALD is a restricting depositing of thin films approach centered around surface-saturated interactions among compounds, having numerous benefits over previous mounting techniques (such as flaring and CVD). The benefits include the capacity to manage atomic-level dimensions, high conformality, and large-area homogeneity [10]. Metallic oxides and nitride substances have received the greatest attention among the known ALD-grown substances. Coating of metallic oxides and nitrites by ALD is thus among the most prevalent, which might be attributable to the necessity of these substances in the creation of Si-based gadgets, notably circuit dielectrics [14–16]. The restrictive tendency of surface interactions gives ALD its exceptional conformity along with the capacity to adhere to substantial aspect ratio geometries [17]. Detrimental reactions on the exterior also cause coating accumulation that cannot be estimated by averages. Other physical as well as chemical vapor condensation methods also have an unforeseen initial flow [18].

4.2 Thin Film Deposition Methods

4.2.1 Chemical Vapor Deposition

Chemical vapor deposition is a process that uses chemical processes involving substrates in the deposition chamber. CVD is frequently implemented to produce silicon thin films, and it demands elevated temperatures to commence the entire process.

4.2.2 Physical Vapor Deposition

This process incorporates the tangible transport of substances under vacuum, with vaporized components condensing on the substrate. PVD is suitable for depositing metals and alloys.

4.2.3 Pulsed Laser Deposition

This approach employs an intensely powerful laser to vaporize the target material and then implant it upon the backing material.

4.2.4 Molecular Beam Epitaxy

Molecular beam epitaxy is a thin-film deposition process that produces superior, high-purity epitaxial structures on appropriate surfaces. It is very beneficial for coating semiconductor substances such as III-V compounds, silicon, germanium, and silicon carbide, as well as oxides and organic molecules.

4.2.5 Lithography

This is the process of transferring patterns onto a substrate. The most popular method is optical lithography, which employs light to display a photographic substance (photoresist) on the substrate. During exposure, the photoresist is produced and the pattern is etched into the underlying material.

4.2.6 *Ball Milling*

Ball milling is a mechanical technique that reduces the dimensions and homogenies of materials. It can be conducted in either humid or dry situations. Planetary ball milling (PBM) is a type of ball milling in which the grinding jars move around both their axes and the axis of the circulating bearing disk.

4.2.7 *Dry Etching*

This is a method in which a substrate is exposed to a plasma or reactive gas so that material can be etched off in a pattern. It often is used in combination with lithography to attach patterns to thin films. Dry etching can etch materials that are either not volatile, or volatile at reduced pressures.

4.3 Chemical Methods Involve

4.3.1 *Chemical Etching*

Chemical etching, also referred to as photochemical processing, acid polishing, or chemical milling, refers to the subtractive manufacturing method carried out based on immersion or showering of a substrate in temperature-controlled chemicals for the purpose of removing a specific section and creating a complex shape or geometric features.

4.3.2 *Solgel Method*

The sol–gel technique forms a sol by reacting metallic alkoxides with water, which is then applied on the substrate. This sol is normally deposited using various techniques including dip coating, spin coating, and spraying. The sol will then go through hydrolysis and condensation reactions that transform it into a gel-like material, which is dried and calcined into a thin coating with particular properties.

4.3.3 *Spray Pyrolysis*

Spray pyrolysis is a process that includes spraying a precursor solution or suspension into a high-temperature furnace or another hot environment. The starting material decays and interacts with the external environment, resulting in a thin coating. This approach is especially useful for manufacturing nanoparticles. Figure 4.1. illustrates of thin film deposition through various processes.

4.4 History of ALD

Tuomo Suntola was born in 1943 in Tampere, Finland. He finished school in Turku, Finland, and passed his initial college exams (national exams that allow for higher education) in 1962. Suntola participated in electrical engineering at Helsinki University of Technology (HUT-Teknillinen korkeakoulu, TKK, now Aalto University) in 1963. In 1967, he defended his final thesis ("diplomityö" in Finnish) on Schottky diodes for radar monitors [19]. Atomic layer deposition (ALD), a gas-phase deposition of thin films methodology that utilizes recycled, self-terminating gas–solid behaviors, has grown into the norm in the fabrication of semiconductors and numerous additional sectors of technology for dumping fragile uniform inorganic material layers for a variety of programs [20]. ALD was initially identified and developed autonomously at least twice, beneath distinct designations: molecular layer (ML) in the Soviet Union in the 1960s and atomic layers epitaxy (ALE) in

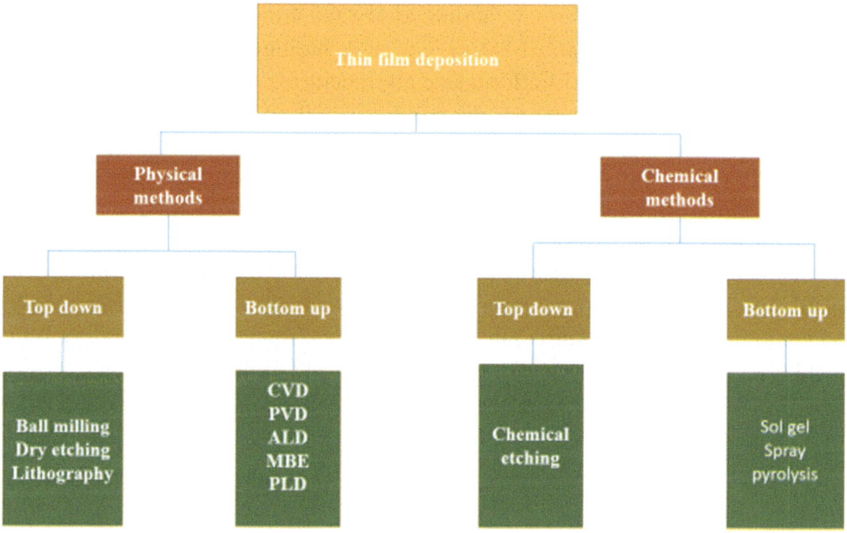

Fig. 4.1 Thin film deposition through various processes

Finland in 1974. ALE is widely recognized as the genesis of ALD [21, 22]. Particle ALD was discovered at the University of Colorado in the late 1990s 990 and it has been distinguished by its capacity to cover primary particles (without aggregating them) with conventional films, incorporating nanomaterials. The primary objectives of granular ALD were to develop pinhole-free films as thin as possible to build external barrier coatings (EBC) or enhance the outer layer of particles for a particular purpose. A lot of work was done to define a minimal coating thickness for an obstacle to achieve the desired outcome [23–25].

4.5 ALD Processes

There are several ALD production methods, but among those most commonly utilized are thermal ALD and plasma-enhanced ALD (PEALD). Furthermore, plasma-assisted or radical-enhanced ALD is also known as PEALD. The term thermal ALD is primarily a contact-triggered procedure that happens solely primarily surface effects, allowing for precise thickness control and conformality regardless of platform configuration or furnace architecture. Thermo ALD techniques necessitate rather extreme temperatures (usually 150–350 °C), which limits their potential uses but can be rectified by PEALD [26]. PEALD, formerly known as plasma-assisted ALD (PA-ALD), has resilient interactivity of the ionized species, allowing for lower deposition rates without affecting the film's performance. It is notably preferred when fragile films are applied to temperature-dependent substances. PEALD may additionally result in superior material characteristics. Furthermore, the sensitivity of the plasma promotes the implementation of more types of progenitors and the placement of elements that are problematic or intractable by thermo ALD [27]. Boosting the surface dimension and overall volume of an ALD furnace causes prolonged burst and flush times, leading to reduced capacity. Performance enhancement is achieved by employing spatial ALD (SALD) that bypasses the pulse-driven flush chamber with the spatial specified tip by treating the backing material to a distinct progenitor dependent on locality. However, the durability of the thin layer would have been degraded [28]. PEALD enables production at comparatively low temperatures, and the resulting film characteristics are superior to those of thermal ALD [29]. PEALD also delivers extremely active molecules or catalysts that improve the reaction, allowing for the use of fewer degrees Celsius, in addition to a wider spectrum of substrates and precursors for deposition [30]. Thus, PEALD is advantageous for building films susceptible to heat substances. PEALD can also yield films that are more pristine than those formed by thermal ALD. Additional benefits of the PEALD include a wider range of predecessors, a faster pace of advancement, along with operational adaptability.

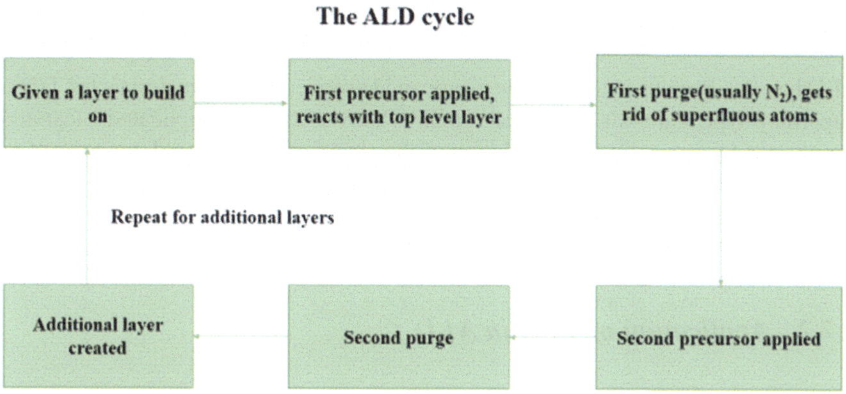

Fig. 4.2 Schematic representation of ALD cycle

4.6 Working Principal

The essential premise of ALD is to expose an absorbent material to a variety of gaseous progenitor substances in a periodic fashion. The method usually comprises the subsequent steps.

The initial reagent is injected and reacts with the material's interface. This process is permitted to continue unless the outer layer is coated with chemisorbed substances. The leftover reacting agents and explosive combustion outcomes are subsequently eliminated to prevent unregulated adsorption. The additional reagent is free to interact with the outermost species that were removed from the first reacting agent, forming an "atomic layer" on the surface and preserving sites of reaction for the previous reactant. An expulsion/purge concludes the ALD reactivity chain. Reaction repetitions are performed to improve the film [31]. Figure 4.2. describes the schematic representation of the ALD cycle.

4.7 Application of ALD

4.7.1 Application of TFTs in ALD

4.7.1.1 N-Type Oxide TFTs

ALD methods have successfully generated high-performance n-type oxide semiconductor TFTs. When coated by ALD, materials such as zinc oxide (ZnO), indium gallium zinc oxide (IGZO), and indium zinc oxide (IZO) exhibit outstanding flexibility and rigidity.

4.7.1.2 P-Type Oxide TFTs

ALD has also been used to generate p-type oxide semiconductor materials for TFTs. Superior p-type materials are critical for CMOS inverters and complicated integrated circuits. ALD allows for fine regulation of the chemical makeup and diameter of p-type materials, facilitating the creation of lightweight device circuits. Figure 4.3 describes the application of ALD in various fields.

4.7.2 Application of TCO in ALD

In ALD, Transparent Conductive Oxides (TCOs) are used to create semiconductors. TCOs, like as Indium Tin Oxide (ITO), are frequently employed for their transparent electrodes in TFT. ALD methods can be used to develop TCO films with desired features such as strong electrical conductivity and transparent optical properties. These films are essential for producing transparent gate TFTs along with additional optical-electronic components.

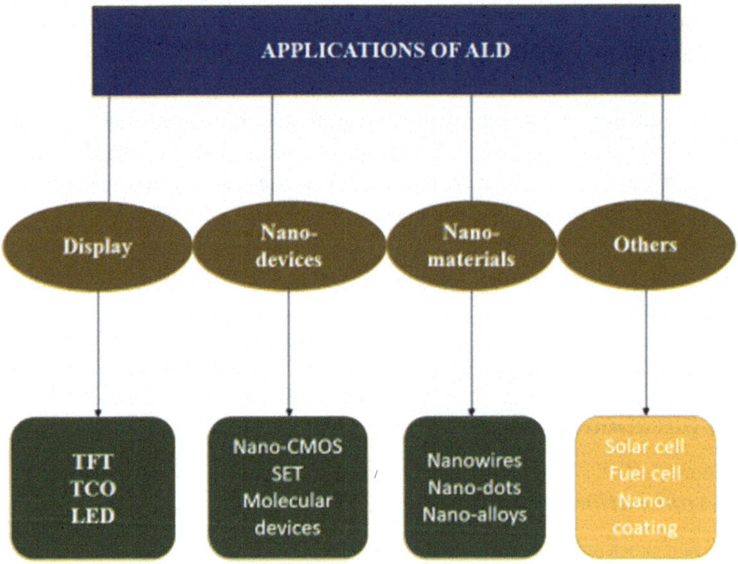

Fig. 4.3 Applications of ALD in various fields

4.7.3 Application of LED in ALD

The use of LEDs in Atomic Layer Deposition (ALD) entails employing LEDs as a source of illumination to accelerate the process of deposition. ALD is a flexible film deposition technology that employs chemical processes to deposit substances layer by layer. LEDs are employed to offer the light required for these processes to proceed effectively and precisely.

4.7.4 ALD Applications for SCs

Due to the benefits of rapid CD interest rates, a substantial power supply, excellent durability, and cost-effective functioning, electromechanical SCs have attracted a lot of interest being a possible option for the fabrication of top-performance energy-storage platforms in the not-too-distant future. Large particular areas of surface and high conductivity for electricity are typically needed for higher-capacity electrodes to promote redox kinetics and/or enable double-layer development [32–34]. To create SC electrode substrates with the desired shape, appropriate charge transmission actions precise thickness, and surface chemical composition, appropriate manufacturing methods will soon be needed. ALD has been the most popular method for creating advanced-generation SC electrode constituents with outstanding macro-/nano scale architectures and exterior characteristics by utilizing its special operating conductions. ALD is especially well-suited for use in battery packs due to its benefits, which include more extensive manipulation of the generated layer, evenly distributed deposition over a broad surface area, compliance of formation over complicated amenities, and straightforwardness in constructing hybrid nanoscale [35–37].

4.7.5 Latest Developments of Atomic Layer Deposition in Nanotechnology

Atomic Layer Deposition is an approach appropriate for nanotechnology. Nanotechnology fabrication benefits from precise layer depth surveillance, consistency across wide regions, exceptional conformance and repeatability, multilayer manufacturing, and exceptional layer generation at extremely low temperatures. Atomic Layer Deposition (ALD) techniques sustain depth precision up to 100 nm and are ubiquitous and homogenous over broad regions, making them suitable for deposition on complicated substrates like those used in nanotechnology. ALD's features make it a promising process for producing nanotechnology. As technology gets smaller, accuracy at the atomic level becomes more important. Because of the chemistry behind ALD procedures, they may assemble themselves layers and respond without any outside stimulus. This simplifies nanotechnology applications and strengthens the case for ALD as an effective thin film manufacturing approach.

4.7.6 ALD in Microelectronics

ALD is being investigated as a possible method in semiconductors for coating ferroelectrics in, high-k memory capacitor dielectrics, high-k gate oxides, and metals and nitrides for electrodes and interconnects. ALD is probably likely to become more widely used at 45 nm technology in high-k gate oxides, where oversight over incredibly thin layers is crucial. Conformal films are needed in metallization; at the 65-nm node, ALD is currently anticipated to be employed in standard production. The conformity requirements in dynamic random access memory (DRAMs) are much greater, and ALD is the only technique that can be applied when Characteristic diameters drop below 100 nm [46].

4.7.7 Power Lithium-Ion Batteries Modified by ALD

Lithium-ion batteries are currently powering an unprecedented global development in alternative energy transportation. While the lithium-ion cells used in modern electrically powered cars are capable of meeting the fundamental efficiency criteria, issues such as inadequate cycle integrity and security persist. These issues are intimately linked to the malfunction of the battery's components. Atomic layer deposition (ALD) on the cathode, anode, and isolator can boost battery longevity by guaranteeing the movement of Li ions as well as electrons while simultaneously enhancing the stability of the electrochemical cycle. Surface nanotechnology coatings on both anode and cathode materials can reduce adverse reactions between both electrodes and electrolytes during cycles of charging and discharging and enhance the inherent strength of the materials [38].

4.7.8 Application of ALD in PSCs

ALD is currently gaining popularity as a means of increasing the reliability and efficacy of power supply units. ALD can be employed to create an ongoing, thick, and nearly pinhole-free film as an inhibitor or encapsulated layer, preventing material and humidity migration through the device and unfavorable charge recombination between surfaces. Furthermore, the noninvasive approach to ALD manufacturing does not cause damage to the manufactured devices. Furthermore, outstanding lightweight, and functional ALD layers improve carrier extraction efficiency, increasing device performance. The utilization of ALD in the manufacture of PSCs yielded excellent results in recent years. It is interesting to note that ALD methodology is suitable to several PSC topologies, including planar, reversed, and mesoporous structures, which led to high PCEs of 20.00%, 18.30%, and 16.20%, respectively. Consequently, the stability and effectiveness of PSCs can be improved by sophisticated processing methods such as ALD.

4.8 Types of ALD

4.8.1 Thermal ALD

Thermal ALD demands temperatures between room temperature (~ 20 °C) to 350 °C for ligand swapping or combustion interface processes. It happens through surface responses, allowing precise depth modulation regardless of material morphology (depending on aspect ratio) or reactor layout. One of the most widely recognized instances of thermal ALD is the manufacturing of Al_2O_3 using trimethylaluminum (TMA) and water. Following TMA exposure, TMA dissociatively chemisorbs on the material's surface, and any residual TMA is flushed from the compartment. TMA's dissociative chemisorption results in a surface coated with $AlCH_3$. The exterior is then exposed to H_2O vapor, which interacts with the surface $-CH_3$, producing CH_4 as a reaction byproduct and leaving a hydroxylated Al_2O_3 surface [41].

4.8.2 Plasma ALD

Plasma-Enhanced Atomic Layer Deposition (PE-ALD) is an improved thin-film manufacturing technology that employs plasma as the reactant substance in a single phase of the ALD phase. In PE-ALD, the substrate is subjected to a plasma rich in reactive substances such as free radicals and ions. These species add energy to the outermost layer, allowing interactions at cooler temperatures. Plasma can be created either manually or locally on the backing material. PE-ALD provides more versatility in precursor selection and manufacturing parameters than thermal ALD. Nevertheless, it may additionally trigger issues like as diminished conformality and plasma ablation. To achieve optimal film quality, plasma characteristics must be carefully controlled [41].

4.8.3 Spatial ALD

Spatial Atomic Layer Deposition (SALD) is a novel version of Atomic Layer Deposition (ALD) that provides distinct benefits for thin-film fabrication. SALD incorporates continually providing substrates in multiple areas inside the unit, buffered via a neutral gas, resulting in quicker deposition rates than standard ALD. This approach retains the primary advantages of ALD, which include exquisite width manipulation at the nanoscale scale, great conformity, and the capacity to layer superior materials at ambient temperatures. SALD methodology is distinguished by its form-following, and excellent finishing at its atomic level, which results in enhanced production at reduced expenses, enhanced performance, effectiveness, renewable material elimination, and a greater lifespan for plated substances. The technique's capacity to easily transition from laboratory settings to high-volume manufacturing offers substantial prospects in diverse growing areas.

4.8.4 Photo-Assisted ALD

Photo-assisted Atomic Layer Deposition (Photo-ALD) is a specialized technology that uses photons to provide energy throughout accumulation reactions on the substrate, hence accelerating layer formation. This approach enables extremely material-selective and self-aligned deposition, which is particularly advantageous for purposes that need fine monitoring of film characteristics and width at an atomic level. Photo-ALD has benefits in regards to the consistent application of conformal films with adjustable depth, regardless of complicated three-dimensional surfaces, rendering it advantageous for different industries involving microchips, battery storage, catalytic processes, and therapeutics [41].

4.9 ALD Precursor

A metal encased in organic functional groups and kept in a bubbler is commonly referred to as an ALD precursor. The characteristics of the precursor have an impact on the bubbler temperature in an ALD system. ALD precursors have high reactivity, are volatile, and are thermally stable [40]. Rapid reactions between precursors and other precursor molecules as well as the active sites of substrates are desirable. Precursor chemistry, therefore, plays a crucial role in an ALD synthesis and influences the development principles. The emergence of novel antecedents has garnered significant interest in the past few decades. When creating new materials, mathematical techniques are useful for forecasting their physical characteristics before synthesis and for redesigning them based on available evidence from experiments [41].

4.9.1 ALD for Porous Membrane Preparation

The ALD coatings were previously used on a range of cell substrates, including porous polymers and inorganic ceramic template substrates. While the majority of published research focuses on modifying the physicochemical qualities of the pores' surface (hydrophilicity, organophilicity, catalytic activity, etc.) and decreasing their diameter, some studies concentrate on creating separate layers with adjustable characteristics. To the greatest extent of our understanding, Kim and Gavalas' 1995 work is the first work revealing ALD on porous membranes. They lined up H2-specific silica membranes by depositing a small amount of SiO_2 on a mesoporous Vicar Glass tubular substrate through switching loops of up to 35 cycles of $SiCl_4$ and $H2_O$ vapor reactants. This initial investigation of gas penetration via ALD-modified membranes revealed that their H2-modified membranes demonstrated a 20% drop in H_2 permeance and a rise in specificity to values above 2000 (at 600 °C) [42].In 1996, Dücsö et al. conducted the first conformity investigation in porous membranes, reporting the implantation of conical ALD SnO2 layers into the 65 nm broad holes of porous silicon with an aspect ratio (AR) of 1:140 [43].

4.10 ALD Advantages

The great benefits of ALD enable atomic-level pore size modifications and transformation of porous materials. Additionally, the proliferation of publicly available ALD reactors enables ALD accessible to scientists across a wide range of scientific domains, notably the membrane community. Nevertheless, before ALD reactors can be used in commercial membrane systems, there are still problems and technological obstacles that need to be overcome to produce membranes. In fact, by adjusting the ALD process (e.g., with plasmas or brief precursor contact), even the degree of intrusion within the framework might be regulated. It may also be possible to precisely tune (or even repair) membranes using the newly developed specific ALD methods for Nanopatterning [44] and atomic layer etching (ALE) [45].

4.11 ALD Limitations

ALD's primary drawback is its latency; typically, only a small portion of a monolayer is laid down in a single cycle. Accumulation rates are typically 100–300 nm h^{-1}. Therefore, using ALD to create micrometer-thick coatings is not particularly realistic. Manufacturing big quantities of substrate utilizing a single procedure, as in TFEL manufacturing, can make up for the slow growth rate. However, single-wafer processing is preferred in microelectronics. Thankfully, the films required for next-generation integrated circuits (ICs) are extremely thin, making the gradual pace of ALD less of a concern. Despite the chemical nature of ALD, leftovers from the substrates could always remain. Depending on how comprehensive the reactions are, the films' impurities level will vary [47]. The deposition reactor's sophistication and expenditure, as well as the ensuing metal precursors, are the route's constraints. These factors significantly affect the method's appeal to membrane makers. The price of this vacuum-based technology is the first problem. At the moment, membrane production is thought to have relatively high capital and operating expenses. Since the semiconductor industry is the driving force behind this technology, another challenge is the layout of ALD reactors that are conventionally oriented for thin film depositions on wafers. On the other hand, novel possibilities are being created by developing unique designs for ALD reactors that will boost the deposition throughput and expand the potential uses of ALD. For instance, conformal coatings of hydrophilic substances on or in traditional tubular ceramic membranes are made possible by the emergence of cylindrical and circulation reactors [48].

4.12 Financial Sustainability

Equipment for atomic layer deposition can cost about \$200,000 and \$800,000, depending on their quality and effectiveness. Administering an entire cycle of this instrumentation has no fixed expense; instead, the expenditure is determined by the temperature and timing of the equipment's operation, alongside the purity and effectiveness of the substances utilized. Because certain substrates are highly susceptible to oxygen and may accelerate their rate of decomposition, many are less easily accessible than others and call for particular circumstances. The majority of multicomponent oxides and other metals that are historically required in microelectronics manufacturing are not cost-effective.

4.13 Chemical Restriction

The main factors influencing Atomic Layer Deposition's (ALD) chemical constraints are the properties of the substrates utilized in the procedure. To enable gas-phase transport, ALD precursors need to be flexible. However, they also need to be stable, because the precursors are air- and oxygen-sensitive, which may restrict the spectrum of materials that can be employed. The selection of substances for films may be limited by their susceptibility to oxygen or air, particularly in the case of organism-based substrates that are particularly hypersensitive to heat and vulnerable to elevated decay rates. The need for precursors that can detrimentally chemisorb on their surfaces is another of ALD's chemical constraints. The regulated formation of light films with exact depth and compositional regulation at the microscopic level depends on this self-destructive chemisorption. To achieve consistent and conformal film advancement, chemical reactions must stop once all of the surface's accessible sites have been used, hence selecting the right kind of precursors is crucial to the methodology's effectiveness as well as quality.

4.13.1 Manageable Layering for Enhanced Efficiency and Scalability by Using ALD

In electrochemical processes, the degree of thickness and quantity of the active ingredients have a significant impact on the ion's dispersion range and the charge transmission path. Generally speaking, a more dense coating of active ingredients boosts your electrode's energy content at the expense of its electrical concentration. Thus, to maximize power and energy density, having control over the quantity of energetic material is crucial. ALD allows for easy manipulation of the deposited material width by adjusting the number of sessions performed. This allows for well-tuned active material laying with optimal efficiency. Furthermore, electrodes with a tiny amount of ALD provide a special advantage for studying and comprehending the active materials' charge transportation and retention process [39].

4.14 Corrosion

- **Prevention of Corrosion**

Corrosion refers to the degradation of substances caused by chemical processes. It is typically oxidation with molecules of air, usually in the presence of water. Corrosion can also occur when an acidic or basic corrosive chemical comes into contact with another substance. Corrosion alters a material's physical properties. Corrosion issues are most common with metals, although other materials can also corrode. Corrosion is a sort of erosion. Certain materials, such as stainless steel, are extremely resistant to corrosion. Metals that may erode can be maintained through plating, coating, and other methods. One type of severe temperature corrosion can result in the creation of condensed oxide layer glazes, which, according to particular conditions, minimize wear.

- **Alloying**

An alloy is a combination of one metal and another substance or metal. Alloying the exterior of metal prevents rust from forming, and metal combinations can lower the total expense of the substance.

- **Tinning**

Tin is a malleable, flexible, silver-colored metal. It's utilized to cover other metal surfaces to create alloys. Tin is not reactive to acidic meals, is hypoallergenic, does

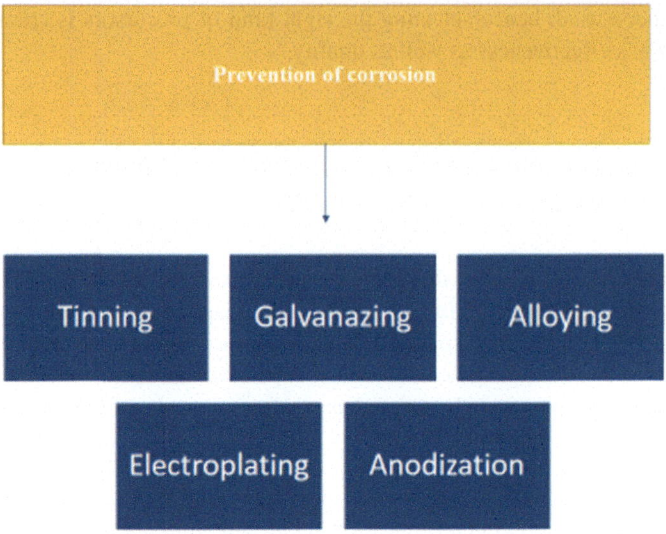

Fig. 4.4 Schematic representation for prevention of corrosion

not rust, and can prevent corrosion. It is also used to coat steel cookware to keep it from rusting.

- **Galvanizing**

Galvanization is the process of treating iron and steel surfaces with zinc to prevent rusting. It is among the most extensively utilized.

- **Anodization**

Anodizing is an electrochemical conversion procedure that produces an oxide coating, often on aluminum, in an electrolyte (chemical solution). The aluminum pieces serve as the anode (thus the word "anodizing"), and current flows between them and a cathode, typically flat aluminum bars, via the aforementioned electrolyte (sulfuric acid is most commonly employed) [49]. Figure 4.4 illustrates the representation of steps to prevent the surface from corrosion.

4.15 History of Electroplating

It's unclear when exactly the very initial electroplating experiment took place, although most people believe that Beccaria was the person who first effectively deposited metal in 1772 by discharging a Leyden bottle and dissolving metallic salts with the flame. Galvani's discovery in 1791 of the physiological effects of electricity on frog legs, which he named "an own fluid identical with electricity," was a significant advancement in the field [49]. Brugnatelli was a close companion of Alessandro

Volta, the man who gave his name to the electric unit "volt." Volta had just discovered the Chemical theories that permit the fabrication of "voltaic" electrical cells. Volta's first real instance of proving that was dubbed the "Voltaic Pile." This advancement made it possible for Brugnatelli to play around with different plating solutions thanks to his initial research with voltaic power. He had sufficiently perfected his method by 1805 to cover huge silver metal objects with a thin layer of gold [50].

4.15.1 Electroplating

The procedure of electroplating a metal onto an object to create an artifact is referred to as electroplating [51]. An essential method for creating functional, attractive, and/or repellent coatings on surfaces is electroplating. Reducing metal cations and then depositing them on the substrate is a crucial stage in the electroplating process. Using electrochemical theory as a guide, Brugnatelli demonstrated silver electroplating in 1805. Elkington submitted an initial patent request for cyanide silver plating in 1840, and it later became utilized in production, marking the beginning of the electroplating business [52]. An extensive selection of metals and their alloys are used in the multifaceted sector of electroplating for coating on various kinds of substrates for an extensive spectrum of practical and ornamental uses. Plastic is being coated in more and greater quantities, despite metals and alloys making up the vast majority of substrates. Operational qualities are more often utilized in plated materials than aesthetic ones [53]. The approach of electroplating has advanced from a sophisticated science to an art in the past few years. This discipline of applied science and engineering is thought to have an ever-growing quantity and variety of potential uses because of this progress. Among the technical domains where electroplating techniques and processes are crucial are electronic devices, including both micro and macro electronics, optics, optoelectronics, and a wide range of detectors [54]. The past twenty years have seen remarkable economic and industrial development in Asia, and this has greatly aided the growth of the engineering sector in the electroplating along metal-finishing industries. The rise in consumer goods consumption has led to the establishment of numerous small-to-medium-sized electroplating plants. As a result of the electroplating industry's significantly lower water usage compared with rival sectors, less wastewater is generated overall. Nevertheless, the existence of metals like these as cyanides, nickel, copper, zinc, cadmium, and chromium makes the effluent extremely dangerous [55]. The amalgamated sewage evacuated from zinc and copper plating facilities contained cyanide in amounts ranging from 1 to 6 mg/l. Given that these two plating procedures' baths contained the origins of cyanide found in the wastewater ought to come from cleaning the drag-out solution that was still adhered to the plated objects' surface. Metallic elements like copper, nickel, zinc, aluminum, and silver were also found in the effluent, depending on the type of plating technique (Fig. 4.5) [56].

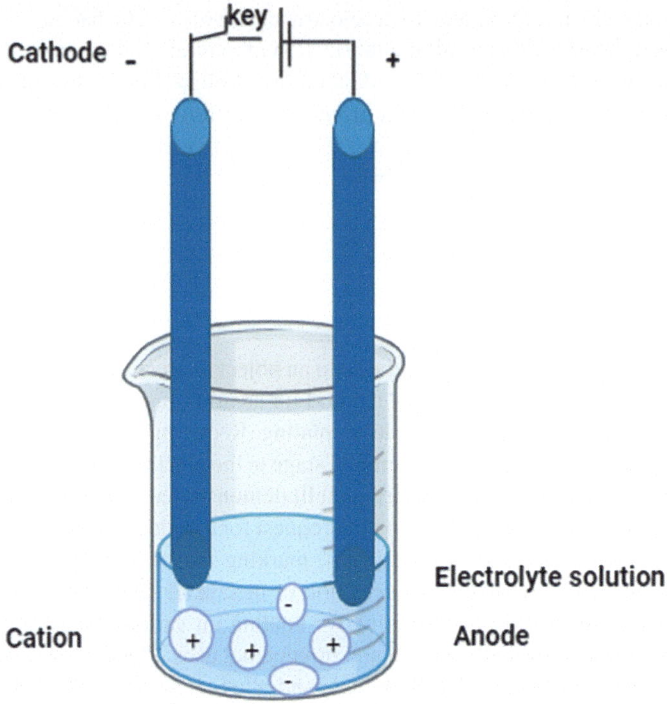

Fig. 4.5 Schematic representation of Electroplating

4.15.2 Working Principle of Electroplating

The term "Electroplating" describes a material coating technique that makes use of the electrochemical metal deposition mechanism. The substrate covering the sample is applied to serve as the anode, while the sample to be coated serves as a cathode. The two electrodes are submerged in an electrolyte mixture and further linked in a single orientation (DC). Nickel, chrome, copper, and zinc are among the metallic coatings that are frequently utilized. In essence, the purpose of the metal electroplating technique is to slow down the pace of corrosion while leaving an ornamental imprint on the metal surface. These electroplating techniques are used in the motor vehicle development and manufacturing industries.

4.15.3 Types of Electroplating

There are various methods for electroplating metals, including continuous plating, rack plating, line plating, and mass plating (also known as barrel plating).

4.15.3.1 Line Plating

Line plating is less expensive because an automated process is utilized to plate the pieces and fewer substances are utilized.

4.15.3.2 Rack Plating

More costly than mass plating, yet appropriate for intricate as well as huge objects. Components that are commonly submerged in solutions with "racks".

4.15.3.3 Continuous Plating

Anodes are regularly circulated via items such as cables and valves at a predetermined pace. The total expenditure of this operation is slightly lower.

4.15.3.4 Mass Plating

Considering it is inadequate at preventing scrapes and tangles, it is not the ideal choice for intricate items. However, this technique encompasses an extensive amount of components efficiently.

4.15.4 Purpose of Electroplating

- Presentation.
- Barrier.
- Unique characteristics of the surface.
- Mechanical or technical characteristics.

4.15.5 Plating Metals

- **Sacrificial Coating**

The sacrificial coating is mostly utilized for safety reasons. The metal utilized for the covering serves as a sacrificial element, being consumed during the process of reactions. Common metals include zinc and cadmium, which are now banned in many nations.

- **Minor Metals**

Although they are relatively infrequently employed in plating, minor metals often include iron, cobalt, and indium since they are simple to mount.

- **Decorative Coating**

The main uses of decorative coating are aesthetic and appealing ones. Typical metals consist of zinc, tin, nickel, copper, and chromium.

- **Functional Coatings**

Functional coatings are applied to metals according to their utility and requirement. Gold, silver, platinum, tin, lead, ruthenium, rhodium, palladium, osmium, and iridium are examples of frequently encountered metals.

- **Unusual Metal Coating**

Even less frequently than minor metals, unique metal covering refers to metals. As, Sb, Bi, Mn, Re, Al, Zr, Ti, Hf, V, Nb, Ta, W, and Mo are a few of them.

- **Alloy Coating**

A substance composed of multiple elements that exhibit metallic characteristics is called an alloy. Two distinct metals are plated in an identical cell to create these coatings. Gold–copper–cadmium, zinc–cobalt, zinc–iron, zinc–nickel, brass (a copper-zinc alloy), bronze (copper-tin), tin–zinc, tin–nickel, and tin–cobalt are examples of frequently occurring pairings.

4.16 Preparation for Plating

Cleaning, or getting rid of grease or other "foreign" impurities, and pickling, or getting rid of oxides or other metallic compounds, are the two main steps in the plating preparation process for metals. Modifications must be pursued by speeding up or lowering the expenditures of the current scrubbing techniques, which mostly include solvents, emulsions, and electrolysis of alkaline solutions. These techniques are generally satisfactory [57].

4.17 Electroplating of Various Metals

4.17.1 Nickel Electroplating

There are similarities between nickel plating and various other electroplating techniques using aqueous metal anodes. It needs two electrodes submerged in a conducting, water-based solution of nickel salts to transmit a direct current across them. One electrode (the anode) dissolves when direct current flows and the other electrode (the cathode) gets wrapped in nickel. Divalent positively charged ions (Ni^+) are a sort of nickel that appears in suspension. The positively charged ions react with two electrons ($2e^-$) whenever the current passes, turning them into metallic nickel (Ni^0) at the cathode surface. The opposite happens at the anode, in which the metallic nickel dissolves and releases positively charged divalent ions into the solution. The anode produces new nickel ions that compensate for the ones released at the cathode [58].

4.17.2 Copper Electroplating

Substantial mechanical strength copper films having a high volume of twin barriers are well-known for having minimally sacrificed electrical conductivity. Dual lamellae and separation will be dependent on the nanoscale to reach such a high ratio. Pulse electro-deposition was used to create 10 μm copper films with varying imposed pulse maximum current densities and pulse times. It was discovered that the pulse-based plating characteristics affect the microscopic characteristics of the coatings. Although lower energy pulses resulted in a more chaotic microstructure in the deposits and rougher surface topography, stronger energy pulses led to bigger self-annealing impacts on crystal re-crystallization and expansion, resulting in better fiber patterns [59].

4.17.3 Gold Electroplating

Of all metals, gold is ranked third best at transferring heat and electricity at room temperature. Upon its surface, dielectric surface oxide cannot develop because of its great durability, outstanding resistance to wear, and stability. Gold and wire made of aluminum coupling work quite well on this surface. The majority of wire bond and surface mount applications that require a lot of bump bonding on a chip employ soft gold electroplating. To allow for slight variances in its thickness, the bumps on the gold should be reasonably pliable and flexible. Although gold electroplating creates deposits with qualities that meet most of the unique requirements for these applications, it is significant to the electronic devices sector. Technologies such as wire bonding and wafer bump hyperlinks are made possible by the development of pure gold electrodeposition. The majority of traditional methods for pure gold electroplating utilized in the field of electronics are cyanide-based. Wafer technologies have historically relied on sulfite-based gold methods because of the cyanide's inconsistency with favorable photoresists and its hazardous nature in the shape of free cyanide that results through electrolysis [60].

4.17.4 Platinum Black Electroplating

Microelectrodes made of platinum black are a popular and efficient way to capture brain signals. The key benefits of platinum black microelectrodes are their appropriate biological compatibility, superior signal-capturing performance, and ease of manufacture. Different physical stimuli are delivered to microelectrodes when they come in contact with a realistic biological structure. Despite the exceptional electronic characteristics of platinum black microelectrodes as neurological sensors, their porous exterior is so sensitive that external forces can easily break it [61].

4.17.5 Tin Lead Electroplating

Tin and tin/lead alloy deposits are helpful to the electronics sector, especially for making boards for printed circuits, electrical conduits, semiconductors, and other related items when these coatings' intrinsic qualities are required. Any plating electrolyte used to deposit tin and tin/lead alloy deposits can attain a certain level of high-performance plating if the amount of agitation, electrolyte concentration, and/or solution temperature is increased [62].

4.17.6 Silver Electroplating

A product composed of a different kind of metal can have a tiny layer of silver applied to it via a chemical technique called silver electroplating. To create a solid coating of silver which is chemically linked to the opposite substance, this procedure employs an electric current to deposit silver ions from an electrolyte solution onto the object. Many advantages come with silver electroplating, such as antibacterial qualities, good thermal and electrical conductivity, resistance to corrosion, and affordability. Since it has extensive applications, silver is utilized in a wide range of sectors, notably telecommuting, automobiles aircraft, gadgets, cutlery, medical treatment, jewelry, and military equipment. To produce different textures and quality, the procedure uses several techniques such as inner anode plating, rack plating, and barrel plating. The methodology of silver electroplating uses silver as the anode material and the target object as the cathode. It is an electrolytic method. When an electrical current is applied to the electrolyte, the silver ions attach immediately to the metallic surface and form a film of pure silver. Figure 4.6 describes the schematic representation of silver electroplating.

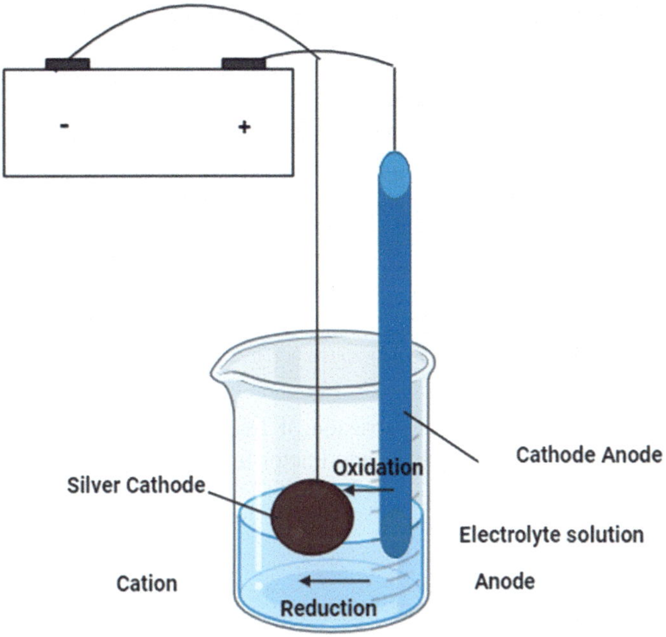

Fig. 4.6 Schematic representation of silver electroplating

4.17.7 Cadmium Electroplating

Cadmium plating, alternatively referred to as cadmium electroplating, is a metallic coating type possessing unique and practical characteristics. Steel, copper, iron alloys of all kinds, and powdered metals are common substrate elements for cadmium plating, which is mostly utilized in the aerospace sector and other locations subject to elevated salinities. The properties of cadmium electroplating is

- Strong defense against corrosion from saltwater.
- Gives adhesives a remarkable bonding surface.
- It serves as an excellent electrical conductor.
- Exhibits a strong solder ability.
- Inbuilt immunity against germs and mold.

4.18 Application of Cadmium Electroplating

4.18.1 Aeronautics and Defense

Because of its inherent lubrication and anti-corrosive properties, cadmium electroplating is perfect for critical in-flight applications. Because of its adhesive qualities, it is particularly helpful for attaching things to aircraft aluminum frames.

4.18.2 Transportation

Because cadmium plating resists corrosion, it frequently appears in disk brakes and hoses of outstanding durability cars, tractor-trailers, and various other forms of transportation.

4.18.3 Marine

Cadmium plating is extremely beneficial in maritime applications because it provides defense against corrosion from seawater.

4.19 Types of Cadmium Plating

Class I. This kind of plating, often known "as plated," is the standard color palette for cadmium electroplating; no further color processing is used.

Class II. This kind of yellow-colored cadmium plating is iridescent. The external surface of the plated element is sprayed with salt to evaluate the type II treatment's performance. After 96 h, the component is examined for corrosion.

Class III. This kind of cadmium plating is typically used as a paint base and has an additional pigmentation with phosphate.

4.19.1 Zinc Electroplating

Rusting and corrosion are all kinds of metal's worst enemies. Rust's corrosive action can be quite damaging. According to statistics from multiple organizations, the annual cost of damage due to rusting is approximately $276 billion in the

United States and approximately $1 trillion worldwide. Rusting has a disastrous effect on several businesses, including shipment, manufacturing, and production. The application of zinc electroplating halts corrosion and rust.

- **Colors and types of zinc coating**

Zinc electroplating comes in a variety of colors, however, each color is a product of the manufacturing techniques employed, meaning that each color has a unique rusting and corrosion impact.

- **Yellow Zinc Coating**

This zinc coating provides an excellent degree of resistance against rusting and is typically employed in the automotive sector.

- **Black Zinc Coating**

This is a naturally occurring black coating, or it might be a gray-green oxide. Generally speaking, the grayish-green shade will result in marginally improved rust resistance.

- **Blue zinc**

Blue zinc sometimes known as clear zinc, is said to be less hazardous to the environment than other coating formulations, but it still aims to provide some defense against corrosion.

- **Acid Zinc**

Since cyanide is not added to the coating fluid, acid zinc coating differs from various kinds of zinc coating. One important advantage of zinc acid is that it works well on challenging surfaces.

4.19.2 Brass Electroplating

To achieve perfectly polished iron or, even more desirable, bright nickel-plated iron, brass plating is either directly applied, or it serves as a barrier for later nickel plating (brass plating of die-cast alloys composed of aluminum or zinc, for example), or it can be employed to help rubber adhere to steel, most likely through a chemical bond with sulfur or sulfur compounds found in rubber. However, beautiful ornaments for the home, apparel, hinges, handles, lamps, toys, etc. are the main applications for brass plating. The copper and zinc in the anodes (which in turn are made of brass) must disintegrate in the same amount as they are deposited at the cathode for the brass plating baths to function properly; if not, the mixture needs to be adjusted with the proper additions. As a result, the composition of the two metals in the mixture could vary even though the anodes must be constructed of brass that is identical to, or very similar to, the brass that needs to be deposited. A specific proportion of the elements are present in brass of that particular color. Nevertheless, the percentage in electrolytic brass and molded or laminated brass is different, even though the shade is the same [63].

4.20 Treatment of Electroplating Industry Wastewater

Improper electroplating effluent release may cause hazardous metal–organic compounds that are bad for the ecosystem. Various methods have been documented for the treatment of wastewater resulting from electroplating, including chemical precipitation, adsorption, coagulation and flocculation, ion exchange, electrocoagulation as well membrane filtering, improved oxidation procedure, and bio-recovery. These procedures offer the chance to extract organic contaminants as well as separate various harmful elements. The entire sanitation process, from freshwater extraction to the last step of effluent discharge to the ecosystem, should be included in the evaluation of costs [63].

4.21 Application of Electroplating

4.21.1 Application of Electroplating in the Aerospace Industry

- **Prevention against Corrosion**

Electroplating is employed to prevent corrosion, a key aerospace sector significance, as on parts of spacecraft and fighter jets. Section durability can be compromised by corrosion, increasing expenses for upkeep, and posing potential hazards to the safety of citizens. The service life of these elements is increased by electroplating them with materials such as nickel and zinc-nickel alloys, which contribute to avoiding corrosion.

- **Heat Resistance**

Aeronautical components' heat resistance can be increased through electroplating, which is essential in high-temperature uses including radiation shields and parts for engines. Because of extremely excellent conductivity and heat resistance, materials like gold and silver are frequently utilized.

4.21.2 Application of Electroplating in Electronic Devices

- **Conductivity**

Connectors, toggle switches, and other electronic gadgets can have their electrical conductivity improved by electroplating them. Because of their exceptional conductivity, materials like copper, gold, and silver are frequently employed. Electroplating is a technique used to prevent corrosion on gadgets, which can shorten their service life and cause breakdowns in devices. Because they repel corrosion, materials like nickel and zinc-nickel alloys are frequently utilized.

- **Aesthetics**

Electroplating improves the visual appeal of electrical parts, particularly consumer gadgets' ornamental embellishments. Because of their appealing aesthetic and long endurance, materials including copper, silver, and gold are frequently employed [75].

4.21.3 Application of ALD in Jewelry Manufacturing

- **Improving Appearance**

During the production of jewelry process, electroplating is frequently utilized to adhere a thin layer of valuable metals, such as gold or silver, over less costly metals, improving the jeweler's items' look and worth.

- **Customization and Layout**

Electroplating gives jewelry a wide range of textures and shapes to suit a range of preferences and aesthetics. It also offers creative craftsmanship and flexible customization options.

- **Tarnish Resistance**

Despite its appearance, electroplating improves tarnishing resistance in jewelry by shielding the underneath metal from deterioration and wear, hence enhancing the item's functional life.

4.21.4 Application of ALD in Home Decoration

- **Ornamental Finishing**

Electroplating is a home décor technique that applies metals such as copper, gold, silver, or chrome to a variety of items to create artistic finishes that change their look and give them a sleek, contemporary feel.

- **Protection and Longevity**

By providing safeguards against corrosion and damage, electroplating guarantees that ornamental products maintain their original splendor across time as well as are resilient and long-lasting when used for home décor purposes.

4.21.5 Application of Electroplating in Dentistry and Medical Field

Electroplating is a technique employed to give various dentistry and medical components durable exterior surfaces. Tooth fillings and other dental operations are frequently made possible by the use of gold plating. Electroplating is a common technique used on embedded parts, which include replacement joints, screws, and plates, to increase their corrosion resistance and compatibility with pre-insertion sterilization. Electroplating is also frequently used for medical and surgical equipment, such as forceps and radiological components [75].

4.22 Advantages of Electroplating

The following are some advantages of electroplating,

- **Heat resistance**

Several metals, like zinc-nickel and gold, might withstand heat better than other metals, which helps the substrate resist heat-related deterioration. This might therefore increase how long plated products last.

- **Electrical Conductivity**

Copper and silver plating enhances electrical characteristics in equipment and offers an effective, reasonably priced way to increase conductivity in electronics and electrical parts.

- **Shield Measure**

When electroplating, an obstacle is applied to the base to keep contaminants out. There's a chance that this shield may occasionally impede atmospheric deterioration. Because they can endure rougher settings for an extended amount of time and need fewer instances of replacement, this feature is very helpful for equipment.

- **Hardness Enhancement**

Electroplating is often used to fortify substrate substances, lowering their vulnerability to deterioration from abuse or stressful employment. This feature can reduce the need for refurbishment and increase the lifespan of gilded systems [75].

4.22.1 *The Positive Impact of Electroplating: Dazzling Surfaces*

- **Beautiful Looks**

Electroplating gives everyday items a stunning sheen, transforming them into intriguing beauty.

- **Fight Rust**

Objects with plated surfaces have an extended lifespan because they are resistant to corrosion.

- **Extra Toughness**

Surfaces can be made stronger through electroplating, offering a further layer of defense.

- **Electrically Astute**

Coated surfaces have improved electrical conductivity, which is advantageous for electronic devices.

- **Artful Precision**

Electroplating allows for the creation of complex motifs by focusing on particular locations.

4.22.2 *Analyzing the Drawbacks of Electroplating: The Obstacles*

- **Environmental Concerns**

Electroplating generates rubbish and uses several unfriendly chemicals.

- **It's Expensive**

The expenditures of materials, energy, and environmental concerns make the entire procedure far from inexpensive.

- **Difficult Technique**

Electroplating excellence demands exact monitoring and is not something that can be easily attained.

- **Coating Depth Problems**

Achieving the ideal coating thickness can be challenging, similar to trying to strike an object that is shifting.

- **Extended Plating Time**

In certain instances, the plating process may take unduly long. Elevating the power source or levels of electrolytes can accelerate the formation rate; nevertheless, this may result in inconsistent deposition.

- **Intricate Procedure**

For the sake of satisfactory electroplating is a challenging endeavor that necessitates exact monitoring of factors such as current, voltage, solution composition, and temperature. The preliminary treatment and preparatory errors might result in flaws and inferior products [75].

4.23 Molecular Beam Epitaxy

The technique known as molecular beam epitaxy (MBE) is used to create thin, epitaxial Coating on an extensive variety of substrates including metals, oxides, and semiconductors. In the beginning compound semiconductor fabrication was the focus of its application. In this method, elevated crystals that have been treated previously to create a surface that is almost precisely pure are impacted by a stream of atomic particles or molecules in an ultrahigh vacuum phase. As the individual atoms arrive, they arrange themselves into a crystalline layer, the network layer or epitaxial film, in alignment with the surface [64]. Using molecular beam epitaxy, semiconductor crystals can be grown with exceptionally thin epitaxial layers at extremely high vacuum conditions. Extensive dimensional supervision of contamination inclusion and large compositional alterations can be accomplished because the procedure is intrinsically sluggish growing [65]. The technique of growing compound semiconductor films epitaxially by reacting one or hotter molecule beams with a crystal's interface in an ultra-high vacuum is known as molecular beam epitaxy, or MBE. Similar to vacuum evaporation, MBE enables fast switching of beam species and provides considerably greater supervision over the incident atomic or molecule energies to allow adhering coefficient differences to be taken into consideration [66]. Naturally, today's materials systems include elemental and compound semiconductors, metals, magnetic materials, and minerals, oxides, and additionally organic systems, however the III–Vs saw the majority of progress. A reasonable definition of MBE would be the formation of epitaxial films as a result of incident molecular beams on crystalline surfaces. It is important to note that one of the forerunners of the current method grew Si from SiH4 as an extraordinarily early illustration of the creation of epitaxial films using MBE [67]. When it comes to controlling the thickness and homogeneity of epitaxial layers across a sizable substrate region, molecular beam epitaxy (MBE) remains the gold norm. We can now manufacture epitaxial materials free of oval flaws with electrical and spectral characteristics that are either superior to or equal to those of any other crystal growth technology because of current advances in expansion circumstances. We can investigate the chemical and physical features of materials

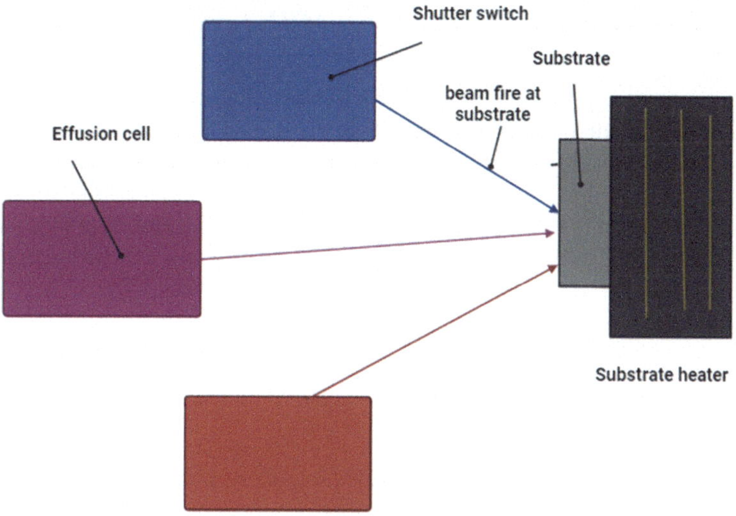

Fig. 4.7 Working diagram of molecular beam epitaxy

that are substantially organized and that are not found in ecology [68]. Because of its powerful sticking power, reduced weight, and increased rigidity, epoxy resins are a viable glue substance for transport systems like cars and airplanes. MBE is essentially a more sophisticated type of vacuum evaporation that happens when heated substrates are impacted upon by prescribed, impartial thermal atomic and molecule beams operating in ultra-high vacuum (UHV). Large sections of liquid nitrogen-cooled panels are incorporated into the two- or three-chamber stainless-steel UHV system, which is often ion- or cryo-pumped. To produce films with superior electrical and optical qualities, a latching system must be included that allows the substrate to be added to the growing chamber without rupturing the vacuum [64]. Figure 4.7 describes the working diagram of molecular beam epitaxy (MBE).

4.23.1 Vacuum Requirement for MBE

Vacuums offer a special environment that allows the material to be geared up, described, and altered; as a result, they are crucial to a variety of innovations. The desired vacuum level is dependent on how leftover gases affect the process. The kinetic theory of gases describes how gases behave as an outcome of pressure. This approach yields a minor correlation between the molecular density, gas pressure, mean free path, and surface molecular impingement rate [69].

4.23.2 Fundamental Component of MBE Growth Chamber

The fundamental components of the MBE growth chamber consist of a stainless steel funnel with a diameter of less than 20 inches, a cell that generates a molecular beam, a shutter that turns the molecular beam on and off, a substrate holder that can be heated to temperatures of several hundred degrees Celsius, and a load lock system that permits the entrance of wafers into the chamber and their removal without disrupting the UHV proliferation environment. To analyze electron diffraction from growing surfaces in real life, an electron gun is usually offered [70].

4.23.3 Epitaxial Films of SCs

Semiconductor epitaxial films have remained important in instrument analysis due to the fact they can be manufactured with envisioned electrical characteristics that differ in comparison to those of the substrates such as higher utmost purity less imperfections, or a distinct amount of electrically active impurities. It is possible to generate unique device designs without requiring manufacturing processes that include the diffusion of impurities to make doped coatings by depositing a series of epitaxial layers with certain features. Vapor-phase epitaxy (VPE), which involves reacting an appropriate combination of gaseous vapors comprising the individual components with a warmed seed or substrate crystal, is a method that can produce various semiconductor materials epitaxially [71].

4.23.4 MBE Growth Devices

The creation of cutting-edge electrical and optoelectronic devices is the main inspiration behind MBE development right now, even though a large portion of the field's early work focused on the kinetics of surface film formation. As the film growth capabilities of MBE became apparent, scientists produced a wide range of previously unthinkable and inconceivable gadgets. While many of these devices were originally designed for use in MBE systems, others have since been manufactured using alternative epitaxial processes. MBE can create precisely discontinuous heterojunctions and doping profiles because of the benefits of typically lower growing temperature and growth rate. Extremely rapid devices have been made possible by the capacity to create these compositional variations with material systems that have high electron mobility by nature, such as GaAs, In, and InGaAs. It has been possible to create microwave devices with operating frequencies close to 100 GHz and fast digital switching times of about 5 picoseconds. Additionally, by controlling the contact layers, parasitic resistances have been decreased, greatly enhancing FET performance (MAG and NF) [72].

4.23.5 MBE Characteristics

For the MBE procedure, vacuum parameters are usually greater than 10^{-10} Torr. This allows for the growth of epitaxial films at comparatively low underlying temperatures that have good crystal clarity along with elevated purity. Furthermore, by using a range of structural and quantitative sensors, the ultrahigh vacuum environment enables immediate measurement of the bulk, interface, and surface features of the expanding coating.

4.23.6 MBE History

Despite Günther's 1958 proposal, the MBE deposition technique was not put into practice until the advancement of ultrahigh vacuum technology. Epitaxial GaAs films were successfully grown by Davey and Pankey in 1968 using the MBE technique. Concurrently, Arthur's research on the kinetics of GaAs production established the foundation for Arthur, LePore, and Cho to create superior MBE films of GaAs and other III-V compounds [73].

4.23.7 Organic Thin Film Growth by MBE

Molecular beam epitaxy has been used to successfully generate high-quality, ultrathin organic films on a variety of substrates. Despite the significant lattice discrepancy among the newly formed organic crystals and the substrates, several distinct types of organic films have been epitaxially synthesized on such materials, using van der Waals surfaces as the fractured faces of stacked substances. Additionally, it has been discovered that the epitaxial development of organic films can occur on economically significant substrates like silicon and GaAs, provided that the appropriate termination of surface dangling bonds with appropriate atoms transforms those substrates' surfaces into quasi-van der Waals ones. Ionic crystals like alkali halides, which have an asymmetrical charge distribution, can also be employed as suitable substrates for organic molecules; epitaxial growth is achieved with the help of electrostatic interaction [74]

4.23.8 Molecular Beam Epitaxy for Fabrication of Quantum Wires and Quantum Dots

Molecular beam epitaxy technology provides substantial benefits in semiconductor technology due to its high flexibility, particularly when preparing materials such as quantum wires and quantum dots. Chemical beam epitaxy (CBE), metal–organic compound molecular beam epitaxy (MOMBE), and laser molecular beam epitaxy (L-MBE), which merge molecular beam epitaxy with pulsed light and other technologies, have all permitted the development of novel sorts of quantum wires. Quantum dot materials have become increasingly diversified, and the prospect of combining laser and molecular beam epitaxy innovation has piqued physicists' interest. In addition, the created quantum wires and quantum dots have numerous uses in superfast optical technology, energy, and micro-nano optoelectronic systems [75].

4.23.9 Characteristics of MBE

1. Molecular Beam Epitaxy (MBE) is a thin film deposition mechanism that grows exceptional epitaxial layers with perfect regulation of their chemical makeup and physical makeup. Some characteristics of MBE are as follows,
2. Controlling surface composition and morphology.
3. Altering chemical composition at interfaces.
4. Controlling crystal formation at the atomic level.
5. Low development at a pace of roughly a single monolayer (lattice structure plane) per second.
6. Insufficient expansion temperature, usually about 550 °C for GaAs.
7. Linear growing surface, with atomic-height steps and vast plain slopes.
8. Stoichiometry secured by excess group V flux impinging on the growth surface.
9. Layer-by-layer growth.
10. Unity sticking coefficient for many group III elements like Al or Ga, but significantly different for group V elements.
11. Requires ultra-high vacuum (UHV) conditions, typically in the range of 10^{-9} to 10^{-11} Torr.
12. Supports the emergence of several materials, notably metals, oxides, and semiconductors.

MBE offers exceptional oversight over the formation of complex lattice-matched heterostructures and superlattices, allowing the manufacture of technically sophisticated semiconductor devices such as high electron mobility transistors, quantum cascade lasers, and quantum well lasers.

4.23.10 Advantages of MBE

Varying the sort of expansion, there are benefits to preferring particular growth circumstances. When developing III-Sb semiconductors, it is critical to keep substrate temperatures low since antimony is easily eliminated from the surface at higher temperatures. This is particularly relevant for materials with low, non-congruent temperatures. The non-congruent temperature is the temperature at which one atomic species is predominantly vaporized, leaving behind non-stoichiometric material. To decrease atom diffusion on the surface, strained layer development necessitates lower substrate temperatures. This lessens the likelihood of the layer relaxing, thereby negating the positive effects of injecting tension into the framework.

- **Comprehensive surveillance over the composition and layering of the film**
- By carefully regulating the beam fluxes and growth time, MBE enables atomic-scale layer-by-layer growth.
- This makes it possible to create intricate heterostructures and quantum wells.

Exceptional, faultless flicks

- MBE produces high-purity, single-crystal films with extremely low defect densities because of its slow development rate and ultra-high vacuum atmosphere.

Materials flexibility

- Metals, oxides, and semiconductors in grades III-V and II-VI can all be deposited by MBE.
- This makes it possible to create cutting-edge gadgets including quantum-well lasers and high-electron mobility transistors (HEMTs).

In situ oversight and management

- Methods such as reflection high-energy electron diffraction (RHEED) allow the surface's makeup as well as development mechanism to be tracked in continuous time.
- This enables exact monitoring of the development and enhancement of the film's characteristics.

Lower Development Temperature

- In comparison to other epitaxial growth methods, MBE functions at temperatures between 600 and 900 °C. This reduces diffusion and the heat of the allocated funds, which is good for VLSI technology.

4.23.11 Disadvantages of MBE

The drawback is that additional defects are formed during development when deposited atoms' mobility decreases, requiring them to migrate along the substrate to identify the optimum 'epitaxial' position. There are several drawbacks with MBE. Being a very slow process with crystal growth rates of only a few microns per hour, it is a very laborious and tedious operation. Because of its slow speed, MBE operates under conditions favorable to research rather than high-volume production. In addition, since neat high-vacuum conditions can be difficult to establish and maintain, the equipment that is necessary for MBE is complex and expensive.

4.24 Conclusion

Atomic Layer Deposition (ALD) is a highly accurate and versatile thin-film deposition technique that offers almost unprecedented control over film thickness and composition at the atomic level. Its ability to produce uniform and conformal coatings on complex three-dimensional substrates has made it crucial for advancing nanotechnology, electronics, energy conversion, and medical applications. The self-limiting surface reactions in ALD allow for very precise deposition of material at a sub-monolayer level; no other deposition technique can rival the film uniformity and conformity on substrates with high aspect ratios, where failure is experienced with other methods. The inherent ability to control thickness at an atomic level makes the technique best suited for materials and devices fabricated with critical dimensions and properties required. Atomic layer epitaxy was first reported as a concept in the 1970s, and the development of ALD thus has its historical roots on this basic concept. ALD emerged with the quest for further miniaturization of electronics; the developments realized for ALD were largely because ALD could easily accommodate materials of a very wide range, including metal oxides, nitrides, and other complex compounds. Thus, ALD has emerged as an indispensable tool for the fabrication of next-generation electronic devices, nanostructures, and high-performance coatings. Although ALD is very widely spread and has been used for a tremendous amount of benefits, there are still some physical and chemical mechanisms inside the ALD processes that are not wholly understood. It is still in need of more research and development of ALD, using both theoretical calculations and multistate modeling, to improve the fundamental knowledge and widen its applications further. Conclusion In a nutshell, ALD is one of the most revolutionary technologies in the realms of material science and engineering-a material that can be controlled very precisely and facilitated with new solutions of various high-tech industries. Further evolutions, in all likelihood, will serve as a stepping stone toward future technological breakthroughs—atomic and near-atomic scale manufacturing. Electroplating is one among electrochemical processes more or less widely used for depositing thin layers of metals onto substrates with the help of an electric current. With this method, corrosion resistance improves upon it, and aesthetics are also enhanced for the better, and while not as ideally important, specific surface properties tailored to a particular application are also achieved. The process of electroplating involves immersing two electrodes into an electrolyte solution containing an anode, which is the material to be plated, and a cathode, which is the substrate to be coated, under a direct current. Electroplating is very much used across industries ranging from automotive to aerospace, electronics, and manufacturing because it can produce wear-resistant and aesthetically pleasing metal coatings. Varieties of metals such as nickel, copper, gold, platinum, tin–lead, silver, and cadmium are electroplated, each with specific purposes and benefits. Molecular Beam Epitaxy: This is a quite intricate technique used in the formation of thin, epitaxial coatings on different substrates consisting of metals, oxides, and semiconductors. First developed to make compound semiconductors, MBE has come to play a dominant role in the manufacture of state-of-the-art electronic and optoelectronic devices because it allows the deposition of structures with controlled thickness at an atomic level, as well as homogeneous composition. MBE works at UHV leading thusly to preparations with minimal contamination and in highly crystalline quality. The process involves directing beams of atomic or molecular species onto a heated substrate where they condense to form a crystalline layer aligned with the lattice structure of the substrate. Such control over the parameters of growth permits the growth of very pure and almost defect-free films. MBE remains, despite its limitations, still the cornerstone in materials science as well as semiconductor technology. Much of the recent progress in electronic and optoelectronic devices has been due to the ability of MBE to grow highquality epitaxial films with controlled composition and structure. Flexibility, high accuracy, and the ability to make complex heterostructures make MBE indispensable in stateoftheart research and nextgeneration technologies. Advanced technology in MBE must play a pivotal role in the continued evolution of material science and device engineering.

References

1. Johnson, R.W., A. Hultqvist, and S.F. Bent. 2014. A brief review of atomic layer deposition: From fundamentals to applications. *Materials Today* 17 (5): 236–246.
2. Cremers, V., R.L. Puurunen, and J. Dendooven. 2019. Conformality in atomic layer deposition: Current status overview of analysis and modelling. *Applied Physics Reviews* 6 (2).
3. Oviroh, P.O., et al. 2019. New development of atomic layer deposition: Processes, methods and applications. *Science and Technology of Advanced Materials* 20 (1): 465–496.
4. Knoops, H.C., et al. 2015. Atomic layer deposition. In *Handbook of Crystal Growth*, 1101–1134. Elsevier.
5. Muneshwar, T., et al. 2018. Atomic layer deposition: Fundamentals, practice, and challenges. In *Handbook of Thin Film Deposition*, 359–377. Elsevier.
6. Zhang, J., et al. 2022. Advances in atomic layer deposition. *Nanomanufacturing and Metrology* 5 (3): 191–208.
7. Kang, S., et al. 2024. Toward fast and accurate machine learning interatomic potentials for atomic layer deposition precursors. *Materials Today Advances* 21: 100474.
8. Puurunen, R.L. 2005. Surface chemistry of atomic layer deposition: A case study for the trimethylaluminum/water process. *Journal of Applied Physics* **97** (12).
9. Niinistö, L., M. Ritala, and M. Leskelä. 1996. Synthesis of oxide thin films and overlayers by atomic layer epitaxy for advanced applications. *Materials Science and Engineering: B* 41 (1): 23–29.
10. George, S.M. 2010. Atomic layer deposition: An overview. *Chemical Reviews* 110 (1): 111–131.
11. Meng, X., et al. 2017. Atomic layer deposition for nanomaterial synthesis and functionalization in energy technology. *Materials Horizons* 4 (2): 133–154.
12. Zhang, Z., et al. 2020. Atomic layer deposition-derived nanomaterials: Oxides, transition metal dichalcogenides, and metal–organic frameworks. *Chemistry of Materials* 32 (21): 9056–9077.
13. Lu, J., J.W. Elam, and P.C. Stair. 2016. Atomic layer deposition—Sequential self-limiting surface reactions for advanced catalyst "bottom-up" synthesis. *Surface Science Reports* 71 (2): 410–472.
14. Maeng, W., S.-J. Park, and H. Kim. 2006. Atomic layer deposition of Ta-based thin films: Reactions of alkylamide precursor with various reactants. *Journal of Vacuum Science & Technology B: Microelectronics and Nanometer Structures Processing, Measurement, and Phenomena* 24 (5): 2276–2281.

15. Ye, P., et al. 2003. GaAs MOSFET with oxide gate dielectric grown by atomic layer deposition. *IEEE Electron Device Letters* 24 (4): 209–211.
16. Gusev, E.P., et al. 2003. Ultrathin HfO2 films grown on silicon by atomic layer deposition for advanced gate dielectrics applications. *Microelectronic Engineering* 69 (2–4): 145–151.
17. Elam, J., et al. 2003. Conformal coating on ultrahigh-aspect-ratio nanopores of anodic alumina by atomic layer deposition. *Chemistry of Materials* 15 (18): 3507–3517.
18. Collins, R., and A. Ferlauto. 2002. Advances in plasma-enhanced chemical vapor deposition of silicon films at low temperatures. *Current Opinion in Solid State and Materials Science* 6 (5): 425–437.
19. Puurunen, R.L. 2014. A short history of atomic layer deposition: Tuomo Suntola's atomic layer epitaxy. *Chemical Vapor Deposition* 20 (10–11–12), 332–344.
20. Ritala, M., and M. Leskelä. 2002. Atomic layer deposition. In *Handbook of Thin Films*, 103–159. Elsevier.
21. Puurunen, R.L., H. Kattelus, and T. Suntola. 2010. Atomic layer deposition in MEMS technology. In *Handbook of Silicon Based MEMS Materials and Technologies*, 433–446. Amsterdam: Elsevier.
22. Parsons, G.N., et al. 2013. History of atomic layer deposition and its relationship with the American Vacuum Society. *Journal of Vacuum Science & Technology A* 31 (5).
23. George, S.M., J.D. Ferguson, and A.W. Weimer. 2003. *Atomic layer controlled deposition on particle surfaces.* Google Patents.
24. Allen, N., et al. 1992. Electrophotographic and microwave photodielectric studies (I) effect of various transition metal dopants on titanium dioxide pigments in the solid state. *Dyes and Pigments* 20 (3): 211–226.
25. George, S.M., et al. 2005. *Nanocoated primary particles and method for their manufacture.* Google Patents.
26. Van Bui, H., F. Grillo, and J. Van Ommen. 2017. Atomic and molecular layer deposition: Off the beaten track. *Chemical Communications* 53 (1): 45–71.
27. Napari, M. 2017. *Low-temperature thermal and plasma-enhanced atomic layer deposition of metal oxide thin films.* University of Jyväskylä.
28. Munoz-Rojas, D., and J. MacManus-Driscoll. 2014. Spatial atmospheric atomic layer deposition: A new laboratory and industrial tool for low-cost photovoltaics. *Materials Horizons* 1 (3): 314–320.
29. Musschoot, J. 2011. *Advantages and challenges of plasma enhanced atomic layer deposition.* Ghent University
30. Kim, H., and I.-K. Oh. 2014. Review of plasma-enhanced atomic layer deposition: Technical enabler of nanoscale device fabrication. *Japanese Journal of Applied Physics* 53 (3S2): 03DA01.
31. Noh, S., et al. 2006. Construction of an atomic layer deposition system for nano-device applications. *Current Applied Physics* 6 (2): 171–173.
32. Aneke, M., and M. Wang. 2016. Energy storage technologies and real life applications—A state of the art review. *Applied Energy* 179: 350–377.
33. Bokhari, S.W., et al. 2020. Advances in graphene-based supercapacitor electrodes. *Energy Reports* 6: 2768–2784.
34. Li, X., and B. Wei. 2013. Supercapacitors based on nanostructured carbon. *Nano Energy* 2 (2): 159–173.
35. Kim, Y., et al. 2021. Atomic-layer-deposition-based 2D transition metal chalcogenides: Synthesis, modulation, and applications. *Advanced Materials* 33 (47): 2005907.
36. Peng, Q., et al. 2012. Atomic layer deposition for electrochemical energy generation and storage systems. *Journal of Vacuum Science & Technology A* 30 (1).
37. Marichy, C., M. Bechelany, and N. Pinna. 2012. Atomic layer deposition of nanostructured materials for energy and environmental applications. *Advanced Materials* 24 (8): 1017–1032.
38. Liu, X., Y. Su, and R. Chen. 2023. Atomic-scale engineering of advanced catalytic and energy materials via atomic layer deposition for eco-friendly vehicles. *International Journal of Extreme Manufacturing* 5 (2): 022005.

39. Guan, C., and J. Wang. 2016. Recent development of advanced electrode materials by atomic layer deposition for electrochemical energy storage. *Advanced Science* 3 (10): 1500405.
40. Leskelä, M., and M. Ritala. 2002. Atomic layer deposition (ALD): From precursors to thin film structures. *Thin Solid Films* 409 (1): 138–146.
41. Shahmohammadi, M., et al. 2020. Enhancing the surface properties and functionalization of polymethyl methacrylate with atomic layer-deposited titanium (IV) oxide. *Journal of Materials Science* 55: 17151–17169.
42. Kim, S., and G.R. Gavalas. 1995. Preparation of H2 permselective silica membranes by alternating reactant vapor deposition. *Industrial & Engineering Chemistry Research* 34 (1): 168–176.
43. Dücsö, C., et al. 1996. Deposition of tin oxide into porous silicon by atomic layer epitaxy. *Journal of the Electrochemical Society* 143 (2): 683.
44. Mackus, A., A. Bol, and W. Kessels. 2014. The use of atomic layer deposition in advanced nanopatterning. *Nanoscale* 6 (19): 10941–10960.
45. Oehrlein, G., D. Metzler, and C. Li. 2015. Atomic layer etching at the tipping point: An overview. *ECS Journal of Solid State Science and Technology* 4 (6): N5041.
46. Zaera, F. 2008. The surface chemistry of thin film atomic layer deposition (ALD) processes for electronic device manufacturing. *Journal of Materials Chemistry* 18 (30): 3521–3526.
47. Haukka, S. 2007. ALD technology-present and future challenges. *ECS Transactions* 3 (15): 15.
48. Kemell, M., et al. 2008. Coating of highly porous fiber matrices by atomic layer deposition. *Chemical Vapor Deposition* 14 (11–12): 347–352.
49. Raub, C. 1993. The history of electroplating. In *Metal Plating and Patination*, 284–290. Elsevier.
50. Schlesinger, M. 2000. Electroplating. In *Kirk-Othmer Encyclopedia of Chemical Technology*.
51. Viegas, L.S., A. Oberdiek, and J.Z. Ferreira. 2023. Funerary art and electroplating: History, present and future. *CONTRIBUCIONES A LAS CIENCIAS SOCIALES* 16 (10): 21294–21309.
52. Sun, X., et al. 2020. Revisiting the electroplating process for lithium-metal anodes for lithium-metal batteries. *Angewandte Chemie International Edition* 59 (17): 6665–6674.
53. Horner, J. 2000. Electroplating. In *Kirk-Othmer Encyclopedia of Chemical Technology*.
54. Schlesinger, M., and M. Paunovic. 2014. *Modern Electroplating*, vol. 52. London: Wiley
55. Tay, J.-H. 1996. Cleaner Production for Electroplating Industry. In *Clean Production: Environmental and Economic Perspectives*, 255–277. Springer.
56. Chiu, H., K. Tsang, and R. Lee. 1987. Treatment of electroplating wastes. *Journal of Institute of Water Pollution Control* 86 (1): 12–19.
57. Blum, W. 1953. The Future of Electroplating. *Transactions of the IMF* 30 (1): 182–201.
58. Plating, D., et al. 1994. Nickel Electroplating. *Journal of Nanoscience and Nanotechnology* 5 (8): 1–26.
59. Zhang, X., et al. 2008. Pulse electroplating of copper film: A study of process and microstructure. *Journal of nanoscience and nanotechnology* 8 (5): 2568–2574.
60. Wang, K., R. Beica, and N. Brown. 2004. *Soft gold electroplating from a non-cyanide bath for electronic applications*. In *IEEE/CPMT/SEMI 29th International Electronics Manufacturing Technology Symposium (IEEE Cat. No. 04CH37585)*. IEEE.
61. Kim, R., and Y. Nam. 2013. Novel platinum black electroplating technique improving mechanical stability. In *2013 35th Annual International Conference of the IEEE Engineering in Medicine and Biology Society (EMBC)*. IEEE.
62. Nobel, F.I., B.D. Ostrow, and D.N. Schram. 1988. *Tin lead electroplating solutions*. Google Patents.
63. Rajoria, S., M. Vashishtha, and V.K. Sangal. 2022. Treatment of electroplating industry wastewater: A review on the various techniques. *Environmental Science and Pollution Research* 29 (48): 72196–72246.
64. Joyce, B. 1985. Molecular beam epitaxy. *Reports on Progress in Physics* 48 (12): 1637.
65. Panish, M.B. 1980. Molecular beam epitaxy. *Science* 208 (4446): 916–922.

66. Cho, A.Y., and J. Arthur. 1975. Molecular beam epitaxy. *Progress in Solid State Chemistry* 10: 157–191.
67. Foxon, C. 2003. Three decades of molecular beam epitaxy. *Journal of Crystal Growth* 251 (1–4): 1–8.
68. Cho, A. 1991. Advances in molecular beam epitaxy (MBE). *Journal of Crystal Growth* 111 (1–4): 1–13.
69. Farrow, R.F. 1995. *Molecular Beam Epitaxy: Applications to Key Materials*. Amsterdam: Elsevier.
70. Henini, M. 2012. *Molecular Beam Epitaxy: From Research to Mass Production*. Newnes.
71. Arthur, J.R. 2002. Molecular beam epitaxy. *Surface Science* 500 (1–3): 189–217.
72. Knodle, W.S., and R. Chow. 2001. Molecular beam epitaxy: Equipment and practice. In *Handbook of Thin Film Deposition Processes and Techniques*, 381–461. Elsevier.
73. Moustakas, T.D. 1988. Molecular beam epitaxy: Thin film growth and surface studies. *MRS Bulletin* 13 (11): 29–36.
74. Koma, A. 1995. Molecular beam epitaxial growth of organic thin films. *Progress in Crystal Growth and Characterization of Materials* 30 (2–3): 129–152.
75. Li, X., Q. Xu, and Z. Zhang. 2023. *Molecular beam epitaxy growth of quantum wires and quantum dots*, 960. MDPI.

Chapter 5
Laser Ablation and Spin Coating

Abstract In this chapter, some specific types of CVD techniques have been described such as Laser Ablation CVD and the Spin Coating CVD process in every aspect. The introduction and historical aspect of the CVD processes has also been explained in detail. The working mechanism of the Laser Ablation CVD and Spin Coating CVD is explained with the help of working diagrams that illustrate the internal steps of working the CVD methods. This chapter, the advantages, and disadvantages and the wide range of applications of Laser Ablation CVD and Spin Coating CVD. These two processes performed the deposition of thin film layers at the range of high temperature and high pressure. Materials used for the Laser Ablation CVD have also been illustrated and various properties of the Laser Ablation CVD and Spin Coating CVD have been described.

5.1 Introduction

Diamond surface processing may be achieved simply and effectively with laser irradiation, which can increase the chemical reactivity of the diamond surface and generate surface graphite formation. Thus, laser irradiation may be used to alter the surface removal of diamonds. Ammonia gas and laser irradiation have previously been utilized to insert amino groups into diamond surfaces [1]. In recent times, a lot of work has been done to modify the surface morphology or add chemical ornamental qualities to change the water absorption of CVD diamond surfaces. By imperfections of the smooth diamond surface, its shape may be modified to generate needle-like or micro/nanocone arrays using maskless technology. Such technologies can be costly and sophisticated, such as lithography, inverse template, selective plasma etching, etc. Another approach is chemical decorating [2]. The process of oxidizing diamonds, or producing oxygen suspensions on their surface, is a relatively straightforward way to make diamonds super-hydrophilic. However, the methods described in published literature such as anodic polarization in an acidic solution or ion implantation require sophisticated equipment and processing. There exists additional work to be done on finding a straightforward and trustworthy method for

T. I. Awan et al., *Thin Film Deposition Techniques*,
https://doi.org/10.1007/978-981-96-1364-9_5

oxygen-terminating diamonds. Micromachining components are extensively utilized in the aerospace, defense, medical, and other industries as a result of scientific and technological advancements [3]. The widespread use of micromachining components is closely linked to their ease of manufacture, affordability, superior quality, and high production efficiency achievable through the use of molds and dies. A good method for machining micro dies and molds is micro-milling. The widely utilized cemented carbide end mills wear out severely during the die and mold final finishing process. This lowers the external excellence of these micro expires then molds, which is detrimental to the large-scale manufacturing of micromachining components [4].

Diamond processing can be efficiently carried out with lasers. A comparison of data from various researchers, namely in the areas of cutting, boring, modeling, and then shining, showed that diamond pictures had an ablation fluence threshold that increased as the laser wavelength increased after UV. 193 nm pulsed laser energy to 10.6 mm infrared [5]. On the other hand, material ejection from the beam point during ablation tests is desirable for more accurate analysis that might arise from the existence fascinated in a diamond external coating. By carefully choosing the laser ablation, it is possible to accurately regulate the beat thickness, ray spot size, then radiated strength supply in both the region treated with a laser at identical irradiation parameters then the width of the detached coating [6]. For many years, people have been aware of the distinct mechanical, thermal, and electrical characteristics of diamonds. These material characteristics have traditionally drawn a lot of interest to diamond processing. But, it's still very difficult to machine diamonds into precise geometric shapes because of their high hardness. Despite extensive research, no technique possesses the necessary speed, resolution, and affordability for industrial manufacturing. Although both electric discharge machining (EDM) and electric discharge crushing of diamond apparatuses have been studied, they are limited by the necessity for post-processing, sluggish speed, and quick electrode wear [7].

5.1.1 *Laser Ablation*

Implementing laser energy to ablate hard mark resources is identified as laser ablation. Laser is a shortening for sunlit intensification by inspired release of energy. This method requires focused incredibly tall energy at a solo spot on a hard external to vaporize material that absorbs light. The spot on the substrate that is being irradiated by a laser during LCVD is heated to a temperature of at least 1200 K. Natural convection will occur in the chamber as a result of temperature gradients in the source gases. The chemical reaction occurring on the substrate has an impact on the concentration of the gas mixture adjacent to the hot spot [8].

Equally LCVD and LIFT have been cast off to manufacture 3D microstructures; though LCVD's output rate is incomplete to 0.1 mm/s then necessity occurs in a space to prevent pollution and thinning of the basis forerunner. As a result, LCVD is unable to make designs in the current building [8, 9]. This is due to the limited quantity of solids that can be transported to the place of notice through the usage of vaporous metal–organic forerunners. In LCVD, the precursor gas can break down by thermal (pyrolytic LCVD) or photoactivated (photolytic) routes. Subministered 3D buildings of an extensive variety of resources (mostly alloys, semiconductor devices, and then certain oxides) on several substrata may be achieved in either configuration. LCVD has been used commercially in the preservative restoration of flat panel display junction transistor arrays; nevertheless, due to its slow programming speed. At best, its applicability to additional micro-AM submissions has been restricted [10].

5.1.2 History

Laser-induced deposition has been the focus of a lot of research over the past ten years, with the majority of this work centered on applications in the microelectronics industry. Big processer producers for example IBM and Hitachi have benefited from the high resolution, speed, flexibility, and simplicity of use of laser-assisted deposition technologies in a function construction cycle, thanks to research efforts conducted in the last several years. The overview of laser deposition provided by Ehrlich and Tsao is excellent [11].

5.1.3 Process

The process of laser chemical vapor deposition, or LCDD, typically involves focusing a laser beam into a region that contains organometallic species vapors. Several different materials may be laser deposited, including nearly every metal that is often employed in the microelectronics field. The statement procedure can be skillful thermally, by heating a substrate surface locally with a laser or extra focused energy root, or photochemically, by utilizing the laser to stimulate photochemical responses of the forerunners for the deposition [12]. Typically, the scheme is uncovered to the laser ray after the substrate has been deep in a gas atmosphere covering the organometallic parts that need to dissociate. The parts in the air stage or a fascinating coating can be activated through the beam through pyrolysis, photolysis, or frequently a mixture of the dual [13].

The pyrolytic technique is almost similar to traditional chemical air deposition (CVD), with a few exceptions that the laser beam imparts localized heat. When parent molecules break down due to collisional stimulation with the heated surface, reactant atoms are released. The organometallic particles need to be suitably clear for enough heating to occur, and the substrate or a film covering it needs to be an effective laser light absorber [8]. The optical maser may be used to separate molecules that have been adsorbed on surfaces or in the gas phase in a complimentary manner without significantly or at all heating the surface. The laser may pass through the substrate, and the reactant absorbs the light immediately. When this happens without the gas being heated up excessively, the process is called photochemical or photolytic deposition [14]. The process of photochemical confession ray can be useful equivalent to the external meanwhile it doesn't need to be heated directly. In certain instances, obtaining pure deposits may need further conventional heating to break down the photo-deposition products. Typical LCVD techniques require precursor chemistry that varies substantially. Because of their comfort of photo-dissociation at 257 nm, a wavelength readily achieved by an argon ion laser with twice the frequency metallic alkyls remained between the initial resources to be examined. It has been possible to deposit Cd, Zn, AI, Pb, Ga, In, then Sn using different laser parameters and employed in several different applications [1].

Using concentrated or wide-area laser beams, direct writing from the carbonyls established deposits of Ni and Fe as well as refractory elements including Cr (chromium), Mo (molybdenum), and W (tungsten). Other precursors were discovered since the majority of noble metals are incapable of producing stable, volatile alkyls or carbonyls [15]. In addition to having properties called beta-detonates or acetyl acetonates, 2, 4 pentane-donates allow for the photolytic and pyrolytic deposition of alloys for example Pt, Au, Ag, Cu, and Pd. There are other chemistries as well; trimethylamine aluminum hydride, for instance, can be used to deposit aluminum. The primary focus of IBM has been on beta-diketones. This work has demonstrated that the pyrolytic gas phase deposition technique may provide exceptionally pure metal films (up to 95%), with film resistances usually three to five times that of the bulk material, in conjunction with particular process optimization measures [16]. This is caused by trapped porosity and structural flaws in the film; annealing can increase the reduced conductivity. (Photolytically formed films, which often have large concentrations of carbon, oxygen, and other contaminants, rarely reach this purity.) A variety of metal and dielectric surfaces are well adhered to the recently deposited layer, and cross sections reveal good intimate contact [17].

Materials Used for Laser Ablation

- The process of laser ablation is versatile and may be used in an extensive variety of resources, for example, composites, metals, and polymers. Here are several examples:
- Polystyrene (PS) and Polymethyl methacrylate (PMMA) (acrylic)
- Polyimide (PI)
- PET, or polyethylene terephthalate
- Polycarbonate (PC)
- Polythene (PE)
- PTFE, or polytetrafluoroethylene (Teflon)
- Polyetherketone, or PEEK
- PDMS, or polydimethylsiloxane
- Stainless steel
- Titanium
- Copper
- Carbon fiber bound with resin
- Kevlar bound with resin

In general, metals react favorably to laser ablation. Surface coatings can be removed differently by adjusting the energy levels. This melts with almost no removal, erodes the metal itself, or modifies the surface characteristics without affecting the mass. Conversely, composite materials can have their surfaces roughened and surface energy increased by laser ablation, which will improve bonding.

5.1.4 Equipment

Japanese and American businesses are now promoting fully integrated, automated LCVD systems. The vapor phase process of LCVD has certain restrictions. It requires that the deposition occur in a vacuum chamber or pressure-controlled vessel, or it can use specially made gas systems that produce primarily focused attention of the organometallic type in specific areas while keeping the ability of the optical maser ray to pass through. Furthermore, because many organometallics are poisonous, corrosive, and/or pyrophoric, handling them requires precaution [18]. The main parts of a deposition system, like those seen in commercial products, often include the following:

Of course, the energy source is the main distinction between a laser tool and a photo exposure tool. The growth of a consistent manufacturing-mark excimer optical maser and the mask remain two essential components for the effective use of laser technology. Today's lasers can have 200 W of power, 500 Hz repetition rates, and more than 10’ beats among significant facility interferences of dependability [19]. Next key apparatuses:

- Optical maser through optical transfer structure with influence-reducing structure.
- Influence monitoring.
- Ray focusing size controller eyes.
- Motor phases through produce fixturing for *X*–*Y* position.
- Space structure containing roughing and turbomolecular drives, a procedure space through the gap, and suitable vapor treatment various; ray gesture structure, also through a touching phase or touching eyes; watching microscope with locked route tube; operative comfort through shows then wheels; computer controller of serious device/procedure purposes; procedure analytic and size ability [20].

Reparation of faulty tinny-film electrodes to enable tall-performance gate array connectivity.

Repairing faulty tinny-film electrodes in parcels like the ones cast off in the ES9000 structure is IBM's main application. Reliable, tall-value maintenances are often found. These are sequential procedures that are virtually completely applied to brief new material segments. To write whole microcircuit layouts using LCVD, laser writing speeds must be increased or alternate methods of processing data must be used to reduce processing time. Large-area patterning has been demonstrated to be supported by both laser projection deposition and optical maser-encouraged pyrolytic confession of seed coatings for later electro-less electroplating [21].

5.1.5 Working Principle and Construction

To generate localized hyperthermia by a specified action, optical maser ablation (LA), a transdermal growth removal method, delivers optical maser bright into living tissue through an intermediary channel. The laser light may be sent across optical fibers with minimal energy loss from the source to the target [22]. It is coherent, monochromatic, highly collimated, and focused. The type of effects that laser light has on structures depends on a variety of factors, the most important of which are the optical maser wavelength, control, experience period, pulse length and duplication incidence (if the emission is pulsed), beam characteristics, applicator optical characteristics, and connective tissue physical properties. Light may be absorbed, dispersed, transmitted, and reflected inside biological tissue [23]. The shape, extension, and location of the heated volume may be impacted by the other but only absorbed energy can cause biological impact. Depending on the speed of heating, coagulation becomes evident during the ablation process at temperatures between 54 and 60 °C. Larger structural proteins and biological components denaturate more quickly over 60 °C, resulting in rapid cell death and widespread coagulation in less than a second. Presently, Nd: YAG ($\lambda = 1064$ nm) or semiconductor device diode optical maser ($\lambda = 800$–980 nm) with 2–40 W of power is used in the majority of LA treatments [24].

If the optical maser apparatus has numerous optical maser causes within, then numerous laser fibers may be inserted into the tissue and triggered concurrently to treat a big volume of tissue quickly. These days, cooled lumens represent a technological advancement in ablative methods. The effectiveness of the removal procedure for the transmission of energy to the tissue the situation is limited by the overly fast drying, carbonization, and sublimation of the tissue that may be avoided using these cooled systems [25]. The maximum popular leadership system for placing the applier in the area of tissue to be ablated is ultrasonic imagination; CT imaging systems are the least used; attractive quality imagination systems are intriguing, then they are also actually costly, difficult to use, and uncomfortable for the patient. On the other hand, they provide real-time control over every stage of the ablative process, from planning to the final evaluation. Now that we have a better grasp of the components of a PLD apparatus, let's look more closely at the physical and chemical processes involved in PLD. The secret of engineering thin films is found in the basic processes that comprise PLD [26]. Figure 5.1 describes the working mechanism of the Laser Ablation CVD. There are three categories for these processes:

(1) Ablation by laser.
(2) Plasma creation and target material expansion.
(3) Target material nucleation and growth on substrate.

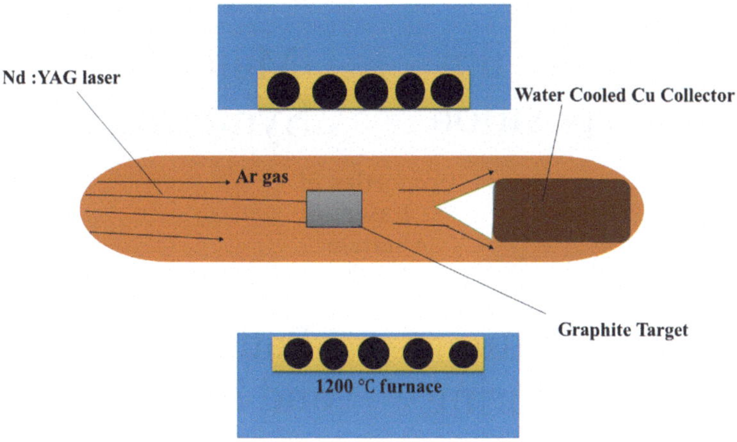

Fig. 5.1 Graphical mechanism of laser ablation CVD

5.1.6 Laser Absorption

In essence, the mark solid needs to be talented to engage the practical wavelength to be physically removed. But when it comes to material removal using ultra-short laser pulses, this reasoning falls short. In this instance, current and non-thermal procedures frequently a mixture of equally are used to further laser ablation. Although these processes occur on distinct time scales, their combined effect aids in the generation of a plasma plume. During the initial phase of optical maser removal, which is dominated by non-thermal interactions, happening photons run sufficient energy to cause group changes in the mark solid as it engages light [27]. An incoming photon excites an electron, which is the primary non-thermal absorption mechanism. The happy electron reduces over electron–phonon connection after undergoing an intra-band change after the pounded state to a stable state. It is also feasible for two incoming photons to excite one molecule. The two photons' combined energy is used to stimulate the molecule when this process takes place on a timescale of picoseconds or less. This process, known as multiphoton excitation, can be sequential or coherent. The absorption process can be influenced by the characteristics of the material; for example, the existence of flaws can modify the band gap, which is the property that determines how much light is absorbed [28]. The Beer-Lambert law (Eq. (5.1)), where A is the optical density, I is the bright strength, I_0 is the happening bright strength, ε is the molar death constant, c is the grinder attention, and t is the width, can be cast-off to determine the extent to which photons spread done the solid. Thus, the dispersion penetration of the captivated laser light and the level of the non-thermal procedures can be estimated using Eq. (5.1).

$$A = \frac{I}{I_0} = \varepsilon ct \tag{5.1}$$

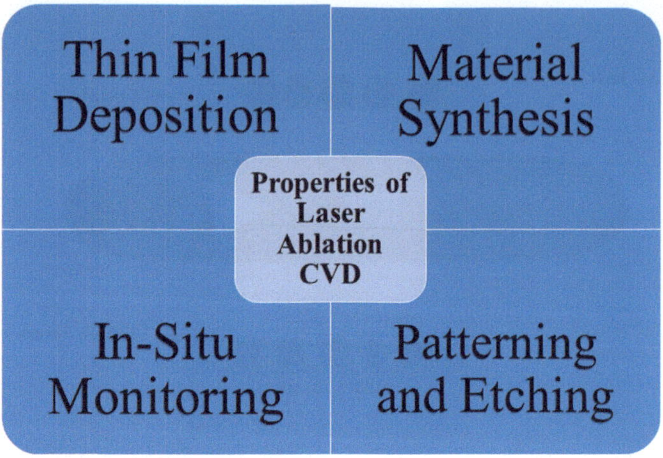

Fig. 5.2 Some specific properties of laser ablation CVD

Thermal excitation is another kind of communication. Laser-induced current innervation is characterized by its ability to rapidly heat a small part to temperatures higher than numerous thousand Kelvin [29]. The system receives heat on a picosecond timescale through the vibrational reduction of happy electrons, often known as an electron–phonon connection. Thus, temperature reactions can predominate because the average PLD pulse lasts for around 20 ns. The surface heat will spread throughout the target as a result of these activities, which may have unanticipated side consequences (such as mechanical stress, the development of microcracks, solidification, and re-melting, among others). Equation (5.2) can be used to compute the thermal diffusion length L_T while accounting for the optical maser beat distance τ_L and the thermal diffusivity D [30].

$$L_T = (2DT_L)^{1/2} \tag{5.2}$$

Its ionization, disintegration, or desorption from the mark external. The current expulsion that occurs after optical maser contact might therefore be attributed to temperature-encouraged promise cleavage, either by promise detachment or evaporation. Intermolecular collisions facilitate this process further by causing excitations and ionization in the extremely dense vapor while allowing the vapor to continue interacting with the incoming laser pulse [31]. Figure 5.2 explains some specific properties of Laser Ablation CVD.

5.1.7 Advantages and Disadvantages

Some significant benefits of pulsed laser ablation are as follows,

- **Multi-component Films**

PLD's ability to easily produce multi-component films that maintain the required stoichiometric ratios is one of its primary features. This feature makes PLD a desirable technique for generating cutting-edge functional materials, especially in applications requiring exact chemical compositions.

- **Elevated Deposit Rate and Adaptability**

Because of PLD's rapid deposition rate, testing times are short, and substrate temperature requirements are low. This efficiency makes PLD a very successful thin film deposition technique, especially when combined with the produced films' consistency. Furthermore, the process's innate flexibility and simplicity indicate its enormous development potential and suitability for a wide range of materials science projects [32].

- **Adaptable Parts of the Process**

PLD also features the flexibility to change process parameters at will. This adaptability permits the creation of heterojunctions and multilayer films with ease and does not impose restrictions on the kinds of PLD targets that can be used. The flexibility in selecting components with multiple targets highlights PLD's ability to respond to a variety of material preparation requirements.

- **Non-Polluting and Simple to Handle**

PLD is unique in that it is a non-polluting method of plasma generation that uses UV-pulsed lasers as its energy source. This feature, along with the method's controllability, guarantees that PLD complies with the increasing need for thin-film material production techniques that are both easily managed and ecologically friendly [33].

The following are some drawbacks of pulsed laser ablation,

- PLD has many benefits, but it is not without difficulties. One important problem that can seriously impair the excellence of the placed film is the existence of molten tiny particles or target pieces. These particles introduce imperfections that degrade the integrity of the film as they scatter during the explosion caused by the laser.
- PLD's applicability to businesses requiring extensive film coverage is limited by its insufficiently tested viability for large-area deposition. These drawbacks, along with the average deposition rate being moderate in comparison to alternative techniques, point to potential areas for PLD innovation and improvement [34].
- Furthermore, PLD's current effectiveness appears limited to high-tech industries like microelectronics, device expertise, visual expertise, then novel solid films, given the expense then size of optical maser film grounding tackle. This focus on a specific market segment emphasizes the necessity for ongoing research and development to increase PLD's usefulness in more applications [35].

Examples of Laser Ablation CVD

- A variety of high- and low-precision applications are served by laser ablation, a scientific and industrial instrument that is being employed increasingly. Here are a few instances:
- Generation of microfluidic channels in plastics, glass, and metals for use in medication administration systems.
- Diamond micro-drilling for precise gas leakage routes in equipment.
- Brand promotion, barcode, and serial number on plastic and metal parts.
- Adhesion to dental implants and growth of bones are greatly enhanced by laser ablation.
- Laser ablation may be used to precisely and repeatedly generate this level of texturing.
- The adherence of bacterial contaminants can be decreased by laser ablation using precisely regulated nano-structure feature sizes. Adhesion to patient tissues is improved concurrently.
- Improved pressing surfaces can be produced via laser ablation.

5.1.8 Applications

The visual characteristics of the solid, the optical maser wavelength, and the beat duration all influence the quantity of solid detached by a solo optical maser beat and the depth across which the optical maser energy is absorbed. Generally speaking, the removal rate refers to the overall form removed from the mark for each optical maser beat. The removal procedure can be greatly influenced by features of optical maser radiation, for example, the coverage of skimming lines and the scanning velocity of the laser beam.

Laser pulses are highly controllable and can range in duration from milliseconds to femtoseconds, as well as in flux. Because of this, laser ablation has significant value in both industrial and research applications. The greatest rudimentary usage of optical maser removal is the measured elimination of solid from a hard external. Examples are laser machining and, in particular, laser drilling, where pulse lasers may penetrate very hard materials to make incredibly tiny, deep holes. Teeth enamel and other gentle or temperature-complex resources can be optical maser drilled on because of the brief optical maser beats that remove solid so rapidly that the nearby solid engrosses actual slight temperature (optical maser dentistry). Numerous individuals have utilized optical maser removal and vapor compression techniques to generate metal nanoparticles, metal oxides, and metal carbides [36].

Furthermore, CO_2 or Nd: YAG beat optical maser can be cast-off to fresh sides, eliminate dye or coverings, or make exteriors for image deprived of harming the fundamental external because optical maser energy can be selectively captivated by coverings, especially on metal. A solo beat from a tall-influence laser can penetrate a big part. Many tiny pulses are used by lower-power lasers, which can be used to scan a large region. In certain sectors of the economy, optical maser removal is also recognized as optical maser washing. The single benefit is that it doesn't require solvents, making it safe for the environment and exposing operators to no chemicals (as long as nothing toxic evaporates). It is not too difficult to automate. Compared to thirsty media or thirsty-ice attacking, the operating prices are less expensive, but the initial expenditure is significantly more. The procedure is less harsh than abrasive methods; for example, carbon fibers in composite materials are not harmed. The target is not very heated. Laser ablation is used in the extra lesson of requests to convert the solid detached into original procedures that are difficult or impossible to manufacture through extra processes [37]. Carbon nanotube synthesis is one recent example. Effective corrosion elimination from iron matter, oil, or fat elimination after a variety of exteriors, and painting, sculpture, and fresco restoration are more applications for laser cleaning. Because it causes the least amount of surface damage to the mold, optical maser removal is single of the recommended methods for cleaning elastic decay. The first reports of ablation of a pure graphite block with a laser date to the study by Guo et al. in March 1995. Later, graphite combined with catalytic metal was also reported. Elements like cobalt, Nb, platinum, Ni, Cu, or a dual mixture of these can make up the catalytic metal. The metal, carbon cement, and powdered graphite are combined to make a paste that is used to create the composite block [29]. After that, the paste is put into a cylinder shape and roasted for a few hours. Following solidification, argon air is pushed in the track of the optical maser point and the graphite block is located in an oven with an optical maser focused on it. The oven has a temperature of about 1200 °C. Carbon nanotubes are produced as the optical maser ablates the mark, and the air movement carries them onto a cold Cu hoarder [38]. Carbon nanotube threads are put in an uneven and warped manner, much to carbon nanotubes created by the electrical-curve release method. While multi-enclosed nanotubes emerge from the clean graphite preliminary solid, solo-enclosed nanotubes form from the block of graphite and metallic substance atoms. An alternative use for this kind of application is the creation of coatings through laser ablation, which involves ablating the covering solid from a basis and then allowing it to sum on the external that needs to be covered. This process is known as pulsed laser deposition (PLD), and it can be used to establish coatings from materials that are difficult to evaporate through alternative methods [39]. Certain kinds of high-temperature superconductors and optical maser gemstones are formed using this way which has significant applications.

In isolated optical maser spectroscopy, the external solid is optical maser ablation to procedure a plasma, and the wavelengths of light emitted by the plasma are analyzed to classify the surface configuration [40]. To produce patterns and remove selective coating from dichroic filters, laser ablation is also utilized. These items are used for mechanism idea equipment correction or as phase light for large-scale forecasts.

5.1.8.1 Propulsion

Furthermore, because the ablated solid transfers a high-compression beat to the external below it as it expands, optical maser extirpation can be used to transmit impetus to an external. It has the same impact as if you were to strike a sharp object on the external. This process is one way an optical maser armament might be damaged and is also used in manufacturing to effort-toughen metallic exteriors. It helps as the basis for spaceship propulsion using throbbed lasers.

5.1.8.2 Manufacturing

- There are now procedures under development to remove the warm air barricade coating after high-compression vapor turbine mechanisms using optical maser extirpation. TBC deletion can be finished with slight damage to the parental solid and fundamental metal coverings because of the short warmth contribution [41].

5.1.8.3 Production of 2D Materials

One effective way to exfoliate unpackaged minerals into their two-dimensional (2D) procedures, for example, dark phosphorus, is via optical maser removal in the fluid stage. The depth and side magnitude of the 2D resources can be manipulated by varying the optical maser energy and solvent [20].

The suitability of a deposit's electric and solid goods, in addition to its moral grip on the substratum, are crucial factors that determine whether or not LCVD can be applied to a particular manufacturing requirement. Additionally, the process as a whole is necessary to be compatible with current buildings. Many microelectronics applications positively utilize laser chemical vapor deposition; the preferred approach is localized, pyrolytic deposition. For some of these applications, deposition resulting from the breakdown of condensed phases is also used [42]. The following are areas of particular emphasis:

- Reorganizing LSI devices,
- Repairing photolithographic,
- *X*-ray lithography masks.

5.1.8.4 Problems with Contaminants in Laser Micromachining

Micrometer-scale element pollution has an important influence on a diversity of businesses in industries like manufacturing. The show sector delivers a supportive demonstration: Through or after the machining procedure, debris from the machining process may be shaped disqualified, then left on the external of the micro-lens system display, which is utilized to evenly diffuse colored light in the direction of the observer. This may disturb visual excellence by intrusive normal machining connections or by causing residual bright attenuation and then refraction during operation. Debris problems can have a significant impact on the manufacture produce number of produced properties later panels through significant remains-interested machining errors are considered to be quality defects. To avoid dirt from absorbing light during operation, post-process cleaning is necessary for all screens that are assessed to demonstrate appropriate geometric quality. The modeling events cast off by semiconductor device manufacturing are additional technical fields anywhere reliability then quality are significant. Since semiconductor devices are electronic goods, they have a superior edge that must be satisfied. If it is surpassed, the creation will malfunction meanwhile there is an arithmetical chance that a single of its many millions of junction transistors will not be electrically remote effectively, requiring its disposal as imperfect. When more, the show subdivision provides a brilliant design.

Dead or trapped pixels, or even more severe, a catastrophic breakdown of the entire device, may arise from this. The majority of semiconductor microprocessors are presently manufactured utilizing light etching techniques like immersion lithography, which makes achieving electrical isolation easy. Nevertheless, the technology is quickly approaching its limits.

Using a shaped excimer beam to etch the resist is a competitive technique, but it has a lot of debris problems. Debris can cause a semiconductor to malfunction by compromising the circuit pathways' electrical isolation.

Debris can also directly impact subsequent product machining by either ranging above the wafer during laser machining and partially absorbing the beam before it reaches the intended surface or by settling back onto a surface to be machined. Debris can indirectly reduce tool linearity and accuracy by accumulating in the supply tubes or by aggregating to cause friction in lubricants and bearings. Research in this field has potential benefits not just for the micro-manufacturing industry but for many other applications as well.

Debris on a machined surface can lead to poor connections or inappropriate fitting of the entire item, which can affect not only the machining of the component but also its installation into a system. A significant amount of money is spent on filtration, pressurization, intricate job routines, costly worker attire, and other means of preserving a low particle count in the working environment in clean surroundings where many micron-sized gadgets are constructed and manufactured. When significant volumes of particle debris are produced during the machining process, which might disastrously pollute other plans throughout machining or meeting, all of this currency is wasted. Research on the containment and prevention of such floating debris is therefore necessary.

The fact that debris is comparatively not expected is single of the belongings that makes it such tricky for manufacture. When weighed against the forces of importance or liquid struggle on an element with such a tiny form, the services associated with submicron-rule elements which are frequently quantified in dynes—come out as the leading power. Flying elements deliver several problems for the production process, facility operating expenses, and the health of the workforce. A lot of time, money, and effort goes into keeping a clean environment for the production of micron-scale goods. Specially designed buildings are used for manufacturing; all points of contact with the unfiltered external atmosphere are sealed with air-tight seals on doors.

The structure of the establishment, the purchase of HEPA filters and staff uniforms, and the training of the workforce to operate appropriately, safely, and environmentally conscious all demand significant capital expenditures. A facility's operating expenses are quite high once it is operational. The huge air conditioning systems use a lot of energy, and HEPA filters have a limited lifespan and need to be maintained. Due to the training that has previously been provided to them and the occasionally challenging working circumstances they experience—which get worse by the merchandise that is required of them, staff members frequently request raises in compensation.

5.2 Spin Coating

A mutual method for putting on thin, same coverings on flat substrates is spin coating. A fluid substance is practical to a rotating substratum, and centrifugal power is used to distribute the substance uniformly throughout the surface. A thin, solid film is left behind as the liquid layer thins and dries while the substrate rotates. Rotation covering is extensively cast off to fabricate thin films with exact thickness and homogeneity in various industries, such as optics, nanotechnology, and microelectronics. Because spin coating can generate thin, homogenous films, it is used in numerous industries [43].

5.2.1 Introduction

Spin coating usually comprises covering (forming) an answer of the wanted solid in a solvent (an "ink") while it is revolving, therefore applying an even thin layer (an insufficient nm to an insufficient um) crossways the external of a substratum. To describe it just, a spinning substrate is covered with a liquid solution to create a slim layer of hard solid, such as a polymer. When the substratum rotates quickly, usually more than 10 revolutions per second, or 600 rpm the solution's surface tension and centripetal force pull the fluid covering into an equivalent coat. The solvent before vanishes throughout this time, leaving the obligatory material in an even coating on the substratum [32]. Functional thin film production has extensively used spin coating, another cost-effective method with exceptional application performance. They have gained a lot of attention because of their outstanding control of the depth of the film. The main parameters influencing the film depth are found using the ideal model and empirical formula, and they may now be used to generate amazing functional films. High-quality thin films are needed to meet the production demand for solar cells in the energy domain. As a result, spin-coating is now an attractive choice. Because there are no high-angle grain boundaries, spin coating of epitaxial films provides a low-cost and easily accessible path to materials that resemble single crystals and should have better electrical and optical characteristics [44]. With remarkable ease, an extensive variety of resources may be put onto various wafer-sized substrates. The films were formed by depositing the material from solutions or precursors that easily transformed into the final product, generating only volatile byproducts. Other metal oxides should be able to be deposited using the same precursor pathway that was utilized to deposit ZnO from an ammine complex. Additionally, the spin coating provides multiple paths to extremely well-ordered semiconductor devices for solar cells, shows, and flexible electronics. The resources can be formed using more traditional gas statement methods onto rotation-covered aquatic-soluble salts, for example, NaCl, which act as sacrificial patterns for the epitaxial boost-off of able-standup semiconductor device foils, or they can be rotation-covered onto elastic metallic foil that resembles a single crystal [28].

5.2.2 *Working Principle and Construction*

Rotation covering is a well-clear process that uses centripetal force to arrange a precursor or gel and evenly disseminate an answer crossway and external to sum an even then homogenous slim layer over the substratum's external. The substratum is covered with a precursor gel that contains a particular solute and solvent, and it is then rapidly spun. Due to the gel's release and vaporization, a solid thin coat covering is developed as an outcome of mutually centripetal power and then the precursor's external tightness. An insufficient nm to insufficient micrometers of homogeneous film depth is formed by the rotation covering technique. Its primary advantage is that it can be deposited quickly and easily, with very uniform deposition of various nanostructured films [45]. Rotation covering allows a very uniform coating through a comparatively controllable and reproducible depth to be applied over a vast area on a flat substrate. Inorganic, organic, and mineral/carbon-based solution combinations can all be covered via rotation covering.

Four basic stages may be used to roughly categorize this process:

- **Deposition**

First, the solution is introduced onto the substrate, usually with a tool called a pipette. The radial wave will push the fluid crossways the substratum whether it is rotating at the moment (energetic rotation covering) or rotating later confession (stationary rotation covering).

- **Spin Up**

After that time, the substratum either arrives at the target rotation rapidity right away or later a dispersal stage at a slower rapidity. The popularity of the solution has now been released from the substratum. The liquid may rotate faster than the substratum initially, but gradually the rotation rates will converge as drag equalizes rotating speeding up, causing the liquid to become level.

- **Spin Off**

At this point, viscous forces start to dominate and the liquid jumps to slim. Due to intrusion properties, the film frequently variations color when the liquid is thrown off. The majority of the film will be thirsty when the color changes cease to exist.

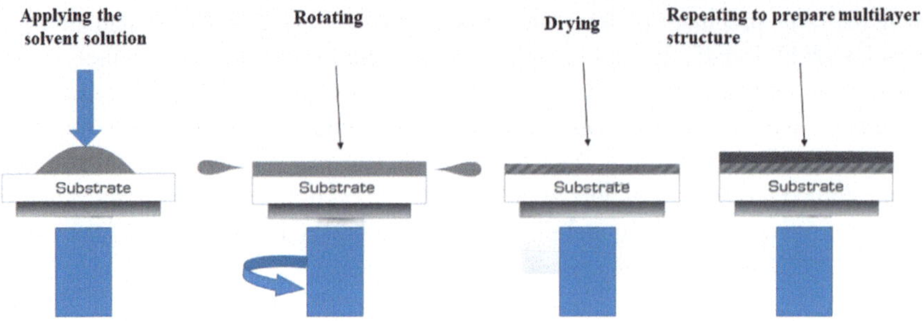

Fig. 5.3 Graphical mechanism of the process of spin coating CVD

Since the liquid necessity procedure precipitations close the edge to be flung off, edge properties can occasionally be observed [46].

- **Evaporation**

At this point, the fluid stops flowing out, and the solvent evaporates to dominate the thinning process. Vapor pressure, atmospheric conditions, and solvent volatility will all affect the rate of solvent evaporation. A substrate's edge or other non-uniformities in the evaporation rate will result in similar non-uniformities in the film. Figure 5.3 illustrates the general mechanism of the process of Spin Coating CVD.

5.2.3 Certain Conditions of Spin Coating

- **Environmental Factors and Variations in Drying Time**

Since the drying time and film qualities are closely related for numerous carbon-based electronics then nanotech requests, external factors can occasionally have an enormous effect. A protective glove box in a professional cleanroom is often filled with pure nitrogen gas, and the temperature and humidity are typically quite tightly regulated [47]. But while this might not matter too much in most situations, dangerous instances can impair constancy and then lead to turn-covering problems. Study laboratories are frequently not that healthy measured. Depending on the weather, the summertime ambient humidity in the UK can vary greatly from less than 20% to almost 100%. Films turned from aqueous explanations may remain wet later the standard 30-additional rotation extent during periods of extremely high humidity, usually after rainfall. This can significantly impact device performance [48]. Because of this, even while most effort is complete in facilities with controlled temperature and moisture, it is still a good idea to monitor the weather outside and maintain a thermometer and hygrometer close to the rotation coater. It is too desirable for the solvent to jump aggregating in the rotation coater's container when coating numerous big surfaces with significant solvent volumes (100 μl or more). This can to have a similar result as raising the ambient vapor heaviness and extending ventilation intervals. If great capacities of solvent are cast off, a simple solution to this issue is to dry the rotation coater container later each rotation gently [49].

- **Steps for Spin Cleaning and Washing**

Although spin coating is a feasible method for substrate cleaning, its efficacy is often lower than that of ultrasonic bath-based cleaning, and its cleaning duration is longer than that of cleaning a batch of substrates at once. Additionally, it frequently leads to the spin coater using huge quantities of solvents, which can be problematic if these processes are carried out in a setting where the solvent requirements to be detached (such as a laminal flow). Rotation cleaning and shower procedures, however, can occasionally be quite beneficial [50]. For instance, HMDS and several additional external alteration actions advantage from a spinning shower. Temporarily, certain additives can also be eliminated by rotation covering a semi-orthogonal solvent. Investigators typically spin wash repeatedly using an active dispensing of an average capacity of solvent (50 μl for normal 20 × 15 mm substrata). However, be aware that if the solvent is not eliminated before (typically by spreading extra solvent from the confidential of the rotation coater through a fresh-area tissue), this may produce a large amount of solvent vapor that may impact the drying dynamics of adhering to films [50]. Figure 5.4 describes the specific properties of the spin coating CVD.

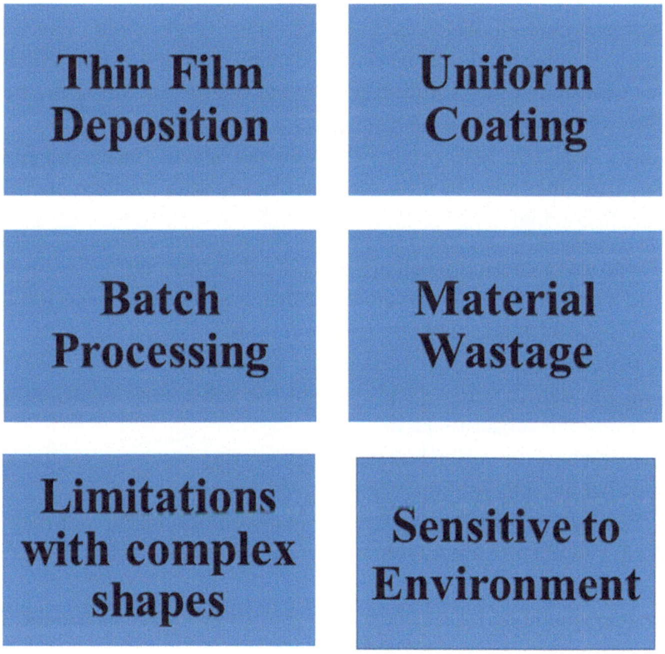

Fig. 5.4 Some specific properties of the spin coating CVD

5.2.4 Preventing Defects

- **Two-Step Rotation Covering and Particle Elimination from Edges and Corners**

The center of the substratum frequently desiccates much faster than the advantage when spin coating at actual small rotation rates (under 500 rpm) using viscid or high hot point solvents like trichlorobenzene. The advantage of the substratum may occasionally take minutes to thirsty, even though the center may be accomplished in just a couple of seconds. Although the substrate's edges are frequently intended to be devoid of active or crucial elements, the film quality might be severely compromised if the spin coater were to stop, causing the edge or corner bead to extend back toward the center [51].

The edge/corner beads can be removed via essentially two methods. One more preferred method is to employ a two-step spin cycle. In the first cycle, the ink is allowed to dry over most of the substratum and the wanted film depth is provided. In the second cycle, the maximum rotation speed is used to remove the turn beads and thirsty the remaining substratum. To rise the consistency of the rotation covering, removing edge beads usually requires two stages of spinning, which is the more convenient method. The substrate may need to be removed from the rotation coater although it is still to some extent wet to proceed with additional dispensation, such as a methanol shower, vacuum drying, or even slower ventilation in an environment drenched with solvents. In these cases, another stage (high-speed) spin is not ideal. In these situations, edge beads can be removed using a second, more delicate method that involves absorbing extra solution with a fine cotton bud while the bead is still spinning [52].

The primary objective is to protect the substrate's active area and keep it from coming apart from the hole. To remove corner beads with a cotton bud effectively, remember the following:

- Proceed with great caution when approaching the substrate.
- Use the lid's center hole to maintain equilibrium and direct the cotton bud inside while keeping your hand stable by resting it on it.
- Hold the fiber bud tip slightly overhead the substratum so that it only comes into interaction through the edge bead and not the substratum itself, attempting to avoid touching the substrate itself.
- Keep an eye on how the cotton bud moves; even a tiny touch against the substratum will cause shaking that you must be talented to sense.
- Keep a close eye on the fiber bud; since many carbon-based and nanotech inks are colored, you must be intelligent enough to determine once it is near sufficient to the ink to absorb it.
- The fiber bud should be kept as near to the substrate's edge as feasible [53].

5.2.5 Low Viscosity Spin Coating Solvents

The viscosity of a solvent can significantly affect the film quality during spin coating. Due to their extremely low viscosities, some solvents that are commonly used in study fields for past or additional explanations can provide serious problems. There are two main problems with this type of solvent, which includes acetone and chloroform as examples. The solution leaked out of the pipette earlier than anticipated, when an even moist layer is formed, the film dries, causing spirals in the substrate [54]. The first thing to do if the ink is not dropping out of the pipette when it should be is to usage the minimum pipette tip existing, as it will have a minor diameter. To minimize the impact of weight on surface tension, it is also helpful to pipette the smallest quantity of ink compulsory for covering (10 μl of solution will consider semi as far as 20 μl while the external tightness leftovers the similar). It can to be beneficial to leave the pipette tip in the ink for a short earlier taking it out if the solvent is halogenated (such as chloroform). This is because the majority of pipette tips are composed of polypropene, which swells a little and decreases the active diameter of the aperture in halogenated liquids instead of dissolving [55]. Since non-halogenated solvents like acetone won't cause the polypropylene to expand, this method is ineffective for them. An approach would be to raise the pipette at a position to lower the heaviness-to-external tightness ratio; the pipette should lone be brought back to vertical right before the solvent is dispensed. It is important to take care not to tilt the pipette to the point where solvent gets inside and breaks it [53].

5.2.6 Benefits and Drawbacks

Among the many benefits of spin coating include its simplicity of use, consistency, and capacity to generate thin, homogeneous layers. It can deposit films over huge areas quickly and affordably, which makes it appropriate for highly efficient production. The spin coating does, however, have several drawbacks. It works best with flat surfaces and can waste material since extra liquid is ejected during the spinning process. Furthermore, the final film qualities may be susceptible to external factors, which require vigilant parameter control throughout production [56].

Some additional benefits of spin-coating deposition techniques are the following,

- Spin coating is easy to apply and handle, so beginners as well as experts can benefit from it.
- Additionally, it is affordable, as the necessary equipment is not too costly, making it more accessible to a wider range of financial situations.
- Its capacity to generate a homogenous thin film is one of its primary characteristics. Because of this, it's perfect for applications that need accuracy and consistency.
- Because of the quick process time, more work may be completed in less time, which boosts productivity.

- Depending on the particular requirements of the project, there is freedom in the coating thickness adjustment [57].

Certain drawbacks of the spin coating deposition technique are listed below,

- Because spin coating can generate thin films, it is not appropriate for applications requiring thicker layers.
- The resultant product's homogeneity is impacted by the process' frequent tendency to produce inconsistent film thickness throughout the surface.
- This is not a highly materials-efficient technique because the spinning process wastes a lot of material.
- Because spin coating is primarily intended for small-scale operations, scaling it to accommodate mass production is a challenge.
- This method's applicability to materials with high viscosity is limited due to its difficult handling of such solutions [58].

5.2.7 Applications

Rotation covering is a popular method for producing homogeneous slim films with nanoscale thicknesses once micro-fabricating useful oxide coatings on glass or solo crystal substrata using sol–gel forerunners. It is frequently cast off in photolithography to credit around single-micron-abundant photoresist coatings. Frequently, the photoresist is turned for 30–60 s at a speed of 20–80 revolutions/s [59]. The generation of polymer-based planar photonic structures is another common application for it. The constancy of the film depth is one positive aspect of rotation-covering slim films. Because of self-leveling, there is only a one percent variation in thickness. The optical characteristics of these materials may also be impacted by the thickness of films made in this manner as well. This is crucial for electrochemical testing because thicker films generally do not allow light to shine through due to their lower optical transmittance, whereas thinner films allow light to pass through before the film's [60]optical density drops too drastically [61]. This is especially legitimate when recording absorbance measurements from spectroscopy using ultraviolet and visible wavelengths. Furthermore, films with a less favorable grade of absorbance are not the best choices for procedures like Cyclic [62].

5.2.7.1 Microelectronics

In the manufacturing process of integrated circuits and solar cells, among other microelectronic devices, spin coating is widely employed. It is used to deposit photoresists, conductive polymers (like PEDOT: PSS), dielectric layers (like silicon dioxide and silicon nitride), and photoresists onto silicon wafers and other substrates. For electrical devices to operate dependably, spin coating's exact thickness control and homogeneity are essential [60].

5.2.7.2 Optics and Photonics

"Spin-optical light sources incorporate photonic modes and electronic transitions and therefore provide a way to study the exchange of spin information between electrons and photons and to develop advanced optoelectronic devices." "Either in their photonic or electronic parts, lifting the spin degeneracy between the two opposite spin states is a prerequisite to constructing these sources" [54, 63].

5.3 Advanced Methods for Spin Coating

Although traditional spin coating is extensively applied, sophisticated variations of the method have been implemented with specific objectives. For instance, spin coating and atomic force microscopy (AFM) are used in dip-pen nanolithography to deposit functional materials in nanoscale configurations. Another version that makes it possible to coat non-flat surfaces is spray coating, which is the process of depositing liquid droplets onto a spinning substrate. These cutting-edge methods increase spin coating's potential for certain uses [64].

5.4 Conclusion

PLD is a very helpful technology for understanding the properties of materials at the nanoscale, which leads to breakthroughs in knowledge that are beneficial for many emerging technologies. The degree to which the growth parameters in PLD can be tuned offers the potential for a fine degree of regulator completing the films' bodily, chemical, physical, and then crystallographic features, all of which can affect their useful attributes. This control makes it possible to conduct sophisticated research that reveals novel phenomena and intricate linkages. This technique's multivariate character, which permits a great degree of tune ability, frequently causes challenges for the development of new material classes. It is common practice to optimize the growth parameters iteratively by analyzing pertinent properties once the film grows, which serves as the foundation for subsequent optimization. This turns into a laborious, resource- and time-consuming operation, particularly for people who are not familiar with the method. In the fabrication of slim films, pulsed laser deposition (PLD) offers an intriguing method with a special combination of accuracy, adaptability, and environmental friendliness. Although the approach's benefits offer new possibilities, its drawbacks point out areas that need more study and improvement. Collaborations with suppliers of materials such as Stanford Advanced Materials (SAM) will be essential to realizing the full potential of PLD and advancing the technologies that depend on superior thin films as the area develops.

References

1. Russo, R.E., et al. 2002. Laser ablation in analytical chemistry—A review. *Talanta* 57 (3): 425–451.
2. Gray, A.L. 1985. Solid sample introduction by laser ablation for inductively coupled plasma source mass spectrometry. *The Analyst* 110 (5): 551–556.
3. Thompson, M., J.E. Goulter, and F. Sieper. 1981. Laser ablation for the introduction of solid samples into an inductively coupled plasma for atomic-emission spectrometry. *The Analyst* 106 (1258): 32–39.
4. Miller, J.C. 1997. 1. Introduction to laser desorption and ablation. In Experimental Methods in the Physical Sciences, 1–13. Elsevier.
5. Speakman, R.J., and H. Neff. 2005. The application of laser ablation ICP–MS to the study of archaeological materials—an introduction. *Laser Ablation-ICP-MS in Archaeological Research* 1–16.
6. Günther, D., S.E. Jackson, and H.P. Longerich. 1999. Laser ablation and arc/spark solid sample introduction into inductively coupled plasma mass spectrometers. *Spectrochimica Acta Part B: Atomic Spectroscopy* 54 (3–4): 381–409.
7. Zelenyuk, A., and D. Imre. 2005. Single particle laser ablation time-of-flight mass spectrometer: An introduction to SPLAT. *Aerosol Science and Technology* 39 (6): 554–568.
8. Stafe, M., A. Marcu, and N.N. Puscas. 2014. *Pulsed Laser Ablation of Solids*. vol. 10, 978–3. Springer, Berlin
9. Richner, P., et al. 1994. Applications of laser ablation and electrothermal vaporization as sample introduction techniques for ICP-MS. *Fresenius' Journal of Analytical Chemistry* 350: 235–241.
10. Harilal, S.S., et al. 2014. Femtosecond laser ablation: Fundamentals and applications. *Laser-Induced Breakdown Spectroscopy: Theory and Applications* 143–166.
11. Franck, P., P.W. Henderson, and K.O. Rothaus. 2016. Basics of lasers: History, physics, and clinical applications. *Clinics in Plastic Surgery* 43 (3): 505–513.
12. Balling, P., and J. Schou. 2013. Femtosecond-laser ablation dynamics of dielectrics: Basics and applications for thin films. *Reports on Progress in Physics* 76 (3): 036502.
13. Gross, A.J., and T.R. Herrmann. 2007. History of lasers. *World Journal of Urology* 25: 217–220.
14. Mahamood, R.M. and R. Mahamood. 2018. Laser basics and laser material interactions. *Laser Metal Deposition Process of Metals, Alloys, and Composite Materials* 11–35.
15. Miller, J. 1994. History, scope, and the future of laser ablation. In *Laser Ablation: Principles and Applications*, 1–10. Springer.
16. Phipps, C., et al. 2010. A review of laser ablation propulsion. In *AIP Conference Proceedings*. American Institute of Physics.
17. Fan, C., and R. Rox-Anderson. 2008. Endovenous laser ablation: Mechanism of action. *Phlebology* 23 (5): 206–213.
18. Jelínková, H. 2013. Introduction: The history of lasers in medicine. In *Lasers for Medical Applications*, 1–13. Elsevier.
19. Hecht, J. 2010. Short history of laser development. *Optical Engineering* 49 (9): 091002–091002–23.
20. Wei, C., et al. 2019. Study on femtosecond laser processing characteristics of nano-crystalline CVD diamond coating. *Applied Sciences* 9 (20): 4273.
21. Kim, M., et al. 2017. Synthesis of nanoparticles by laser ablation: A review. *KONA Powder and Particle Journal* 34: 80–90.
22. Guo, Z., et al. 2021. Optimisation of spray-mist-assisted laser machining of micro-structures on CVD diamond coating surfaces. *Ceramics International* 47 (15): 22108–22120.
23. Guo, B., et al. 2020. Water assisted pulsed laser machining of micro-structured surface on CVD diamond coating tools. *Journal of Manufacturing Processes* 56: 591–601.
24. Cao, Y., et al. 2010. Laser micro/nano-fabrication techniques and their applications in electronics. In *Advances in Laser Materials Processing*, 629–670. Elsevier.
25. Liu, P., et al. 2010. From nanocrystal synthesis to functional nanostructure fabrication: Laser ablation in liquid. *Physical Chemistry Chemical Physics* 12 (16): 3942–3952.

26. Makarov, G.N. 2013. Laser applications in nanotechnology: Nanofabrication using laser ablation and laser nanolithography. *Physics-Uspekhi* 56 (7): 643.
27. You, R., et al. 2020. Laser fabrication of graphene-based flexible electronics. *Advanced Materials* 32 (15): 1901981.
28. Li, Z., et al. 2020. Influence of the grain size of CVD diamond on the thermal conductivity, material removal depth and surface roughness in nanosecond laser machining. *Ceramics International* 46 (12): 20510–20520.
29. Sahin, R., E. Simsek, and S. Akturk. 2014. Nanoscale patterning of graphene through femtosecond laser ablation. *Applied Physics Letters* 104 (5).
30. Hanabusa, M. 1992. Unique features of laser ablation. *MRS Online Proceedings Library (OPL)* 285: 447.
31. Rao, M. 2013. Pulsed laser deposition—Ablation mechanism and applications. *International Journal of Modern Physics: Conference Series*
32. Mouhamadali, F., et al. 2020. Nanosecond pulsed laser-processing of CVD diamond. *Optics and Lasers in Engineering* 126: 105917.
33. Malshe, A., et al. 1999. A review of techniques for polishing and planarizing chemically vapor-deposited (CVD) diamond films and substrates. *Diamond and related materials* 8 (7): 1198–1213.
34. Diam, W.A. 2019. *Identification of the Advantages of Laser Ablation Method for Preparing the Nanomaterials: Efficient, Reliable and Cost-Effective.*
35. Abid, N., et al. 2022. Synthesis of nanomaterials using various top-down and bottom-up approaches, influencing factors, advantages, and disadvantages: A review. *Advances in Colloid and Interface Science* 300: 102597.
36. Gil-Villalba, A., et al. 2018. Single shot femtosecond laser nano-ablation of CVD monolayer graphene. *Scientific Reports* 8 (1): 14601.
37. Jamkhande, P.G., et al. 2019. Metal nanoparticles synthesis: An overview on methods of preparation, advantages and disadvantages, and applications. *Journal of Drug Delivery Science and Technology* 53: 101174.
38. Stern, E., et al. 2006. Comparison of laser-ablation and hot-wall chemical vapour deposition techniques for nanowire fabrication. *Nanotechnology* 17 (11): S246.
39. Butler-Smith, P., et al. 2013. Micro/nanometric investigations of the effects of laser ablation in the generation of micro-tools from solid CVD diamond structures. *Journal of Materials Processing Technology* 213 (2): 194–200.
40. Ou, Q., et al. 2008. Characteristics of graphene-layer encapsulated nanoparticles fabricated using laser ablation method. *Diamond and Related Materials* 17 (4–5): 664–668.
41. Lade, R., et al. 1999. 193-nm laser ablation of CVD diamond and graphite in vacuum: Plume analysis and film properties. *Applied Physics A* 69: S935–S939.
42. Chen, N., et al. 2022. Investigation of metal-coating-assisted IR nanosecond pulsed laser ablation of CVD diamond. *Journal of Materials Research and Technology* 18: 4114–4129.
43. Chen, N., et al. 2019. Investigating the ablation depth and surface roughness of laser-induced nano-ablation of CVD diamond material. *Precision Engineering* 57: 220–228.
44. Zhai, J., Q. Zhang, and Y. Zhu. 2022. Picosecond laser ablation of polycrystalline CVD diamond. *Optics & Laser Technology* 155: 108403.
45. Tyona, M. 2013. A theoritical study on spin coating technique. *Advances in Materials Research* 2 (4): 195.
46. Mustafa, H.A.M., and D.A. Jameel. 2021. Modeling and the main stages of spin coating process: A review. *Journal of Applied Science and Technology Trends* 2 (02): 119–123.
47. Tyona, M. 2013. A comprehensive study of spin coating as a thin film deposition technique and spin coating equipment. *Advances in Materials Research* 2 (4): 181.
48. Kamaruddin, S.A., et al. 2011. Zinc oxide films prepared by sol–gel spin coating technique. *Applied Physics A* 104: 263–268.
49. Nistico, R., D. Scalarone, and G. Magnacca. 2017. Sol-gel chemistry, templating and spin-coating deposition: A combined approach to control in a simple way the porosity of inorganic thin films/coatings. *Microporous and Mesoporous Materials* 248: 18–29.

50. Habibi, M. and M. Khaledi Sardashti. 2008. Structure and morphology of nanostructured zinc oxide thin films prepared by dip-vs. spin-coating methods. *Journal of the Iranian Chemical Society* 5: 603–609.
51. Tahir, M.B., et al. 2017. Review of morphological, optical and structural characteristics of TiO_2 thin film prepared by sol gel spin-coating technique. *Indian Journal of Pure & Applied Physics (IJPAP)* 55 (10): 716–721.
52. Lukong, V.T., K. Ukoba, and T.-C. Jen. 2022. Review of self-cleaning TiO_2 thin films deposited with spin coating. *The International Journal of Advanced Manufacturing Technology* 122 (9): 3525–3546.
53. Sahu, N., B. Parija, and S. Panigrahi. 2009. Fundamental understanding and modeling of spin coating process: A review. *Indian Journal of Physics* 83 (4): 493–502.
54. Norrman, K., A. Ghanbari-Siahkali, and N. Larsen. 2005. 6 Studies of spin-coated polymer films. *Annual Reports Section" C"(Physical Chemistry).* 101: 174–201.
55. Butt, M.A. 2022. Thin-film coating methods: A successful marriage of high-quality and cost-effectiveness—A brief exploration. *Coatings* 12 (8): 1115.
56. Krebs, F.C. 2009. Fabrication and processing of polymer solar cells: A review of printing and coating techniques. *Solar Energy Materials and Solar Cells* 93 (4): 394–412.
57. Scriven, L. 1988. Physics and applications of dip coating and spin coating. *MRS Online Proceedings Library* 121 (1): 717–729.
58. Kaddachi, Z. et al. 2016. Design and development of spin coating system. In *2016 17th International Conference on Sciences and Techniques of Automatic Control and Computer Engineering (STA).* IEEE.
59. Gupta, J., H. Shaik, and S.A. Sattar. 2024. A review on perspective of spin coating technique towards profound electrochromism of tungsten oxide thin films. *Indian Journal of Physics* 1–15.
60. Chaudhari, M.N., R.B. Ahirrao, and S.D. Bagul. 2021. Thin film deposition methods: A critical review. Int. J. Res. Appl. Sci. Eng. Technol 9 (6): 5215–5232.
61. Hall, D.B., P. Underhill, and J.M. Torkelson. 1998. Spin coating of thin and ultrathin polymer films. *Polymer Engineering & Science* 38 (12): 2039–2045.
62. Mitzi, D.B., et al. 2004. High-mobility ultrathin semiconducting films prepared by spin coating. *Nature* 428 (6980): 299–303.
63. Moreira, J., A.C. Vale, and N.M. Alves. 2021. Spin-coated freestanding films for biomedical applications. *Journal of Materials Chemistry B* 9 (18): 3778–3799.
64. Larson, R.G., and T.J. Rehg. 1997. Spin coating. In *Liquid Film Coating: Scientific Principles and their Technological Implications*, 709–734. Springer.

Chapter 6
Advanced Deposition Techniques

Abstract This chapter gives an in-depth review of advanced deposition techniques focusing on hybrid deposition techniques, such as Ion Beam-Assisted Deposition (IBAD), High-rate Chemical Vapor Deposition (HCPVD), and Plasma-Enhanced Atomic Layer Deposition (PEALD). In the description of each technique, basic ideas will be considered along with important elements related to processing benefits and drawbacks to building an appropriate picture of the potential and constraints of each technique within modern material science. Hybrid deposition techniques incorporate the best features of several techniques of deposition with minimum disadvantages. The following methods allow depositing films of complicated composition with specific desired properties, acting in combinations of PVD and CVD processes. The hybrid methods permit much better control of the film properties and higher uniformity, and superior quality films can be obtained using more than one deposition source simultaneously. Multilayer coatings or sophisticated material features are especially beneficial for applications with the hybrid method. Such hybrid systems can be quite costly and complicated, while the fine-tuning of process parameters required for balancing the contributions coming from each approach needs strict control and calibration. PEALD is a variant of atomic layer deposition, wherein a plasma is employed to enhance the deposition process, offering a fine-grained molecular-level control of film thickness and composition. By contrast with the convention, ALD uses alternating pulses of precursor and co-reactant gases; the plasma activation in the PEALD allows for lower processing temperatures. This approach, in general, yields high-quality thin films with outstanding conformality and uniformity, even on complex geometries. Its advantages are the facts that PEALD can deposit films at low temperatures and that it is apt for high-precision applications. It does feature exceptional thickness control. Unfortunately, there is a few challenges that yet remain to be overcome, like sophisticated plasma equipment use, a limited deposition rate, and some plasma-induced contamination or damages. In the IBAD technique, besides the physical vapor deposition, an ion beam is used to blast the growing layer. The modification of microstructure and improvement in adhesion due to energetic interaction with the ion beam results in better film characteristics. IBAD technique allows for obtaining homogeneous coatings, high-density films, and stress control on complex geometries. While control of stress and modulation

T. I. Awan et al., *Thin Film Deposition Techniques*,
https://doi.org/10.1007/978-981-96-1364-9_6

of film characteristics hold great potential, especially in applications involving high-performance coatings, its drawbacks are serious: expensive equipment, complicated operation, low deposition rates, and the possibility of ion-induced damage to substrate or film. Another factor is material-specific substrate heating, adding more challenges that must be overcome for effective utilization.

6.1 Advanced Deposition Techniques

Hybrid techniques play a vital role in thin film fabrication by incorporating most deposition methods to improve film characteristics and/or add functions. Most of the hybrid techniques take advantage of each deposition method's merits to tailor the thin films' properties toward specific requirements. In the case of a hybrid deposition process, which might combine several techniques such as PVD, CVD, and ALD, the composition, thickness, and structure of the films will be more controlled. In this process, one is highly likely to obtain thin films with specific properties that would not have been achieved had one of these deposition methods been used individually [1]. Some of the performance features of thin films possibly afforded by hybrid procedures include improved adhesion, homogeneity, and usefulness. Hybrid methods may be involved in creating multilayered thin films, where each layer may possess special characteristics, thus enabling the construction of intricate structures for particular uses in sensors, optics, and electronics. Hybrid techniques always combine the advantages of several deposition methods in various ways to tailor the features of thin films flexibly for a variety of applications, from commercial to scientific [2].

Coatings and thin films for a multitude of applications are advanced deposition processes that employ precise characteristics. These range in complexity from molecular beam epitaxy to chemical vapor deposition, and to physical vapor deposition techniques, including sputtering and evaporation. Such PVD processes have been used in controlling the thickness and composition of layers in electronics, spintronic, and solar cells [3]. However, CVD techniques such as atomic layer deposition and plasma-enhanced CVD yield high-quality thin films with outstanding homogeneity and conformality for state-of-the-art semiconductor technologies. These techniques will be at the frontline in the development of oxide films, hybrid structures, and functional thin films, which will power advances at the frontiers of topological insulators, thin film sensors, and spintronic. The deposition processes have to be continuously improved to meet the new challenges imposed by continuously changing needs of materials science and technologies, which current technologies have enabled for new applications and improving performance in a wide range of activities [4].

These hybrid methods open routes in design and production for future-generation thin-film products such as flexible electronics, high-performance coatings, and energy efficient materials. These hybrid methods allow researchers and engineers to venture into and discover the possibilities of developing optically superior, electrically superior, mechanically stable, and superior chemical characteristics thin films.

Advanced hybrid techniques in thin-film fabrication involve versatile and flexible approaches toward depositing material structures with tailored properties. Hybrid techniques combine several deposition processes for the manufacture of thin films into one single process that opens new applications in various technological fields and areas of research. It improves control over the properties of the film, adds new functionalities, and increases the performance characteristics [5].

6.2 Advanced and Hybrid Techniques

Advanced hybrid thin film deposition processes have made possible tremendous and revolutionary progress in the field of materials science, thus allowing modern technologies and applications by combining the best characteristics of several deposition processes, offering new opportunities and improved control of the characteristics of the film [6].

A large number of techniques come under the category of physical vapor deposition, where materials are vaporized and then deposited on a substrate in a vacuum. PVD is the most efficient technique for making deposits of metals, alloys, and ceramics. The great advantage of PVD techniques lies in good control over the purity and thickness of films. Conversely, materials from a vapor phase are deposited through chemical vapor deposition following chemical processes which occur at high temperatures. CVD enables complex chemicals, polymers, and coatings with very good conformity and uniformity even onto harder surfaces. Hybrid deposition techniques apply several deposition processes to achieve synergistic effects and tailor the characteristics of thin films suited for specific application scenarios [7]. For example, hybrid PVD-CVD allows the depositions of highly complex, multilayered structures with excellent properties. Thin films with specified composition, microstructure, and functionalities can be produced by such a method and are suitable for energy storage applications, electronics, and optoelectronics. The advantages of PVD and CVD can further be combined into a hybrid process that is called advanced atomic layer deposition. Sequentially deposited thin films can be deposited with atomic-level accuracy using ALD: by successive exposure of a substrate to precursors. Uniform deposition of highly conformal coatings makes this method exceptionally well suited for microelectronic applications [8]. An additional hybrid process combining the benefits of PVD and CVD is advanced atomic layer deposition, or ALD. The method allows for the use of sequential self-limiting surface reactions, with the alternating sequence of precursors to deposit very thin films. Extremely homogeneous and uniform coatings make ALD an excellent promise in microelectronic and nanotechnology applications. This ever-rising demand for current technology has urged relentless innovation as well as hybridization of the processes of deposition of thin films. Emerging problems create requirements with the advancement of material science, which requires the development of novel deposition techniques and new applications satisfying the demanding specifications [9].

Hybrid processes that combine the different deposition techniques have made possible the introduction toward new thin films opportunities of diverse innovations from all fields and research domains to improve the properties and functions.

6.2.1 Hybrid Physical–Chemical Vapor Deposition Technique

The development of thin film deposition techniques has been a crucial factor in the growth of numerous high-technology industries, particularly in the area of electronics, optics, and energy-related systems. Conventional deposition methods do exist, which up to now, in most cases, cannot easily produce the desired properties in materials, specifically when complicated or even multi-functional materials are required. All the above-mentioned issues have created the driving force for hybrid techniques development, which combines strengths of different deposition methods attempting to avoid deficiencies of traditional processes. One of the most interesting recent developments is Hybrid Physical–Chemical Vapor Deposition, a technique coupling the advantages of both PVD and CVD [10].

HPCVD represents a quantum leap in thin film technology owing to the enhanced control possibilities of film composition, microstructure, and properties. The technique allows for depositing complex material with tailored functionalities, making it an essential technique in advanced material applications. This paper overviews principles, mechanisms, and applications of HPCVD methods, advantages compared to conventional methods, and prospects for further use in future thin film technologies [11]. HPCVD combines the physical and chemical aspects of both PVD and CVD, respectively, into one to make use of the advantages both techniques offer. The basic underlying concept is to make use of simultaneous or successive employment of physical vapor generation in PVD and chemical reactions in CVD for depositing thin films of superior properties.

Most of the time, the process in HPCVD involves the physical generation of a vapor phase from the target material by some physical means, such as sputtering or evaporation. The generated vapor is subsequently transported to the substrate, where interaction with the reactive gases introduced into the system takes place just as it does in CVD. These reactive gases take part in a chemical reaction with the vaporized material on the substrate surface to deposit a thin film. This hybrid technique has immense possibilities for improved control of film composition, structure, and morphology in complicated materials or multilayered structures. HPCVD can be carried out at a lower temperature compared to the conventional CVD processes, and hence it could be applied to a wide variety of substrates [12].

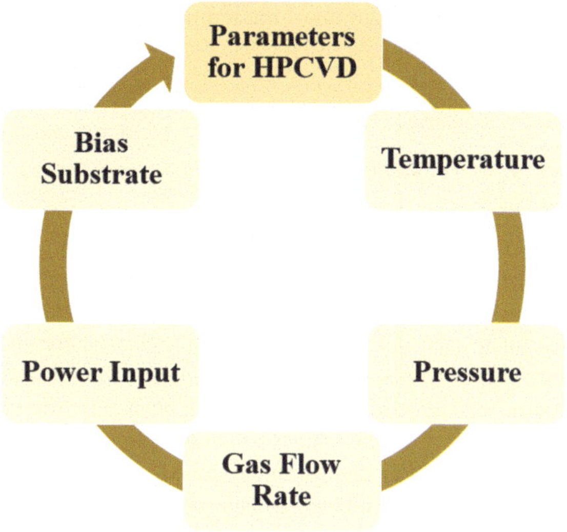

Fig. 6.1 Some parameters that affect the HPCVD

6.2.2 *Process Factors*

Process parameters and their control are basic for obtaining the characteristics of the film, such as thickness, uniformity, crystallinity, and chemical composition required in HPCVD. Appropriate control of the parameters mentioned here ensures reproducibility, efficiency, and quality film deposition as expressed in the Fig. 6.1. Herein, a discussion will be made in detail on the most important process parameters and how these are being controlled in HPCVD.

6.2.3 *Temperature*

Temperature is a key factor in HPCVD since it has an immense effect on the decomposition of precursor gases, film quality, and deposition dynamics in general. Improved kinetic energy with rising temperature improves the decomposition efficiency of precursor gases and hence results in a raised growth rate of the deposited film. The crystallinity and surface morphology of the film are also temperature-dependent: higher temperatures favor crystalline structures and smoother surfaces due to increased adatom mobility. On the other hand, too high temperatures may cause secondary gas-phase reactions possibly leading to the deposition of undesired by-products that can affect film purity. The temperature during deposition, and substrate temperature, will affect adhesion, residual stress, and uniformity of the film. Temperature control is very critical because any deviation from this may lead to the loss of reproducibility and desired properties of the material. This, therefore, means that optimization of the temperature factor will enhance both efficiency and quality in the process of HPCVD [13].

6.2.4 Pressure

Among the parameters critical in the HPCVD process, pressure is one of the most important. It influences not only the kinetics of the chemical reactions involved but also the properties of the deposited film. In general, lower pressure increases the mean free path and reduces the collision frequency of gas molecules, which enhances the transport of precursor gases to the substrate for more uniform and better-controlled film growth sometimes. Conversely, the higher the pressure, the greater the possibility of gas-phase collisions, which may favorably decompose precursor gases but possibly undesirable secondary reactions, thus leading to the formation of by-products that are likely to incorporate into the film. Pressure in addition changes the nucleation and growth modes of the film; low pressures often result in layer-by-layer growth, whereas higher pressures lead to three-dimensional island growth due to increased super saturation. Also, the changes in pressure modify the film's density and porosity; usually, the films are denser at low pressure. All factors become imperative in optimizing pressure to monitor film microstructure, composition, and quality within the HPCVD process [14].

6.2.5 Gas Flow Rate

In the process of HPCVD, gas flow rate is such a critical parameter that can affect film deposition rates, uniformities, and qualities directly. For example, low gas flow rates imply a low concentration of precursor gases near the substrate, hence leading to a slow deposition rate. It can be manifested in thinner films or longer deposition times. However, low flow of the gases can also permit an increase in film uniformity because it allows growth to occur at a very slow rate, thus enabling it to be more controlled without defects such as pinholes or non-uniform thickness. Another point that comes with a low flow rate is increased interaction time between precursor gases and substrate, which may possibly improve adherence of the film to the substrate and reduce unwanted by-products from gas-phase reactions. On the negative side, too low flow rates may lead to incomplete coverage or inhomogeneous films, especially if the supply of precursor becomes poor for the whole substrate. Thus, optimization of the gas flow rate is quite important to improve film properties and sustain consistent high-quality deposition in the HPCVD process [15].

6.2.6 Power Input

Power input rate is an extremely important parameter for HPCVD since this affects the entire deposition process, such as the decomposition rate of precursor gases, film growth rate, and overall film quality. The basic relationship is that the temperature of the hot plate or substrate varies primarily with power input, which is necessary to provide thermal energy that can break down precursor molecules into active species capable of depositing onto the substrate. Generally, the substrate temperature is higher with a higher rate of power input, and thus there is a more rapid decomposition of these gases, which therefore increases the deposition rate [16]. However, this can easily lead to overheating with too high power input, and consequently thermal stresses that could lead to film defects such as crack or delamination. Insufficient input of power would lead to incomplete decomposition, hence lower quality films due to insufficient adhesion or uneven coverage. Higher rates of power input contribute toward the crystallinity and surface morphology of the film-the higher the temperature, the better the crystallinity and smoother the surfaces; excessive power can give rise to a rough, columnar structure as a result of rapid deposition. Another important aspect is that the rate of power input is directly related to the energy consumption, further affecting the overall process cost and efficiency. Since the rate of power input is very important to ensure stability and good reproducibility of the process, fluctuations may occur in changes in film properties from run to run. Thus, optimization of the rate of power input will be one of the most relevant strategies for achieving films with the expected features, which combine quality with energy efficiency and are cost-effective in the HPCVD process [17].

6.2.7 Bias Substrate

Biasing of the substrate during deposition in HPCVD involves the application of some electrical potential to the substrate, and this has been found to have significant effects on the growth, adhesion, and properties of the film being deposited. A substrate bias is a technique that controls the energy and trajectory of ions and charged species in plasma as they interact with the surface of a substrate. A negative bias normally attracts the positively charged ions toward the substrate, hence increasing their impact energy. Such increased ion bombardment may result in a denser film because the high-energy ions remove the atoms that are weakly bound, enhance surface diffusion, and favor compact and uniform film structure. The substrate bias also affects the film microstructure, with finer grain sizes often appearing owing to the enhanced nucleation rate introduced by ion impacts [18]. This also can have some negative consequences: for example, the more increased stress in a film may lead to cracking or delamination of the film, and the probably unwanted incorporation of impurities or defects due to high-energy ion bombardment. Substrate bias also influences the overall film composition, in particular in many compound material processes, where different elements show different responses to ion bombardment. Hence, optimization of substrate bias becomes a key step in striking a balance between the benefit of improving film density and adhesion and the risk of inducing stress and defects in achieving the deposition of high-quality films with favorable mechanical, electrical, and structural properties in the HPCVD process.

The parameters involved in the HPCVD process need to be rigorously controlled to achieve films of high quality, reproducibility, and desired properties. Careful optimization is required for temperature, gas flow rate, pressure, power input, substrate bias, and deposition time. Advanced monitoring and feedback control systems enhance process stability and allow for real-time adjustments to maintain the deposition process within the desired specification and ensure films of the required performance consistently [19].

6.3 Plasma-Enhanced Atomic Layer Deposition

Plasma-enhanced atomic layer deposition has recently emerged as one of the key technologies in the field of thin film depositions, effectively merging well-established principles of Atomic Layer Deposition with dynamic capabilities from plasma processing. The hybrid technique has indeed revolutionized the way thin films are produced by offering unprecedented control of film characteristics, improving deposition rates, and gaining versatility in the materials processed. The underlying principles, benefits, and fields of application, as well as contemporary issues that have motivated the research and development of PEALD, should be an integral part of a discussion on the scope of the technique [20].

Plasma-Enhanced Atomic Layer Deposition, or PEALD, is an advanced modification of ALD, including plasma technology in the deposition process. Plasma usually is considered as a fourth state of matter and represents partially ionized gas that includes ions, electrons, and neutral species. The electric field applied to any gas will cause it to ionize into plasma and form a highly reactive environment. In the context of PEALD, additional energy is brought into the system by plasma that might enhance the chemical reactions of the deposition process [21].

6.3.1 Working Process

The first step of PEALD involves exposure of a substrate surface to a metal precursor within a reaction chamber. The precursor self-limiting adsorbs on the substrate surface, forming only one monolayer, until saturation of the surface is achieved. Once more, this is characteristic of ALD: the ability to exercise this level of control enables the uniform and conformal growth of thin films across complicated substrate geometries. After precursor adsorption, the substrate is exposed to a plasma field. Plasma is a partially ionized gas, consisting of ions, electrons, and neutral species, generated by the application of an electric field to a gas like oxygen or nitrogen. The step of plasma is an essential one in the process of PEALD since either it activates the precursor or cleans and activates the substrate surface, thereby enhancing the reactivity of the successive chemical reactions [22]. In plasma surface processing, high-energy species are formed-such as radicals and ions-that interact with either the precursor monolayer or the substrate surface. Interactions increase chemical reaction rates that result in the desired material, in this case, a metal oxide or nitride. For instance, if one desires to deposit a metal oxide film, the plasma may contain oxygen, which then combines with the precursor of the metal in forming the oxide layer. This plasma can break chemical bonds to form active radicals, and so in PEALD processes, it becomes possible to deposit materials which otherwise would have required a higher temperature or more chemically reactive environment [23].

It's followed by plasma activation and then the introduction of a co-reactant, which normally includes some kind of oxidizing or reducing agent like water vapor or hydrogen. The co-reactant completes the reaction with the adsorbed precursor to finish off the desired thin film. This again is self-limiting in that the reaction stops when all the available precursor sites have reacted to give consistent thickness of the layers. After the reaction, the chamber is purged, which removes residual precursor and coreactant along with the by-products to ensure that only one clean layer of material is deposited per cycle. For the attainment of desired thickness, this cycle of precursor adsorption, plasma activation, coreactant introduction, and purging is repeated to build up the film, adding about one atomic layer per cycle [24].

One of the key advantages of the PEALD is control over the plasma parameters: power, duration, and composition allow a tailoring of properties. For example, higher power would increase the energy of the reactive species, which should lead to more robust reactions or even denser films. Of course, all these parameters need proper optimization in order not to create unwanted side effects, such as substrate damage or unwanted etching. Besides, the use of plasma allows running PEALD at lower substrate temperatures compared with conventional thermal ALD. That provides a method for temperature-sensitive substrates, including polymers and some electronic materials [25].

6.3.2 Efficient Factors

Plasma-enhanced atomic layer deposition is a very sophisticated technique; it requires the optimization of many parameters for getting the right film properties. Several factors make a difference in the scope of PEALD effectiveness and efficiency in yielding films into desired quality, uniformity, and characteristics as shown in Fig. 6.2. We shall look into each of these factors in detail in the following section [26].

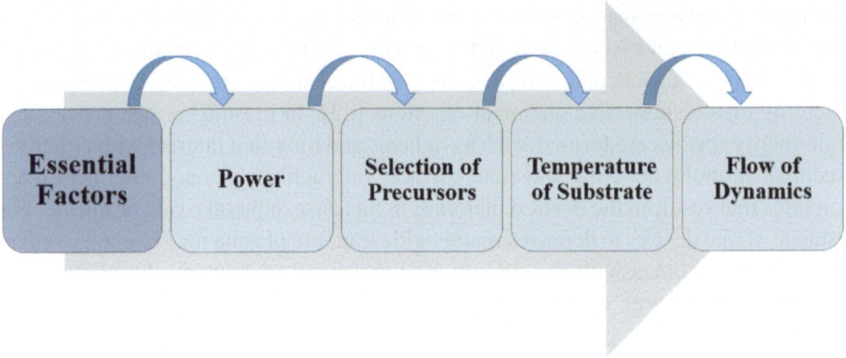

Fig. 6.2 Essential factors of PEALD which determines its working

6.3.3 Power

Another important factor is power in the plasma, which affects the energy level of the plasma and consequently the nature of the reactive species produced. An applied electric field is what normally creates plasma through the ionization of gas molecules into a mixture of ions, electrons, and neutral species. Power fed into the plasma regulates the intensity of such an ionization process. Higher plasma power means higher energy for electrons within the plasma, which in turn increases the ionization and dissociation of gas molecules. What happens here is that more reactive radicals and ions are produced. These can further break the chemical bonds in the precursor molecules more effectively, helping to form a desired thin film on the substrate. More importantly, higher plasma power used for metal oxide deposition can produce more oxygen radicals, which would react more effectively with the precursor of interest and provide a much denser oxide layer. Another factor affecting the choice of plasma power is the nature of the substrate. For substrate materials that are easily damaged by high-energy plasma, such as polymers or organic substrates, one would not want it to be etched, degraded, or otherwise unwantedly changed. In these cases, lower plasma power might be required to protect the substrate while still achieving adequate precursor activation [27].

6.3.4 Selection of Precursors

The chemical nature of the precursor concerns volatility, thermal stability, and reactivity and therefore plays an important role during deposition. Ideal precursors in PEALD should possess adequate volatility for controlled delivery into the reaction chamber, sufficient thermal stability against decomposition before reaching the substrate, and sufficient reactivity toward forming a self-limiting monolayer on the substrate surface [28]. For example, TMA is one of the most widely used precursors for Al_2O_3 deposition because of its high volatility and strong reactivity toward oxidizing co-reactants. In addition, it is important to carefully control the interaction of the precursor with the co-reactant in order not to provoke some side reactions that could lead to the introduction of impurities or defects within the film. For instance, plasmas can be utilized to decompose complicated precursors that may not react well in a purely thermal ALD process. It is also crucial that precursor and co-reactant selection be made taking into consideration the by-products from the reaction because these too may impact film purity and, once more, increase the requirement for effective purging between cycles [29].

6.3.5 *Temperature of Substrate*

The temperature of the substrate will act to affect precursor adsorption on the surface. If the temperature is too low, the precursors may not adsorb well, leading to incomplete or non-uniform coverage. However, if the temperature is too high, it could cause the precursor to desorb from the surface or result in decomposition before film formation, hence yielding a film of poor quality or undesired reactions. Hence, an optimum temperature must be applied which permits efficient adsorption without the decomposition of the precursor. Generally speaking, increasing the substrate temperature provides higher reaction rates between the precursor and co-reactant in such a process, accelerating the film growth rate. However in PEALD, the possible compensation of lower thermal energy by additional activation energy due to high-energy species could be done by using plasma. So, in the process of PEALD, the substrate temperature can be lower than that in thermal ALD, which is very favorable for temperature-sensitive substrates such as organic materials or some kinds of polymers [30].

6.3.6 *Flow of Dynamics*

Chamber pressure generally controls the mean free path of the gas molecules—the average distance a molecule travels before it collides with another molecule. Lower chamber pressures normally lead to longer mean free paths. This lessens the opportunity for gas-phase reactions and gives better conformity to the film. However, very low pressures decrease the density of reactive species in the plasma and can slow the deposition rate. It is, therefore, of paramount importance to determine the value of pressure in an optimum state for a balance of the mentioned effects for efficient film growth. It is known that the precursor and co-reactant gas flow rates, as well as the purge gas flow rate, are factors that influence deposition rate and film uniformity. Adequate gas flow control ensures efficient delivery of the precursors and co-reactants onto the substrate and minimizes unwanted reactions in the gas phase. Efficient purging between cycles ensures the residue of precursor and by-products are removed avoids contamination and affects the purity of the deposited film. Poor purification could result in a film problem, such as CVD-like growth or incorporation of impurities. Depositing films consistently across the entire substrate requires stringent control over the chamber pressure and dynamics of gas flow. Gradients in pressure or non-uniform flow conditions from gases can easily produce film thickness and composition differences across large substrates and even within high-aspect-ratio features. Complex chamber designs and optimization of process parameters in these geometries are required many times to achieve consistent depositions [31].

6.4 Ion Beam-Assisted Deposition

The PVD process known as Ion Beam Sputtering (IBS), often referred to as Ion Beam Deposition or Ion Beam Sputter Deposition (IBSD), generates thin layers on a substrate that are very thick, homogeneous, and lack defects. Noble gas ions, such as nitrogen or argon, are blasted into the target material by a concentrated wide ion beam. High kinetic velocity ions strike the target's surface, ejecting tiny material particles atoms, or molecules that condense on a substrate surface to produce a thin coating. Multiple stacked layers, or multilayer coatings, can be created with good quality by utilizing different targets in a single process chamber.

Ion beam sputtering has some benefits over other PVD processes like magnetron sputtering or evaporation. Particles pushed out impacted the substrate with great force. This gives particles, creating notably thick and flawless layers. Low process temperatures and low sputtering pressure make excellent film characteristics and dense film growth possible. Furthermore, the energy of the ion beam may be accurately adjusted, providing more control over the deposited layer's density, stickiness, and composition. Dual ion beam sputtering, also known as dual ion beam deposition, is a unique type of ion beam sputtering. An additional ion bombardment is added using an assist ion beam source to affect the growth of the film [32].

6.4.1 Introduction

A systematic approach to preparing high-quality films with a certain desired feature is simultaneously depositing atoms and ions on a substrate, where the incoming ions interact with the growing film to alter composition and characteristics. The interaction could give the film more desirable density, adhesion, and mechanical qualities, thus making it suitable for a range of applications that include electronics, optics, and energy storage. A substance is first allowed to evaporate, most often in a vacuum, and then it is bombarded by intense ions. The ions, usually nitrogen or argon, are made to impinge on it to deposit energy and thereby modify the structure of the growing film.

This energy deposition may be due to elastic collisions with film atoms, or compound formation reactions with the deposited atoms. In addition, the ion beam is able to sputter the substrate so that a gradual transition between the substrate material and the coating enhances the interface between the layers. Compared to the traditional PVD methods, some advantages of IBAD concern the possibility of producing films with controlled mechanical strength, adhesion, and density, yielding higher quality [33]. Besides, the technique allows for film depositions at lower temperatures that are appropriate for sensitive materials. Moreover, compressive stress can be introduced to the film by IBAD that enhances its mechanical properties. The process of IBAD is typically conducted in a vacuum chamber where the substrate temperature, ion energy, and beam current density are varied with the view of controlling the ion beam. The ion beam current density is an important parameter in that it controls the amount of energy deposited per atom and final film characteristics. In comparison, the process is slower when compared with the conventional PVD techniques. Though the process is slower in comparison with traditional methods under PVD techniques, the benefit derived is that high-quality films with customized characteristics may be produced. In sum, ion beam-assisted deposition represents a flexible technique for thin film deposition because it integrates precision due to ion bombardment with physical vapor deposition. It allows for the preparation of high-quality films with prescribed properties, making them suitable in diverse applications from different fields. Bombardment by energetic particles in the case of a growing film has been shown to improve adhesion, cause densification of films grown at low substrate temperature, modify residual stresses, control texture—orientation, modify grain size and morphology, alter optical properties, and modify hardness and ductility [34]. These properties and characteristics are critical to the performance of thin films and coatings. Numerous terms have been used to describe the simultaneous thin-film deposition and directed ion bombardment from an ion source: dynamic recoil mixing (DRM) at high energies; ion-assisted coating (IAC); ion-assisted deposition (IAD); ion vapor deposition (IVD); ion beam enhanced deposition (IBED); and ion beam assisted deposition (IBAD). Important processing factors in ion plating include the energy, flow, and mass of the bombardment species as well as the ratio of bombarding to depositing particles. Evaporation, sputtering (bias sputtering), arc vaporization, or the breakdown of a chemical vapor precursor chemical vapor deposition (CVD) are the possible methods for vaporizing the depositing material. Usually, ions of inert or reactive gas or, in certain instances, ions of the condensing film material are the energetic particles utilized in bombardment ("film ions"). Ion plating can be carried out in a vacuum setting, where ions for bombardment are created in a separate ion cannon, or it can be carried out in a plasma environment, where ions for bombardment are extracted from the plasma [7, 35].

6.4.2 Working Principle and Construction

A materials engineering approach called ion beam-assisted deposition (IBAD) combines ion implantation with simultaneous physical vapor deposition (PVD) procedures including sputtering or evaporation. The simultaneous arrival of energetic ions (usually in the range of 100 eV to a few keV) and vaporized atoms to the substrate surface is the fundamental idea behind IBAD.

The vaporized atoms are deposited onto the substrate during the IBAD process, and the growing thin layer is targeted by a high-energy ion beam. The film's atomic layers are only partially penetrated by the ions, but if the deposition process is continued, the IBAD film can thicken to several microns [36]. The two main ways the characteristics of the film can be changed by the ion beam are.

(1) Elastic collisions with the atoms in the film, which deposit energy therein.
(2) By forming compounds with the deposited atoms.

One of the biggest advantages of the IBAD technique is independent control of such parameters as ion energy, substrate temperature, and atomic species arrival rate during deposition. Such control may result in films that are more strongly linked to the substrate by allowing for a gradual transition between the deposited film and the substrate material. Moreover, ion beam energy deposition can contribute to the residual stress reduction in the film, which prevents it from cracking or delaminating. The ion bombardment during IBAD is also able to suppress the columnar microstructure development that is often noticed in thin films deposited by the low-temperature PVD processes. Ion beam-induced collision cascades favor atomic rearrangement and film densification, even for low deposition temperatures, yielding very dense thin films. This may be particularly important for situations in which higher temperatures would affect the material surface characteristics [37].

6.4.3 Working Process

In the combined methods of IBAD, a thin film is generated by deposition on a substrate, onto which a focused beam of energetic ions is continuously bombarded. Some characteristics of the film that might vary because of interaction between the ion beam and the outermost few atomic layers in growth conditions include hardness, adherence to the substrate, density, microstructure, and composition. Despite some interaction within the ion-beam assisted deposition principle. At the same time, energetic ions (100 eV to a few keV) and evaporated atoms at thermal energy reach the substrate surface. The IBAD coating may be several microns thick despite the ion beam only penetrating a few atomic layers. This is due to the currently in-progress vapor deposition. By interacting with the vapor atoms to generate compounds or by depositing energy in elastic collisions with the film atoms, the ions modify the characteristics of the film [38]. Film characteristics can also be influenced by reactive background gas. The initial steps of an IBAD process involve coating and substrate; this is not to be linked with the ion-beam mixing process. This involves ion bombardment of the interface layer between a film that has previously been deposited and the substrate with enough energy for the entire film to be penetrated. Low-energy ions (100 eV to a few keV) and evaporated atoms reach the substrate surface at the same time. A few atomic layers may be penetrated by such low-energy particles, but if processing continues, the IBAD film can grow to a thickness of several microns. The ion beam modifies the film's characteristics in two ways: first, by adding energy to the film; and second, by interacting with the atoms deposited in the vapor to produce compounds [32]. The combined impact of reaction with the atoms of a regulated intake or background gas may occasionally be included in that which follows.

6.4.4 Features and Uses of Film Traditional IBAD

- **IBAD**

Film coating techniques are frequently needed to modify the characteristics of a substrate material. There are several thin-film coating methods available, and most entail depositing a substance onto the surface of an optical element. Although thermal, e-beam, and reactive arc evaporation are inexpensive thin film deposition procedures, their film quality is not very good. Although they are more expensive, other methods like ion beam sputtering provide the best environmental stability and longevity. Ion-beam-assisted deposition, or IBAD, offers superior depositions and a high rate of return on investment the best of both worlds.

- Physical vapor deposition of thin films at low temperatures frequently results in a microstructure with columns oriented in the direction of the atoms that are impinging, which causes porosity in the film, which is generally undesired. Surface diffusion levels out this structure when the deposit temperature is raised over a threshold value, usually 0.25 T_m, where T_m is the film melting temperature. With IBAD, the ion beam's energy is mostly deposited by elastic ion-atom collisions, and the collision cascade's atomic motions prevent the columnar structure from forming. Therefore, even at low temperatures, highly thick thin films may be produced using IBAD. This is highly beneficial when surface qualities are altered by high temperatures [39].
- The decrease in residual tension inside the film has a significant impact on the ion beam's energy deposition. The microstructure of the film, contaminants, and the various thermal expansion coefficients of the film and substrate are the main causes of residual stress. If the ion/atom arrival ratio R is high enough to be certain that every atom in the deposited layer is involved in a collision cascade, the stress may be efficiently annealed out. The ion energy determines the crucial R-value, which is usually < 0.01; that is, a beam current intensity of less than 1.6 $\mu A/cm^2$

is enough for a film growth rate of 2 Å/s. One crucial method for preventing the film from cracking or delaminating is stress annealing [40].

By sputter-cleaning the substrate surface with an ion beam and mixing the atoms of the film and substrate (the latter requiring high beam energy to guarantee sufficiently deep penetration into the substrate), film adhesion can be further enhanced. These methods may be used to create MoS_2 sheets for dry lubrication that have incredibly high adhesion.

6.4.5 Reactive IBAD

Ion bombardment and physical vapor deposition (PVD) are used in on-beam-assisted deposition, also known as ion-assisted deposition, a thin-film deposition technique. There is a vacuum during the IBAD process. High-energy nitrogen and argon ions are directed toward the developing thin film by an ion source as the material is evaporated using the PVD process.

Temperature, the arrival rate of substrate atoms, and ion energy are among the deposition parameters that may be independently controlled by an ion-beam-assisted deposition system. The thin film characteristics can therefore be altered.

Ion-assisted e-beam deposition (IBAD) is a thin-film coating technique that combines evaporation with ion support. Reactive ions can be directed to the surface area to affect the shape, bonding, and chemical composition of thin films. In contrast, IBID uses a high-energy ion beam to break down gaseous molecules to deposit pieces onto a substrate close to the coating material [41].

IBID and electron-beam deposition have similarities (EBID). However, an IBID system employs an ion beam in place of an electron beam and a scanning electron microscope.

The ion beam's main function in all of those situations is to impart energy and momentum to the atoms in the substrate and film. In these situations, noble gas ions are typically used in the process, and the typical ion/atom arrival rate ratios are substantially smaller than one. Ion beams with low and medium currents are therefore effective. The ion species is important when the ion beam provides atoms for compound formation (also known as reactive IBAD).

To achieve an appropriate stoichiometry and the required film characteristics, the *R*-value is also a far more important parameter. For this application, high ion currents are necessary since the ion and atom arrival rates have to be comparable in size. Therefore, a flexible ion-beam system that can generate high beam currents of various gaseous (and maybe metallic) ions is needed. By evaporating Ti or Al and then blasting the film with a nitrogen beam with an intensity equal to $R = 1$, one may create thick films of TiN or AlN with high adhesion using the reactive IBAD approach. They may also be created by using nitrogen as a background gas and blasting the metal layer with a beam of noble gas, such as Ar^+. Zr may be vaporized under an oxygen beam bombardment to create very dense oxide coatings for use in optics. Additional IBAD methods that are now in the research stage involve producing extremely durable and chemically inert carbon and cubic boron nitride coatings that resemble diamonds [42].

6.4.6 Advantages and Disadvantages

Advantages

IBAD is a helpful technology in many advanced applications because it has several critical advantages. Probably the most important of the advantages is the improvement in film adherence to the substrate. In cleaning the substrate surface by sputtering away impurities and oxide layers, an ion beam can render the surface more reactive to the incoming atoms or molecules. Improved bonding at the interface will ensure better adhesion and reduce the possibility of film delamination or peeling for applications on difficult substrates, such as polymers or ceramics [43]. One of the major benefits of IBAD involves film density and microstructure enhancement. An ion-assisted deposition process has the growing film's atomic arrangement influenced by the focused ion beam, through which there is improvement in the packing density and reduced voids or defects. These effects culminate in films with higher mechanical strength, enhanced hardness, and improved structural integrity. These qualities are especially desired, as thin films for protective coatings, optical films, and electronic devices need high-quality films with long-term performance [44].

IBAD also provides good control over film stress, which is one of the problems in thin film deposition. As a result, controlling the energy, flux, and angle of incidence of the ion beam may provide control over the intrinsic tension within the film. Especially in very precise uses, as in microelectronics or high-performance optics, stress control is very important in ensuring that the film remains mechanically stable without any defects such as warping, cracking, or even delamination. Another major advantage of the method is the versatility it allows for film quality adjustment. By controlling the ion beam, refractive index, hardness, electrical conductivity, and optical transparency become but a few film characteristics that can be precisely tuned by IBAD to produce either phase changes or create dopants during deposition. This level of control allows the synthesis of advanced materials with niche applications, including flexible electronics, graded-index optical coatings, and conductive films. Further, IBAD works especially effectively for coatings that are uniformly deposited on substrates of complicated or irregular shapes. Conformal coverage can be achieved on high-aspect ratio or complex-geometry surfaces by directing an ion beam at various angles. A very significant advantage arises by coating elements that have to be compatible and of high accuracy-for example, turbine blades, medical implants, or MEMS [45].

Disadvantages

Ion Beam Assisted Deposition, or IBAD in abbreviation, has a great number of advantages, although there are some disadvantages that also should be taken into consideration. Among the major disadvantages of this method are the high cost and the complicatedness of the process. In general, IBAD systems are very complicated, and this already presupposes an accurate control of a great number of parameters, such as the ion beam energy, flux, and angle of incidence. Issues are further complicated by coordinating the deposition source with the ion source. Due to the large degree of precision and control, IBAD equipment is expensive to purchase as well as to operate and maintain; this may preclude its application in high-value manufacturing environments or specialized research establishments.

Another major disadvantage, commonly associated with IBAD, involves limited deposition rates. The very nature of the process makes it such that the use of an ion beam in conjunction with deposition will result in lower deposition rates compared to traditional PVD methods, such as sputtering or evaporation. Such a reduced deposition rate may prove to be a serious disadvantage in applications needing thick films or high-throughput processing since it increases the time needed for the processing and decreases manufacturing efficiency. With IBAD, one added concern is ion-induced damage [46]. Whereas the ion beam may be used to affect film properties, if it is not judiciously controlled, it also will induce defects, dislocations, or amorphous regions within the film. In particular, high-energy ions may create substrate or film damage which will degrade the desired qualities. The growing film may also be sputtered or etched by the ion beam, thus increasing its surface roughness or porosity, even thinning. A further complication arises because the characteristics of the ion beam need to be precisely adjusted to avoid all these negative effects. Other problems that are possible with IBAD include substrate heating due to ion bombardment. When compared with other techniques, the method can work at a lower temperature of the substrate; because the ion beam introduces local heating, though, the substrate may undergo thermal strains or unwanted chemical reactions. This is particularly critical when working with materials sensitive to temperature, like polymers, since overheating may cause warping, deterioration, or outgassing and hence reduce the quality of the film [47].

6.5 Applications

There are several uses for ion beam sputtering. For example, it is utilized to produce piezoelectric coatings for enhanced mobile communications in the MEMS field. In the optical industry, it is also employed to coat lenses or mirrors to maximize their transmittance or reflectance at certain wavelengths. Moreover, ion beam sputtering is employed in several other fields requiring accurate thin films, such as the fabrication of wear-resistant coatings [48].

6.5.1 Deposition of Dielectric Films by Ion Beam Sputtering on Wide Optical Substrates

Due to point-type sources and the lack of an advanced displacement mechanism, homogeneous deposition on large, flat optical substrates utilizing the most sophisticated electron beam evaporation sources is becoming challenging. As an alternative, compared to alternative procedures, the use of ion beam sputtering (IBS) provides the benefit of greater energy of the sputtered material (i.e., temperature) arriving at the substrate. As a result, denser films with characteristics that are comparable to the corresponding bulk materials are frequently produced using IBS. On an optical glass, Al_2O_3 was applied as an anti-reflective coating. The metallic aluminum target served as the reactive source for the process. As a reactive ground gas, oxygen was introduced. The dimensions of the substrate were 500 × 300 mm. Shapers and a constant velocity linear movement of the substrate level were used to control uniformity. Al_2O_3 was deposited at a static rate of 7 nm/min. Using a software-integrated quartz crystal oscillator, the deposited film thickness was tracked in situ and utilized as an endpoint signal. Using a Sentech SE400 Ellipsometer, measurements of film thickness were performed. Five substrates measuring 150 mm each were used to measure the large sample area. It was possible to determine reproducibility with three more process runs [49].

6.5.2 Sputtered Ion Beams for Anti-Reflective and High-Reflective Coatings

For optical coatings, ion beam sputtering is a reliable technique in the fields of research and manufacturing. Coatings of high-accuracy optics, such as beam splitters, mirrors, and filters, are typical uses. The great mobility of the sputtered particles on the substrate surface, and therefore the good mechanical properties of the layers, is advantageous to optical multilayer stacks. Ion beam sputtering works based on target material being ejected by high-intensity ion bombardment [50]. Angstrom precision in material removal is made possible by the regulation of the ion energy. Unlike magnetron sputtering, minimal contamination and high-quality coatings are produced by low base pressure (10^{-4} mbar) and temperatures below 100 °C. The increased energy (mobility) of the materials that arrive on the substrate in comparison to thermal evaporation or magnetron sputtering further supports this. Due to improved self-organization processes, the density of the material that has been deposited is often higher as well. Ion beam sputtering is the recommended deposition technique for premium optical coatings because of these film characteristics. The film characteristics can be further tuned if an extra ion beam source is used to assist the deposition. To eliminate organic contamination, in particular, a second ion irradiation of the substrate surface can be employed as a substrate pre-cleaning procedure [51]. Moreover, reactive sputtering, additional densification of the growing layer, and surface smoothing/structuring may be accomplished using the secondary ion beam source. The first industrial-scale implementation of IBAD technology was in the manufacturing of optical anti-reflective coatings, which greatly benefit from better adherence and density as compared to non-ion deposited coatings. Additional uses, such as wear resistance, corrosion prevention, and diamond-like carbon coatings, are now finding industrial use.

Given the wide range of opportunities for developing novel low-temperature coatings.

The IBAD approach is certain to stimulate more advancement and research efforts in a regulated manner. Consequently, the technology will find new uses in areas where there is a need for the unique benefits of IBAD films [52].

To convert semiconductor injection lasers into various types of active devices, AR coating films are utilized. By certain application requirements, optical coatings on lenses or mirrors modify how an optic reflects light waves. AR and HR coatings are the two main coating alternatives for modifying reflectance. Sunglasses, camera lenses, and solar cells are frequently coated with antireflective coatings (AR), which boost transmission to decrease surface reflections. High-reflecting coatings (HR), also referred to as mirror coatings, improve surface reflectance and enable the fabrication of mirrors that almost entirely reflect light falling within a specific wavelength band. Utilizing sputtered silicon nitride, high-quality antireflection coatings for laser surfaces have been accomplished. Laser applications use coatings with both AR and HR properties [53].

6.5.3 Organic and Inorganic Hybrid Materials Deposition

The possibility of combining unique features of organic and inorganic components inside a single molecular composite has drawn a lot of interest to organic–inorganic hybrids. In addition to their potential for semiconducting and even metallic behavior, organic materials have the advantage of structural flexibility, easy processing, adjustable electrical characteristics, photoconductivity, and efficient luminescence. High carrier mobility, band gap tunability, a variety of magnetic and dielectric characteristics, and mechanical and thermal stability are among the possible benefits of inorganic compounds. The contact between the organic and inorganic components can induce new or amplified phenomena in addition to combining unique properties.

Because of the smaller dielectric constant and less filtration in the organic component of the hybrid, dielectric manipulation in organic–inorganic systems has also resulted in higher binding energies for excitons in the inorganic framework of the structure. The self-assembling organic–inorganic perovskites are a specific group of crystalline hybrids that offer a route to an alternating framework of semiconducting inorganic sheets and organic layers. The natural quantum well structures' attributes may be adjusted by adjusting the thickness, metal/halogen concentration, and chemical/physical features of the organic cations, such as their length, shape, and polarizability. As the thickness of the perovskite sheet (or well-width) increases, members of the hybrid perovskite family show a semiconductor–metal transition [1].

6.6 Merging Chemical and Physical Methods

Two widely recognized thin-film deposition methods, Physical Vapor Deposition (PVD) and Chemical Vapor Deposition (CVD), have unique benefits and procedures that meet various application requirements. Solid materials are vaporized in a vacuum and then condensed onto a substrate to produce a thin film in the physical process of PVD. The process allows you to have total control over the thickness and growth of the thin film while operating within 20–500 °C. Because PVD films are of very high purity, besides being of excellent uniformity and stickiness, this process commands a premium for important applications in optical coatings, semiconductor devices, and tools [54]. Due to line-of-sight deposition, however, it could have trouble providing step coverage on intricate geometries that might result in shadowing effects. In the CVD method, the interaction between gaseous precursors and substrate surface decomposition is a chemical process to form the solid layer. Because this technique can provide excellent step coverage and, in most cases, requires working at higher substrate temperatures (200–1600 °C), it is ideal for coating complex structures and providing conformal layers. Some of the films that can be made by CVD methods with unparalleled uniformity and tailored for specific material qualities are semiconductor dielectric layers. Because of the very high temperatures and reactive gases involved in CVD, careful handling and disposal of byproducts must be performed because of the related safety and environmental hazards. Generally speaking, CVD yields better conformality and is suitable for more intricate geometries, while PVD acts very well in applications involving low temperatures, with excellent purity and control. Which approach to use will be determined, finally, by the application requirements: material qualities, substrate compatibility, and specific processing conditions [55].

6.7 Impact of Temperature on CVD and PVD

The substrate temperature also plays a key factor in the ultimate characteristics of thin films produced by both PVD and CVD techniques. The substrate temperature, in most of the PVD processes, varies between 20 and 500 °C. A columnar grain structure may occur at lower temperature films, producing smaller grain sizes, increased inherent stress, and lower adhesion. Higher substrate temperature will increase the mobility of adatoms—atoms that have fallen on the substrate—enabling better surface diffusion and a more uniform shape of the film. Higher temperatures can improve the mechanical or electrical properties of a film by increasing its crystallinity and reducing defects. However, higher temperatures might cause the substrate to degrade thermally or react undesirably with the film material [3, 45].

In contrast, CVD deals with substrates whose temperature range is from 200 to 1600 °C. Because of chemical interaction at the interface, this relatively higher temperature in CVD results in the dissociation of precursor gases, which enhances the uniformity and adhesion of the layer. High temperatures enhance reaction kinetics; hence, better films with fewer defects and better step coverage, particularly at complex geometries, can be obtained. However, this may be hampered by the required high temperatures for CVD because sometimes substrates cannot bear the temperature without deterioration. In the end, substrate temperature in both the PVD and CVD processes is one of the important conditions, influencing their microstructure and adhesion, hence the general quality of thin films. Generally speaking, the higher the temperature, the better the quality of the film. However, sometimes having high temperatures leads to adverse effects on the characteristics of the substrate material [56].

6.8 Chemical Compared to Physical Crosslinking Techniques

The thin film improvement during deposition can be enhanced by many strategies, including chemical and physical crosslinking. Chemical crosslinking involves the atoms within the film forming long-term covalent bonds in the structure for strength and stability. This technique employs either reactive species or energetic particles to develop interatomic bonds among atoms, thus improving mechanical strength, thermal stability, and resistance to degradation. In applications such as protective coatings, optical devices, and microelectronics, chemically cross-linked films show better performance and reliability because the covalent nature of bonds confers longevity and endurance under operating conditions. In contrast, however, reactive species also have the possibility to introduce defects or contaminants that could affect film purity and optical qualities. In contrast, physical crosslinking relies on non-covalent interactions such as hydrogen bonding, ionic interactions, or van der Waals forces to improve the film properties [57]. In most cases, this involves the introduction of functional groups or modification in deposition parameters that would promote such non-covalent contacts. Physically cross-linked films are of interest in smart coatings, sensors, and biomedical devices owing to their frequent demonstration of flexibility, self-healing, and responsiveness to stimuli. Other than reducing the possibility of contamination, physical crosslinking, without the use of reactive species, does not detract from the inherent qualities of the film, such as electrical conductivity and transparency. Physically cross-linked films may be less heat and mechanically stable compared to their chemically cross-linked counterparts. Whether the cross-linking approach is more chemical or physical in nature depends on the specific needs of the application, including target substrate, required film qualities, and compatibility with the deposition process. Both approaches have special benefits and can be engineered to create specific thin-film properties that enable the creation of state-of-the-art materials with unprecedented functionality and performance [58].

In polymer chemistry, crosslinking is a stabilizing procedure that encourages the multidimensional connectivity of polymer chains, converting liquid polymers into solids or gels. For polymers to grow mechanically stronger and resistant to things like heat, wear, and solvent assault, they must go through this process. Cross-linked networks in hydrogels keep them from disintegrating. To encourage polymer network reticulation, a variety of cross-linking techniques, including chemical and physical ones, can be used. Hydrogels that are physically cross-linked are often stabilized by hydrophobic, hydrogen bonding, or ionic/electrostatic interactions [59]. The most important advantage of the adoption of physical cross-linking methodologies is biomedical safety, since there are no chemical cross-linking agents involved, thus excluding the possibility of cytotoxicity. Photo-polymerization, Schiff base formation, and enzyme-induced crosslinking are some ways in which chemical hydrogels can be realized. Due to permanent covalent bonds being formed, chemically cross-linked hydrogels compared to physically cross-linked hydrogels present superior mechanical properties, higher stability under physiological conditions, and tunable degradation behavior. Moreover, permanent and longer-lasting relations, like permanent covalent bonds, are formed.

6.9 Advantages and Disadvantages

In contrast to the physical cross-linking approach, the methods used in chemical crosslinking are unquestionably superior in some ways, particularly those dealing with film stability and performance. The first advantage of chemical cross-linking is that it establishes permanent covalent connections between the polymer chains, yielding a stronger and permanent film structure. It significantly raises the mechanical strength and especially the heat stability of such films, making them suitable for a variety of applications, including demanding ones like protection coatings and electronic devices. Chemically cross-linked films can also deliver customized deterioration rates, providing precision control over the length of time they last and perform satisfactorily under a given set of conditions [60].

Another advantage is that chemical crosslinking can achieve films with excellent substrate adhesion-quality which is critical for maintaining the integrity of a film during use. Other than adding resistance to the solvents and degradations of the environmental matrix, the covalent bonds attained by chemical crosslinking ensure that during exposure, the characteristics of the films remain invariant in time. The flexibility of other chemical cross-linking methods, such as photopolymerization and enzyme-induced crosslinking, enables films to be tuned to specific application needs. Films resulting from physically cross-linked methods have often lower stability and mechanical quality but are generally safer and more biocompatible as they do not involve reactive chemical agents. The physically cross-linked films involve non-permanent interactions that may render maintaining performance difficult under varying conditions. Hence, chemical cross-linking methods provide considerable benefits in thin film formation for applications needing high durability with customized features [61].

Disadvantages

There are, however, a couple of disadvantages associated with the use of these chemical cross-linking techniques as compared to physical cross-linking techniques in the growth of thin films. The main drawback to this process is the potential addition of residual stresses to the film due to covalent bond formation. These stresses can lead to distortion or breaking of the thin film, especially if the thermal expansion coefficients of the substrate and the film are not matched. Furthermore, the use of most chemical cross-linkers involves specific conditions that further complicate deposition processes and raise the costs. Thirdly, chemical cross-linking usually involves hazardous or toxic agents, with dangerous environmental and safety impacts during the deposition process and in further handling of films. In contrast, physical crosslinking usually involves non-toxic interactions and, hence, the process is non-toxic and friendly to the environment [62]. Moreover, since chemical cross-linking is permanent and the covalent links are hardly broken, once the films are processed, they may hardly be modified or recycled. It can be thought that this rigidity may impose a big drawback in cases when reprocessing or adaptability is needed. Besides, although the chemical cross-linking techniques may give enhanced mechanical and thermal properties to thin films, the methods of physical crosslinking are more intriguing in certain situations. This can be related to complications such as residual stresses, complexities in the process, safety issues, and inflexibility [63].

6.10 Conclusion

In conclusion, deposition techniques introduce the important achievements that have been achieved with the development of hybrid deposition techniques, such as High-rate Chemical Vapor Deposition-HCPVD, Plasma-Enhanced Atomic Layer Deposition-PEALD, and Ion Beam Assisted Deposition-IBAD. All of the above processes have specific advantages for certain material and application requirements. While more involved and expensive, hybrid deposition techniques synthesize the strengths of various deposition techniques to allow for sophisticated film composition and superior uniformity. The extraordinary deposition rates and efficiency of HCPVD make this technique very valuable for large-scale industrial applications; however, the technique requires a high degree of control over precursor stability and temperature conditions. Although there are still problems with sophisticated equipment and rate depositions, PEALD truly excels in terms of providing accurate control over layer thickness and composition, particularly at low temperatures, and is suitable for applications requiring very high precision. IBAD, on the other hand, though improving adhesion and film density by ion bombardment and the possibility of controlling the stress, has the negative features of being expensive and complicated and also allows the introduction of ion-induced damage. Mastery in applying them to such varied applications as protective coatings and microelectronics requires an understanding of the working principles, critical processing parameters, and trade-offs. The in-depth treatment here will provide the researcher and the engineer with a basis for addressing trade-offs like film quality, production efficiency, and technological feasibility when choosing and adapting deposition processes.

References

1. Mitzi, D.B. 2001. Thin-film deposition of organic−inorganic hybrid materials. *Chemistry of Materials* 13 (10): 3283–3298.
2. Lou, Z., et al. 2019. Programmable three-dimensional advanced materials based on nanostructures as building blocks for flexible sensors. *Nano Today* 26: 176–198.
3. Oke, J.A., T.C. Jen. 2022. Atomic layer deposition and other thin film deposition techniques: From principles to film properties. *Journal of Materials Research and Technology* 21: 2481–2514.
4. Zhao, X., et al. 2023. Organic/inorganic hybrids for intelligent sensing and wearable clean energy applications. *Advanced Composites and Hybrid Matrials* 6 (5): 176
5. Petti, L., et al. 2016. Metal oxide semiconductor thin-film transistors for flexible electronics. *Applied Physics Reviews* 3 (2).
6. Mavukkandy, M.O., et al. 2020. Thin film deposition techniques for polymeric membranes–a review. *Journal of Membrane Science* 610: 118258.
7. Sun, L., et al. 2021. Chemical vapour deposition. *Nature Reviews Methods Primers* 1 (1): 5.
8. Schmitz, J. 2018. Low temperature thin films for next-generation microelectronics. *Surface and Coatings Technology* 343: 83–88.
9. Kunene, T.J., et al. 2022. Review of atomic layer deposition process, application and modeling tools. *Materials Today: Proceedings* 62: S95–S109.
10. Liu, G., et al. 2021. Additive manufacturing of structural materials. *Materials Science and Engineering: R: Reports* 145: 100596
11. Liu, X., et al. 2019. A review of perovskite photovoltaic materials' synthesis and applications via chemical vapor deposition method. *Materials* 12 (20): 3304.
12. Pedersen, H., and S.D. Elliott. 2014. Studying chemical vapor deposition processes with theoretical chemistry. *Theoretical Chemistry Accounts* 133: 1–10.
13. Mbam, S.O., et al. 2019. Thin-film coating; historical evolution, conventional deposition technologies, stress-state micro/nano-level measurement/models and prospects projection: A critical review. *Materials Research Express* 6 (12): 122001.
14. Ukoba, K., and T.C. Jen. 2023. *Thin films, atomic layer deposition, and 3D printing: Demystifying the concepts and their relevance in industry 4.0*. CRC Press.
15. Rawat, R.S. 2017. Dense plasma focus—High-energy-density pulsed plasma device based novel facility for controlled material processing and synthesis. *Plasma Science and Technology for Emerging Economies: An AAAPT Experience* 39–112.
16. Safi, I.J.S. 2000. Recent aspects concerning DC reactive magnetron sputtering of thin films: A review. *Surface and Coatings Technology* 127 (2–3): 203–218.
17. Lin, H.C., et al. 2012. Parameter optimization of continuous sputtering process based on Taguchi methods, neural networks, desirability function, and genetic algorithms. *Expert Systems with Applications* 39 (17): 12918–12925.
18. Mozetič, M., et al. 2018. Recent developments in surface science and engineering, thin films, nanoscience, biomaterials, plasma science, and vacuum technology. *Thin Solid Films* 660: 120–160.

19. Kern, W., and K.K. Schuegraf. 2001. Deposition technologies and applications: Introduction and overview. In *Handbook of thin Film Deposition Processes and Techniques*, 11–43. Elsevier.
20. Barrigón, E., et al. 2019. Synthesis and applications of III–V nanowires. *Chemical Reviews* 119 (15): 9170–9220.
21. Banerjee, S., et al. 2020. Atmospheric pressure plasma deposition of TiO_2: A review. *Materials* 13 (13): 2931.
22. Oehrlein, G.S., and S. Hamaguchi. 2018. Foundations of low-temperature plasma enhanced materials synthesis and etching. *Plasma Sources Science and Technology* 27 (2): 023001.
23. van Ommen, J.R., A. Goulas, and R.L. Puurunen. 2021. Kirk-Othmer encyclopedia of chemical technology. In *Atomic layer deposition*, 1–42.
24. Hu, L., W. Qi, and Y. Li. 2017. Coating strategies for atomic layer deposition. *Nanotechnology Reviews* 6 (6): 527–547.
25. Knez, M., K. Nielsch, and L. Niinistö. 2007. Synthesis and surface engineering of complex nanostructures by atomic layer deposition. *Advanced Materials* 19 (21): 3425–3438.
26. Munoz-Rojas, D., et al. 2019. Speeding up the unique assets of atomic layer deposition. *Materials Today Chemistry* 12: 96–120.
27. Lu, X., et al. 2016. Reactive species in non-equilibrium atmospheric-pressure plasmas: Generation, transport, and biological effects. *Physics Reports* 630: 1–84.
28. Li, J., G. Chai, and X. Wang. 2023. Atomic layer deposition of thin films: From a chemistry perspective. *International Journal of Extreme Manufacturing* 5 (3): 032003.
29. Becker, H., et al. 2023. Impact of impurities on water electrolysis: A review. *Sustainable Energy & Fuels* 7 (7): 1565–1603
30. Weber, M., et al. 2019. Atomic layer deposition (ALD) on inorganic or polymeric membranes. *Journal of Applied Physics* 126 (4).
31. Cremers, V., R.L. Puurunen, and J. Dendooven. 2019. Conformality in atomic layer deposition: Current status overview of analysis and modelling. *Applied Physics Reviews* 6 (2).
32. Smidt, F.A. 1990. Use of ion beam assisted deposition to modify the microstructure and properties of thin films. *International Materials Reviews* 35 (1): 61–128.
33. Cotell, C.M., J.K. Hirvonen. 1996. Effect of ion energy on the mechanical properties of ion beam assisted deposition (IBAD) wear resistant coatings. *Surface and Coatings Technology* 81 (1): 118–125.
34. Zhang, L., et al. 2013. Microstructure, residual stress, and fracture of sputtered TiN films. *Surface and Coatings Technology* 224: 120–125.
35. Hou, X. and K.L. Choy. 2006. Processing and applications of aerosol-assisted chemical vapor deposition. *Chemical Vapor Deposition* 12 (10): 583–596.
36. Groves, J.R. 2010. *Ion-Beam Texturing at Nucleation: Investigation of the Fundamental Mechanism of Biaxial Texture Development in Ion Beam Assisted Deposition and Applications*. Stanford University.
37. Rauschenbach, B. 2022. Ion beam-assisted deposition. In *Low-Energy Ion Irradiation of Materials: Fundamentals and Application*, 481–612. Springer.
38. Stock, H., et al. 1995. *Ion Implantation of Metal and Non-metal Ions into Vapour Deposited Coatings*.
39. Gonzalez-elipe, A., J.M. Sanz, and F. Yubero. 2003. *Low Energy Ion Assisted Film Growth*. World Scientific.
40. Gago, R., I. Jiménez, and J.M. Albella. 2006. *Thin Film Growth by Ion-Beam-Assisted Deposition Techniques*, vol. 63. Elsevier Amsterdam, The Netherlands.
41. Dhara, S. 2007. Formation, dynamics, and characterization of nanostructures by ion beam irradiation. *Critical Reviews in Solid State and Materials Sciences* 32(1–2): 1–50.
42. Waters, C.K., et al. 2006. *20 Synthesis, Optimization, and Characterization of AlN/TiN Thin Film Heterostructures*.
43. Marques, A.C., et al. 2020. Review on adhesives and surface treatments for structural applications: Recent developments on sustainability and implementation for metal and composite substrates. *Materials* 13 (24): 5590.

44. Glynn, C. and C. O'Dwyer. 2017. Solution processable metal oxide thin film deposition and material growth for electronic and photonic devices. *Advanced Materials Interfaces* 4 (2): 1600610.
45. Panjan, P., et al. 2020. Review of growth defects in thin films prepared by PVD techniques. *Coatings* 10 (5): 447
46. Chaudhary, K.T. 2021. Thin film deposition: Solution based approach. In *Thin Films*. IntechOpen.
47. Li, P., et al. 2021. Recent advances in focused ion beam nanofabrication for nanostructures and devices: Fundamentals and applications. *Nanoscale* 13 (3): 1529–1565.
48. Lewis, K. 2002. *10 Advances in Thin Films* 5: 187.
49. Grinevičiūtė, L. 2021. *Nanostructured Optical Coatings for the Manipulation of Laser Radiation*. Vilniaus universitetas.
50. Oktik, S., et al. 2021. *Thin-Film Technologies for Glass Surfaces* 1: 763–774.
51. Hsiao, S.H. 2008. *Ion Beam Modification of Thin Film Barrier Layer and Deposition of Transparent Conductive Oxides on Polymer Substrate for Flexible Display*.
52. Brown, I. and C.T. Marsden. 2023. *Regulating Code: Good Governance and Better Regulation in the Information Age*. MIT Press.
53. Martinu, L., et al. 2000. Plasma deposition of optical films and coatings: A review. *Journal of Vacuum Science & Technology A: Vacuum, Surfaces, and Films* 18 (6): 2619–2645
54. Ferguson, S.W. 2016. *Vapour Deposition and Evaluation of Conducting Polymeric Thin Films*. Carleton University.
55. Mathur, S., et al. 2013. *Metal–Organic Chemical Vapor Deposition of Metal Oxide Films and Nanostructures*, 291–336.
56. Guenther, K.H. 1984. Physical and chemical aspects in the application of thin films on optical elements. *Applied Optics* 23 (20): 3612–3632.
57. Teodorescu, M., M. Bercea, and S. Morariu. 2018. Biomaterials of poly (vinyl alcohol) and natural polymers. *Polymer Reviews* 58 (2): 247–287.
58. Al Zoubi, W., et al. 2020. Recent advances in hybrid organic-inorganic materials with spatial architecture for state-of-the-art applications. *Progress in Materials Science* 112: 100663.
59. Wang, W., Y. Zhang, and W. Liu. 2017. Bioinspired fabrication of high strength hydrogels from non-covalent interactions. *Progress in Polymer Science* 71: 1–25.
60. Younes, H.M., et al. 2023. Overview of tissue engineering and drug delivery applications of reactive electrospinning and crosslinking techniques of polymeric nanofibers with highlights on their biocompatibility testing and regulatory aspects. *Pharmaceutics* 16 (1): 32.
61. Xue, X., et al. 2022. Fabrication of physical and chemical crosslinked hydrogels for bone tissue engineering. *Bioactive Materials* 12: 327–339.
62. Mushtaq, F., et al. 2022. Preparation, properties, and applications of gelatin-based hydrogels (GHs) in the environmental, technological, and biomedical sectors. *International Journal of Biological Macromolecules* 218: 601–633.
63. Zak, S. 2024. Thin film nanoindentation and the importance of the tip radius. In # *PLACEHOLDER_PARENT_METADATA_VALUE#*.

Chapter 7
Thin-Film Attributions

Abstract This chapter describes the thin films' characteristics including their thickness, composition, surface morphology, and adhesion. The thin-film thickness measurement is vital for ensuring the desired application performance. Techniques such as gravimetric methods, optical methods, and direct measurement methods. The composition of thin films significantly influences their properties and functionalities. Techniques like energy dispersive *X*-ray spectroscopy and *X*-ray photoelectron spectroscopy are utilized to analyze the elemental composition and detect impurities. The surface morphology of thin films, including roughness and microstructure, plays a critical role in determining their optical and mechanical properties. Enhancing adhesion between thin films and substrates is essential for the durability and effectiveness of coatings. Understanding the mechanism of adhesion at the molecular level can lead to the development of more robust thin-film applications. Proper cleaning of substrates is crucial for achieving high-quality thin films. In summary, the study of thin-film characteristics encompasses a range of measurements and analysis techniques.

7.1 Introduction

Generally, a thin film is a thin, flexible layer of polymer, either with or without a plasticizer. The patient may find it less intrusive and more acceptable because they are naturally thin and flexible. The polymeric matrices in the thin film fulfill numerous criteria for effective use as a drug release substrate. In general, thin films are superior options to pills or liquid formulations when it comes to targeting sensitive sites [1]. Thin films have demonstrated the capability to intensify therapeutic efficacy, diminution prescription occurrence, and accelerate the onset of action. Desirable characteristics such as satisfactory medication packing ability, quick-dissolving amount, or extended habitation period at the location of management, then suitable origination strength must be present in ideal thin films [2]. They must be biocompatible, biodegradable, and non-toxic as well [1]. It is significantly better than the current standard forms of dosage in terms of increased biodegradability, resilient patient

T. I. Awan et al., *Thin Film Deposition Techniques*,
https://doi.org/10.1007/978-981-96-1364-9_7

compliance, and the active therapeutic component's patent extensions. Additionally, thin-film formulations provide several benefits, such as.

- Easy administration via harmless methods.
- Handling convenience during manufacture and shipping.
- Cost-effective formulation creation [1].

The development of a broad range of thin films has been made possible by the availability of a wide range of suitable polymers and the change in the manufacturing paradigm. As a result, thin films are becoming more and more recognized in the therapeutic industry by way of cutting-edge drug conveyance quantity forms [2]. A portion of exertion has vanished into unindustrialized polymeric thin films that are often applied topically, mouth, spoken, eyepiece, and skin. The use of thin films among other methods, to deliver medication to the cheek or lower tissues, has attracted a lot of attention lately. Meanwhile, ocular barrier-crossing eye films are currently being researched to prevent medical field loss via the lacrimal drainage system [1]. Modifying the composition of polymers with varying grades has made it easier to alter important thin-film features like mechanical strength, waterproof qualities, drug release rate, and other associated attributes. Additionally, different inactive ingredients including sugars, polymers, saliva-stirring agents, colorants, and fillers can be added to improve the product's look [3]. Numerous medical firms have already patented several methods for creating thin films because they find the properties of these films so attractive. Even if there are already a lot of unique works and patents available in the literature, comprehensive research is motionless obligatory to accurately improve thin-film presentation [3]. Inadequate guidelines for the production of goods, characterization, and quality oversight of thin-film therapeutics are drawing attention to the need for further investigation in this area from a pharmaceutical perspective.

Consequently, this study will help provide insights into understanding the important features and character development techniques [4]. Figure 7.1 explains the phenomena of thin-film attributions.

7.2 Thin-Film Characteristics

Understanding the mechanism of thin-film formation which results from the condensation of particles commencing the material's vapor phase has been the concentration of constant research since the 1920s in several domains. Previous research at England's Cavendish Laboratory on evaporated thin films advanced the idea of nuclei creation, growth, and island coalescence [5]. Thin-film technology advanced significantly along with vacuum science and technology. In 1953, the American Vacuum Society was founded. Consequently, the field of thin-film technology improved further in terms of finding lattice defects, epitaxy, and segregation of alloying elements and contaminants due to advancements made in the fields of electron microscopy and surface analysis [6]. Specific qualities, such as electrical, optical,

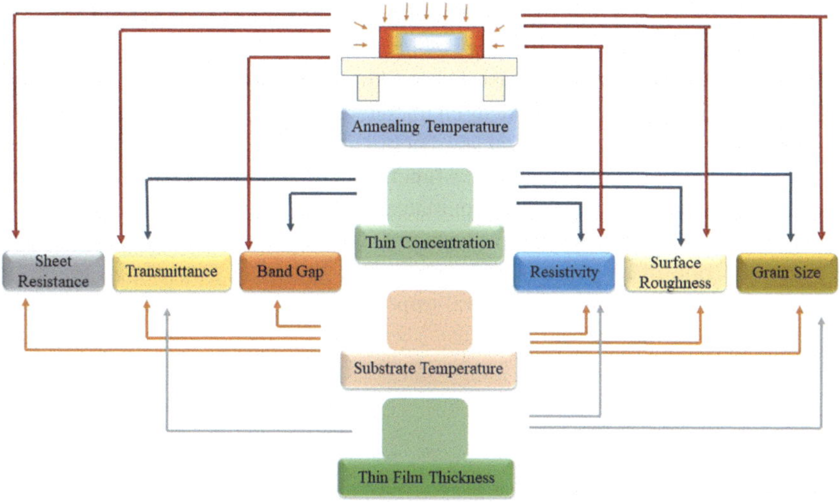

Fig. 7.1 Describes the phenomenon of thin-film attributions

mechanical, chemical, and thermal ones, are created into thin films to encounter the demands of dissimilar solicitations. These layers originate from a variety of deposition methodologies, particularly chemical vapor deposition (CVD) and physical vapor deposition (PVD).

Microstructure, which is governed by deposition parameters such as substrate temperature, deposition rate, and pressure, affects the properties of thin films [7]. Three possible microstructures come from this process: amorphous, polycrystalline, and epitaxial. Each has unique chemical and physical characteristics. The thickness and microstructure of thin films have a significant impact on their optical characteristics [7]. The film's extinction coefficient and refractive index both significantly depend on its thickness, with the latter increasing and the former lowering as thickness increases. The wavelength of the incident beam has an impact on the optical behavior as well, with distinct mechanisms predominating in the visible and infrared areas. Structural flaws that enhance carrier scattering and decrease electrical conductivity, such as voids, unsaturated bonds, and grain boundaries, have an impact on the electrical characteristics of thin films [8].

Various characteristics of thin films set them apart from bulk materials. Among these qualities are:

- Three categories can be used to group thin films according to their thickness: Incredibly thin (50–100 Å).
- Thin (between 100 and 1000 Å).
- In comparison, relatively thick (> 1000 Å).

An important factor in defining the optical, electrical, and structural characteristics of a film is its thickness. Numerous methods, including gravimetric, optical, direct, and methods utilizing electrical or magnetic quantities, can be used to quantify it [8]. The study of thin-film devices and materials requires the measurement of thin film properties. When assessing thin films, one must consider their chemical makeup, crystalline structure, optical characteristics, electrical characteristics, and their mechanical properties. Several approaches are put forth to assess the thin films deliver an impression of the technique's function. The assessment of the surface and thin-film composition of semiconductor materials has advanced quickly among these procedures. Several approaches have been put forth to assess thin films and surfaces [9]. The selection of an instrumental approach to address a particular issue in the surface or thin films is based on several important factors. Figure 7.2 describes the properties of thin-film deposition.

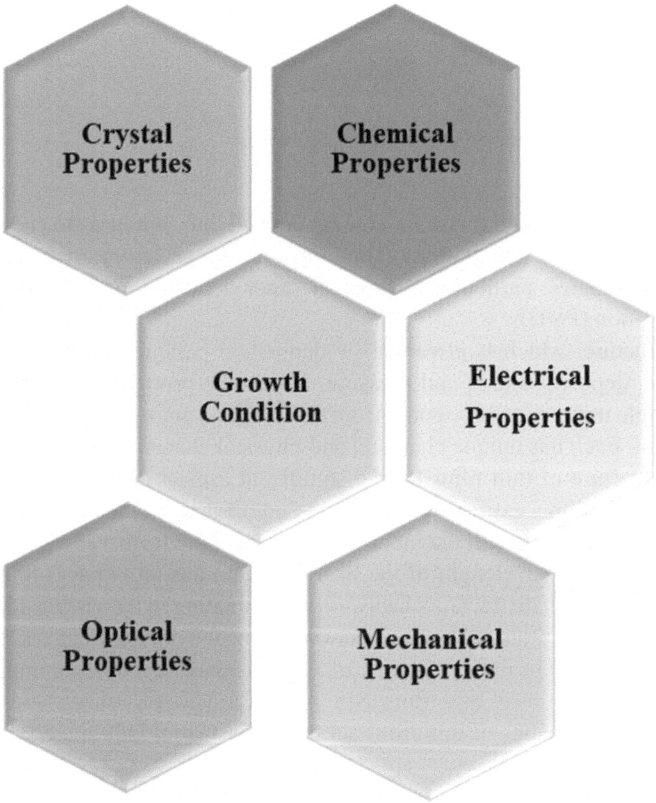

Fig. 7.2 Properties of thin-film deposition

7.2.1 Mechanical Characteristics

The mechanical properties of thin films are different from those of bulk materials. Among these qualities are [7].

Increased yield strength as a result of stress stored during deposition.

- **Hardness**: Enhanced by microstructures such as dislocations, dopants, and grain boundaries.

The thickness and microstructure of thin films affect their mechanical characteristics.

7.2.2 Electrical Characteristics

The substrate and the type of material (insulator, metal, or semiconductor) affect the electrical characteristics of thin films. Important elements influencing electrical conductivity consist of:

- **Effect of Size**: Because charge carriers' mean free pathways are shorter, their conductivity is lower.
- **2-Defects in Structure**: The presence of oxide bonds, voids, and grain boundaries increases dispersion and decreases conductivity [9].

Electrical properties can be characterized by methods such as Rutherford backscattering spectroscopy (RBS) and *X*-ray photoelectron spectroscopy (XPS).

7.2.3 Optical Characteristics

It is possible to build thin films with particular optical characteristics, like:

- **High Reflectivity**: Applied in solar cell and optical detector array applications.
- **Low Reflectivity** (anti-reflective): The film's optical properties are influenced by its viscosity, roughness, and refractive index. Two techniques are utilized to study crystalline structure and thin-film microstructures: *X*-ray diffraction (XRD) and Raman spectroscopy [10].

7.3 Measurement of Thickness

Nowadays, thin films are used at regular intervals by a significant segment of the scientific and industrial communities. Coatings are frequently used to create a synergistic action between the characteristics of the substrate and the covering material.

They can enhance the final product's chemical, physical, and aesthetic properties while lowering its cost. The thickness of composite materials must be measured for each of these reasons. Different methods can be employed to measure the thickness of the layers depending on the substrate's composition and thickness as well as the film's own. The processes can be divided into two categories: destructive and non-destructive methods [11].

$$d = (\Delta m * \lambda_1 * \lambda_2)/(2 * n * \cos(\theta) * (\lambda_2 - \lambda_1)) \tag{7.1}$$

The wavelength range's number of peaks is expressed in Δm. The film material's refractive index is represented by n. θ represents the incidence angle. The start and end wavelengths utilized in the computation are λ_1 and λ_2 [11].

For thinner films, the waveform's pitch in the interference spectrum is longer; for thicker films, it is shorter. By entering the wavelength range and refractive index into film thickness measurement software, the thickness may be simply calculated. This method is effective for film thicknesses between about 0.3 and 60 µm. To determine the thickness with this method, one needs to know the refractive index of the film material. Nowadays, thin films are used daily by a significant segment of the scientific and industrial communities.

To achieve a beneficial action concerning the possessions of the substrate and the wrapper material, coatings are commonly utilized [12]. They can condense the budget of the completed invention and develop its chemical, somatic, and appearance. For all of these motives, measuring the thickness of composite materials is essential to achieving the objective item properties and controlling costs. Films may possess a completely distinct nature because of the use of dielectrics (organic, such as metal oxides and autonomous monolayers (SAM), or organic substances like polymers), semiconductors, and metals [12]. We concentrated happening the description of metal flicks produced by vapor phase deposition or electrodeposition, although the majority of the methods discussed in this article can be utilized for films made of other materials. Regarding the film's size, an atomic monolayer (ML) has the thinnest measured thickness, although electroforming can produce layers as thick as hundreds of microns. As a result, we included a summary of every research technique in this area. Various methods can be employed to investigate the layer thickness depending on the kind, composition, and thickness of the film and substrate. The processes are divided into two categories: destructive and non-destructive methods. A destructive method can be used to measure the thickness directly [13]. This means that the sample must be transformed either macroscopically or microscopically; hence, the inspected sample is no longer useful and must be thrown away. On the other hand, a non-destructive technique facilitates sample measurement without endangering it. Indirect measures are considered non-destructive approaches; specific calculations and assumptions are utilized to arrive at a numerical number for the thickness. The following vital techniques are covered in this review: cross-sectioning, angle lapping, the TEM lamella getting ready, and the conventional Marco test. Before any microscopic analysis, regardless of whether it be using light (optical microscopy, for example), electrons (scanning, transmission, and scanning transmission electron microscopy, STEM), or ions (scanning ion microscopy), these destructive macroscopic techniques include cross-sectioning and the Calo test microscopic techniques include angle lapping and TEM lamella preparation, which is carried out mechanically or with the assistance of a centered ion beam [14].

The most often used techniques for determining thin-film thickness:

- **Stylus Profilometry**

This technique measures the height of a step or groove between the substrate and the film's surface using a stylus. It works well with layers that vary in viscosity from a few nanometers.

- **Interferometry**

This performance measures the breadth of the film by analyzing the interference patterns that are produced when light bounces off the top and bottom surfaces of the film. Films with thicknesses ranging from a few nanometers to microns can also be used.

- **Ellipsometry**

This non-destructive technique examines the polarization state of light reflected from thin films to determine their thickness. It can be applied on films with a maximum thickness range of 1000 Å (0.1 μm).

- ***X*-Ray Reflectivity (XRR)**

This technique measures the surface roughness and thickness of thin films using *X*-rays. Films with a thickness range of up to 100 nm [15].

- **Scanning Electron Microscopy (SEM)**

This technique makes an image of the film by focusing an electron beam, which makes thickness measurements possible. Films with a depth sort of 100 nm to 100 μm are appropriate for it.

- **Transmission Electron Microscopy (TEM)**

This technique allows for thickness measurements by projecting an image of the film onto a focused electron beam. It works well with films that sort in width from insufficient nanometers to one hundred nm.

Since optical interferometry can only detect the optical thickness by nature, the somatic depth of the material must be determined by measuring its refractive index or by knowing it beforehand using a different measurement method [16]. However, many measurement techniques have been put forth in studies of breadth dimensions utilizing SRI to assess the depth and refractive index instantaneously in an actual period. A thick object can be measured with success using a white-light scanning interferometer and the Mirau objective, as illustrated in Fig. 7.3.

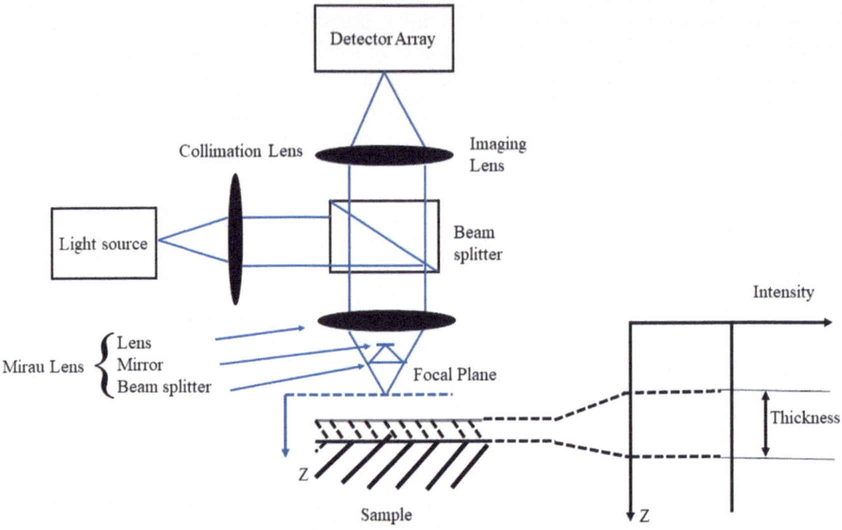

Fig. 7.3 In side parameters, describing measurements of thin-film thickness

7.4 Analysis of Composition

Understanding the composition and performance of the deposited films is a critical first step in thin film deposition analysis. Transmission electron microscopy (TEM), *X*-ray diffraction (XRD), and energy dispersive *X*-ray analysis (EDAX) are approximately of the approaches recycled to ascertain the composition of thin films [17]. Composition analysis of a sample always refers to a quantitative elemental or chemical analysis. In materials science research, compositional analysis is a valuable tool that aids in the identification of unknown materials and the understanding of the structure-property relationships of individual components as well as their synergistic behaviors to characterize notable changes in the distribution of elements and material structure. For example, it has been found that the char yield, char combustion time, and char combustion rate of woody biomass pellets are mostly determined by their composition.

Because of variations in char production, bark pellets had a char combustion duration that was up to 50% longer than that of stem wood pellets [18]. Sometimes, failure analysis, handling, or manufacturing issues with products are investigated, and the assessment of foreign material impurities is done using quantitative chemical analysis techniques. Gravimetric and titrimetric procedures are examples of traditional methods used in wet chemistry. Depending on the kind of material, one or more complementary procedures may be used to complete the analysis [19]. Figure 7.4 illustrates the structure of compositional analysis.

- **Mass Spectroscopy**

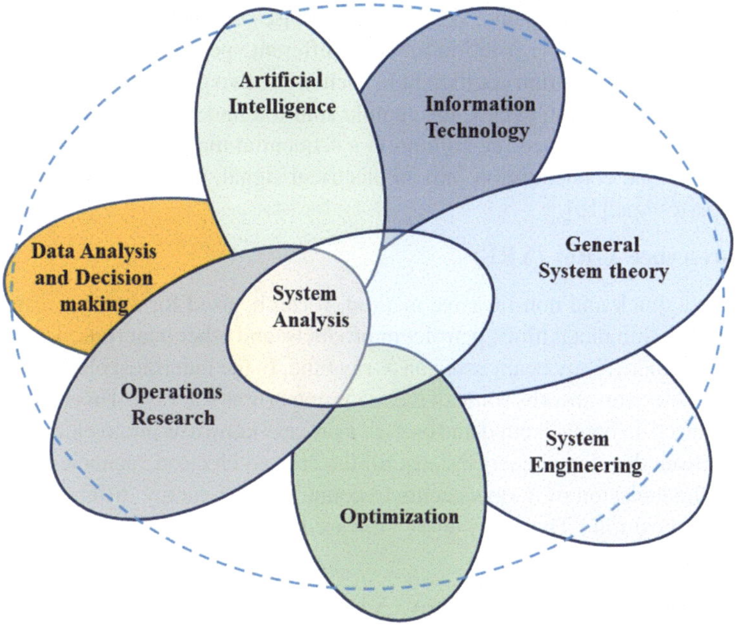

Fig. 7.4 Illustrates the analysis of thin-film composition

Ionizing chemical species and classifying the ions according to their mass-to-charge ratio is the analytical method known as mass spectrometry. In a traditional MS procedure, a sample—which could be solid, liquid, or gas—is ionized by, say, being bombarded with electrons. These ions are then propelled by an electric or magnetic field. For ions having an identical mass-to-charge ratio, there will be an equal amount of deflection. Following the discovery of the ions, spectra display the relative abundance of the discovered ions as a function of the mass-to-charge ratio. The atoms or molecules in the sample can be identified by comparing known masses, such as the masses of a whole molecule, to the reported masses or by spotting a characteristic fragmentation pattern. Mass spectrometry is invariably used in conjunction with other tools. As an illustration, consider several ionization processes including inductively coupled plasma (ICP), matrix-assisted laser desorption/ionization (MALDI), and electron ionization (EI). Additionally, various separation methods including liquid and gas chromatography (GC and LC) are chosen to improve mass spectrometry's capacity for mass resolving and determination. Comparable methods include ICP-MS, Time-of-flight mass spectrometry (TOF–MS), Liquid chromatography-mass spectrometry (LC–MS), Inductively coupled plasma-optical emission spectroscopy (ICP-OES), and Gas chromatography-mass spectrometry (GC–MS) [19].

- **High-Performance Liquid Chromatography (HPLC)**

The solution containing the sample enters the flow component through the sampler and is loaded into the phase that is stationary when the liquid is used as the flow phase under high pressure constraints. When the two phases are shifted in comparison, the components in the sample solution move at different speeds because each component has a different partition coefficient in each of the two phases. Following a series of adsorption–desorption cycles, the sample solution underwent separation into its constituent parts and exited the column in a sequential manner. The detector transforms the sample concentration into an electrical signal, which is then transmitted to the tape-to-tape [19].

- **Fluorescence *X*-Ray (XRF)**

XRF is a quick and non-invasive method. It can be used for direct research into hard samples, thin metal films, petroleum products, and other materials. The tester is showing to a main *X*-ray beam using an *X*-ray tube. In the innermost electron shell of an atom, an electron absorbs some of the incoming primary *X*-rays. This results in the absorbed electron being excited and ejected a process known as photo ejection. Later, electrons from developed energy states fill the created electron vacancies, and fluorescence the emission of *X*-rays occurs to counteract the energy differential among the electron positions. The *X*-ray energy released is indicative of the component that released it [20].

- **Atomic Absorption Spectroscopy (AAS)**

When a component is stimulated and its particles move from the ground state to the first excited state to form a supernatural track with unique properties, this is known as atomic absorption spectroscopy or AAS. AAS uses this function to do our elemental analysis. When the element lamp to be tested passes through the atomic vapor produced by the atomization of the sample, the state of the ground atom containing the element to be tested in the vapor absorbs the distinct path generated by the element lamp. To assess the element's content in the sample, the radiation light's intensity is weakened to a certain degree. Additional comparable methods include Auger electron spectroscopy (AES), Energy dispersive *X*-rays (EDX), *X*-ray photoelectron spectroscopy (XPS), Real-time magnified *X*-ray imaging, and Micro *X*-ray absorption near edge spectroscopy [20]. Dynamic spectroscopy the three primary instruments used in vibrational spectroscopy are Raman, Fourier Transform Infrared (FTIR), and Near-Infrared (NIR). To analyze the functional groups of solid or liquid materials, FTIR is used to calculate the discrete energy levels for the vibrations of the atoms in these groups. Atomic groups inside the material can absorb light of a particular energy when it passes through a very thin sample. This happens when the regularity of the received light and the incidence of atom-to-atom vibrations match. The atoms' masses, chemical environments, and vibration type all affect the vibration energy. The infrared area has a range of 12,800–10 cm^{-1}, which can be further classified as near-infrared (12,800–4000 cm^{-1}). Meanwhile, most organic ingredients and inorganic ions absorb light within this range, 4000–400 cm^{-1} is the commonly used zone for ultraviolet immersion spectroscopy. It is feasible to classify the well-designed group's existence in the material by monitoring the amount of light transmitted at each wavelength while scanning over a range of wavelengths

(400–4000 cm^{-1}). Because each of those three methods has advantages and disadvantages, they are complementary instruments for analyzing different kinds of materials. Similar methods include NanoFTIR/Raman, Micro-FTIR/Raman spectroscopy, and Attenuated Total Reflectance-FTIR (ATR-FTIR) [21].

- **UV Spectrometer**

The absorption of light energy varies among substances due to differences in their molecules, atoms, and molecular space configurations. Each chemical therefore has a distinct, fixed absorption spectrum curve of its own. The foundation for both qualitative and quantitative spectrophotometric analysis is the ability to ascertain a substance's composition based on the absorbance at specific representative wavelengths on the concentration spectrum. The Lambert–Beer law is the cornerstone of quantitative analysis. Benefits high sensitivity, rapid, easy to use, affordable, suitable for a broad range of concentrations, and cheap analysis cost [22].

- **Nuclear Magnetic Resonance (NMR)**

The molecular structure, concentration, and purity of a material can be determined using nuclear magnetic resonance (NMR) spectroscopy, an analytical chemistry technique used in research and quality control. For example, mixes containing known chemicals have been extensively subjected to quantitative NMR analysis. NMR can be used to compare unknown chemicals to spectral libraries or to directly infer their basic structure. According to the NMR principle, every nucleus has an electrical charge and many nuclei have spin. When there is an outside magnetic field, energy can go from the base energy to a higher energy level, which is often a single energy gap.

7.5 Surface Morphology

The stability of the nanoparticle is also influenced by surface morphology, which includes surface roughness and surface of the element. Determining the surface morphology is crucial for identifying the nanoparticles' surface flaws. When gray and gold-leaf nanoparticles were evaluated using UV–visible spectroscopy, it was discovered that the surface morphology was related to the catalytic movement of the nanoparticles; silver nanoparticles had greater catalytic activity than gold nanoparticles. From a formulation perspective, surface morphology is also crucial; silicon absorbent nanoparticles act as a substrate for the creation of silver nanoparticles. Particles with bigger diameters were formed in porous silicon particles with higher surface roughness. Analytical methods including XRD, AFM, and SEM were used to observe the surface morphology. Electron microscopy methods that include transmission electron microscopy (TEM), atomic force microscopy (AFM), scanning tunneling microscopy (STM), and scanning electron microscopy (SEM) can all be used to study surface morphology. In contrast to SPM, which uses a probe to scan a surface and generate a signal that is measured by a sensor, electron microscopy makes use of the communication of electrons with the surface of nanoparticles. This communication can be electrical, chemical, mechanical, magnetic, or optical, depending on the type of SPM [23]. The type and level of surface roughness are known as a surface's morphology. This could apply to specific surface features or the surface as a whole. The micron and submicron surface morphology of the substrate has an impact on the film characteristics, surface coverage, and deposited film morphology [24].

- **Surface Structure**

A key factor in SHSs is surface morphology, which is the decrease in the deceptive connection angle between the surface and droplet as a result of air stuck between an arranged area and liquid. By enhancing the MSHSs' surface structure, artificial durable multifunctional surfaces can be created to address a variety of challenging issues [24].

- **Effect of Surface Morphology**

It is commonly recognized that a material's photocatalytic performance is significantly influenced by its surface shape. Controlling a material's shape is crucial to producing one with high photocatalytic activity [24]. Additionally, the surface shape and surface area of bismuth ferrite nanoparticles are discovered to have a remarkable influence when exposed to visible light. We can deduce that variations in surface distinctions relate to distinct energy levels, such as valence and conduction bands from different mineral lattices, which improves their surface effects and amplifies their photocatalytic reactions about the surface area morphology effect [25].

7.5.1 Methods of Surface Morphology Analysis

Morphology analysis, which is essential for many research projects, comprises both qualitative and quantitative study of a sample. As a result of advancements in micro- and nanofabrication methods, biomolecule-surface interactions can now be uniquely controlled through micro- and nanostructures. SEM is one of the most effective and adaptable methods used in microanalysis, materials research, and investigative applications today. SEM analysis offers elemental mapping, high-resolution imaging, and thorough analysis across a broad range of magnifications for a variety of sample types. We may also easily determine fundamental arrangement with an energy dispersive X-ray spectrometer (EDS) connected to a scanning electron microscope. For the characterization of materials, TEM has been widely used. A very thin sample is exposed to a high-intensity electron beam, allowing characteristics like grain boundaries, dislocations, and crystal structure to be observed through the interactions between electrons and atoms. TEM can be used to observe semiconductor faults, layer composition, and growth. The superiority, figure, size, and density of quantum wells, wires, and dots can all be observed using high-determination transmission electron microscopy (HRTEM) [26]. The surface quality is changed by the thermal and mechanical demands of the production processes, such as vibrations, tool wear, and cutting settings. The machined surface is produced by taking into account various disturbances and utilizing the tool profile and actual tool path [27].

7.6 Preparation of the Substrate and Surface Usage

7.6.1 Substance Pre-Medication

Rubbers, both synthetic and natural, can be bonded to a range of semi-rigid, flexible, and stiff substrates using MEGUMTM and THIXONTM bonding agents. An essential step in the bonding process is cleaning and prepping the substrate surface before applying MEGUMTM or THIXONTM bonding chemicals. Without a physically and chemically clean substrate, it is incredible to obtain a dependable attachment concert between the substrate, the MEGUMTM or THIXONTM bonding agent, and the rubber. After cleaning, the substrate needs to be shielded from further chemical and physical contamination before applying the MEGUMTM or THIXONTM bonding agent [28].

The goal of substrate preparation is to:

- Create and preserve a stable, clean substrate closeness surface before applying cover-cement or MEGUMTM or THIXONTM bonding agent primer in two coats. Systems of THIXONTM single-coat bonding agents.
- Eliminate any oil or grease that might impede the substrate surface from being adequately wetted by the MEGUMTM or THIXONTM bonding agent.
- Eliminate impurities like oxidative rust that could break and result in bond delamination when the bonded component is being applied in service. Prevent re-oxidation or re-infection of the recently prepared substrate bonding surface [29].

7.6.2 Selecting a Preparation Technique

Techniques for preparing substrates might be mechanical, chemical, or a combination of both. The decision to use chemical or mechanical procedures is determined by the following factors:

These include the bonded component's in-service use as well as the substrate's composition, size, quantity, and arrangement of components. The substrate pretreatment method employed may have an impact on the bonded component's durability in harsh conditions and under-bond corrosion. When steel substrates are phosphate using the suggested pre-treatment procedures before the application of MEGUMTM or THIXONTM bonding agents, the bonded components exhibit resistance to harsh in-service environmental conditions. DuPont Specialty Adhesives

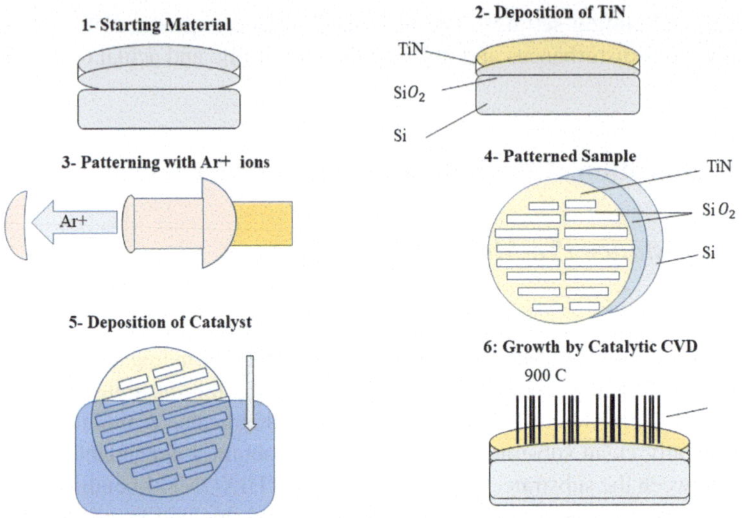

Fig. 7.5 Preparation of the substrate and surface usage

can provide recommendations for appropriate phosphate treatments before and after bonds [30]. The preparation of the substrate is explained in Fig. 7.5.

- Obtaining thermal SiO_2.
- TiN deposition and subsequent patterning of the substrate.
- Catalyst deposition, and
- Growth of SWCNTs by CVD.

7.6.3 Substrate Preparation Methods

- **Elimination of Grease and Oil**

Rubber-to-metal bonded steel apparatuses are typically shipped to the user with a lubricant or oil-defensive coating to stop oxidizing during storage and transit. The protective oil and grease layer must be removed as the first step in the substrate grounding development to guarantee that the MEGUMTM or THIXONTM bonding agent makes good interaction more precisely, that it "moistures" the substrate bonding surface. A vapor degreasing unit is typically used in mechanical surface pre-treatment procedures, where oil and grease are eliminated through contact with the vapor of an appropriate organic solvent. The condensed solvent drips back into the liquid, dissolving and removing the grease and oil, while the solvent vapor condenses on the part. Chemical pre-treatment techniques involve the use of alkali wash tanks to eliminate oil and grease, which are then rinsed with water to stop contamination from leaking into the tanks used for the chemical conversion process [31].

- **Pre-Treatments Using Equipment**

The degreased substrate surface is abrasively polished using mechanical pre-treatment techniques to provide a clean, new surface ready for bonding. The bonding surface is made rougher by the abrasion procedure, which increases giving a key to the MEGUMTM or THIXONTM bonding mediator and revealing its surface area. The exterior that has been automatically worn down will respond with air oxygen very quickly; in the presence of high humidity, this reaction will speed up. After the surface abrasion treatment, it is critical to apply the MEGUMTM or THIXONTM bonding agent as soon as possible to minimize the properties of oxidation and any impurities. Alumina, carborundum grits, cooled iron, and air blast are examples of mechanical pre-treatment techniques. It is not advisable to prepare stainless steel or any other non-ferrous substrate using chilled iron or steel grits. Additional techniques to sand the metal surface consist of emery cloth, wire brushing, and grinding [32].

- **Chemical Preparations**

Alkali wash tanks are typically used in chemical pre-treatment processes to remove oil and grease. To prevent contamination from leaking into the chemical conversion, a water rinse is then performed in tanks used in processes. Alternatives to alkali treatment for some forms of grease (such as animal greases) include solvent degreasing using carbon hydrates or perchloroethylene degreasing in fully enclosed units. Before selecting a degreasing treatment, conduct a laboratory trial. DuPont Specialty Adhesives can provide more detailed advice upon request. Anodization, plating, acid etching, and phosphating are examples of chemical pre-treatment techniques. When phosphorizing is applied to ferrous surfaces, the substrate surface is chemically changed. As a result, the rubber-to-metal bonded component has a longer service life and better corrosion resistance.

- **Prepare and Clean the Metal**

Metal can be cleaned and prepped using mechanical, chemical, or a mix of mechanical and chemical surface pre-treatment techniques. Specific pre-treatment procedures are needed for certain substrate types. Make sure the right pre-treatment technique is being applied, and keep an eye on how well it's working. Non-ferrous grits should be used to blast non-ferrous surfaces, such as plastics and stainless steel. Make sure the substrate surface is prepped and degreased before applying MEGUMTM or THIXONTM bonding agents. It is advised to degrease the substrate bonding surface both before and after grit blasting if pre-treatments such as grit blasting are utilized. The initial degreasing procedure lessens the amount of oil and grease that contaminates the abrasive grit. The second degreasing procedure gets rid of any leftover grease and dust particles from the grit abrasion procedure.

- **Take Carefully**

Before using the MEGUMTM or THIXONTM bonding agent, pre-treated, degreased mechanisms must be held sensibly to prevent recontamination of the bonding surface. It is recommended to wear lint-free, clean gloves when handling degreased components to avoid contaminating the bonding surfaces. Avert the contamination of degreased components with oil, grease, or dust before using MEGUMTM or THIXONTM bonding agents. Keep the components of the prepared metal substrate at room temperature and prevent moisture condensation. Steer clear of moving cold metal parts into warm, steamy environments where a concentration of air humidity vapor is likely to happen [28].

- **Prevent Postponements**

The interval between the conclusion of the pre-treatment procedure and the submission of the MEGUMTM or THIXONTM bonding is one of the most important steps in preparing the components for rubber bonding representative. The ultimate quality of the bonded component is frequently assessed at this point. Metals that are produced chemically or mechanically are susceptible to contamination and oxidation.

7.7 Improvement in Thin-Film Adhesion

An extremely high grip between gold thin films and a polymer based on polysiloxane was established using a unique approach. The polymer's surface was modified using a plasma process. The modification of the wettability and topography of the polymer surface with the plasma conditions were investigated, respectively, using measurements of the water contact angle and scanning electron microscope observations [33]. Realizing the uses of polymers in medical implants, photonics devices, and microelectronic elements requires metallization of polymers. Unfortunately, there is typically relatively little adhesion between metals and polymers, which makes it difficult to investigate the polymer's potential uses. For instance, it is challenging to establish wire bonding for metalized photonics devices based on polymers if there is insufficient adhesion between the metal layer and the polymers. Materials that are quite different from one another are metals and polymers. For example, metals usually have cohesive energies that are two instructions of an amount higher than those of polymers. We created a brand-new technique to achieve incredibly strong adhesion at low temperatures between Au films and a polymer based on polysiloxane. This method involved firstly modifying the polymer surface using a plasma of oxygen and tetrafluoromethane (CF_4) gas mixture (O_2/CF_4). Next, two adhesive layers were deposited on the polymer, one being Al_2O_3 and the other being Al. Finally, an Au film was deposited on the Al layer, and the polymer/Al_2O_3/Al/Au system was annealed at a temperature lower than 160 °C [34]. This is important because poor adhesion and prevented inter-diffusion can result from surface pollution. Moreover, thin film nucleation and development can be further influenced by glow discharge and ion bombardment cleaning, which subsequently impacts adhesion and cohesive failure mechanisms. The temperature of the substrate is another crucial element. Heating the substrate can frequently increase adhesion in the event of inter-diffusion or complex contact [33]. Moreover, adhesion may be impacted by electrostatic charges present during deposition. For instance, the development of an electric. An electrostatic force of adhesion across the contact, for instance, may arise from the creation of an electric double layer as an outcome of charge transmission. The durability and lifespan of the thin-film device depend on these variables guaranteeing strong adhesion between the thin film and the substrate [35]. For the deposition of thin films on a variety of products, vacuum arc evaporation and electron beams are frequently employed at gas pressures between 0.01 and 1 Pa. When the deposition rate is high enough and the plasma density is close enough to the product surface, the arc evaporation is impressive [36]. By doing this, a sufficient flux of ions is guaranteed to bombard the negatively biased products. This makes it possible, for instance, to alter the structure and surface characteristics of wounding tools made of high-speed steel and other manufacturing materials. The two main disadvantages of the vacuum arc are the arrival of metal droplets from the arc cathode spots to the thin films and the no uniformity of the discharge plasma in the process chamber. A filtered vacuum is used for the deposition of thin films that do not include droplets [36]. Magnetron sputtering is another widely used method for thin-film deposition. It makes it possible to deposit fine films devoid of droplets. Unfortunately, a planar magnetron's target only sputters when exposed to an arc-shaped magnetic field on its surface. The target material's utilization efficiency is therefore extremely poor. It was demonstrated that when the target is positioned at the bottom of a hollow cathode, the material is employed more effectively. The energy of condensing atoms determines the qualities of the film. Ion-assisted film deposition occurs when this energy is transferred to the film surface by accelerated ions. In this instance, the substrate temperature has no bearing on the energy delivered to the condensing film atoms. It permits the film to be deposited on the substrate. Film adherence is improved by the ion bombardment [36]. The latter increases monotonically with ion energy as a result of accelerated particles penetrating the film at a deeper rate. Nevertheless, the expanding film sputters as a result of the rise in ion energy. A rise in ion energy to 1–3 keV can cause all atoms that have been deposited to sputter. This implies that pulsed bombardment is the best method for modifying the growing films with high-energy ions. A titanium nitride film developing on a high-speed steel substrate, e.g., was pulsed and bombarded by 45 keV ions, producing an nc-Ti2N/nc-TiN nanocomposite with micro-hardness up to 5000 $HV_{0.04}$. The interface width rises to one because of the high-energy particles that combine the atoms of the substrate and film [37]. It guarantees flawless adherence of coatings made of titanium nitride up to 50 μm in thickness. The same patterns apply to thin films formed at layer thicknesses between 50 and 1000 nm on a smaller scale. Applying a negative voltage to the substrate is required to bombard films developing on electrically conductive substrates with energetic ions. Ions will therefore accelerate out of the gas discharge. However, when magnetron sputtering is employed, the plasma density close to the substrate may be rather low. The plasma density can only increase to a point where there is enough current density of ions assaulting the substrate surface by using an unbalanced magnetron [38]. Because the fragile film depends on the underlying substrate and its stickiness for endurance, adhesion is a critical topic in thin-film science. Furthermore, the adhesive contact between the depositing material and the substrate determines the creation and structure of the film. Adhesion is one of the areas of thin-film research that is poorly understood despite its broad applicability [33].

7.7.1 *Post-Deposition Processes*

Additionally, post-deposition processes can change the thin-film microstructure. Grain growth is encouraged by the heat treatment, which may result in fewer grain boundaries. This may improve the surface zone in interaction with the substrate and lead to a stronger bond between the film and the substrate.

7.7.2 *Scratch Test*

A contemporary destructive testing technique for evaluating the cohesive and adhesive qualities of coatings is scratch testing. The degree to which the substrate adheres to the coating at the interface is referred to as its adhesive characteristics failure within the coated zone is referred to as its cohesive properties.

7.7.3 *Indentation and Imaging*

One simple test is the Rockwell indentation technique, which can be used to assess the performance of coating adhesion to the substrate, causing minimal damage. This can also provide additional information such as whether the coating will fail in a brittle or ductile manner when indented. Description of the properties of thin-film adhesion is described in Fig. 7.6.

To expand the bond between Au films and the polymer, it was discovered that the O2/CF_4 plasma was more efficient than the Ar or O_2 plasma at altering the polysiloxane-based polymer surface. This was achieved by collecting the wettability and irregularity of the polymer surface. It was also demonstrated that increasing the adhesion between the polymer and the Au film could be accomplished by inserting two adhesion layers, Al_2O_3/Al, between them [39].

The following are typical obstacles to strong thin-film adhesion:

- **Control of Uniformity and Thickness**

For many applications, a uniform coating's thickness is essential. Uneven or non-uniform film thickness can have an impact on the properties of the material and the final product's performance.

- **Delamination and Adhesion**

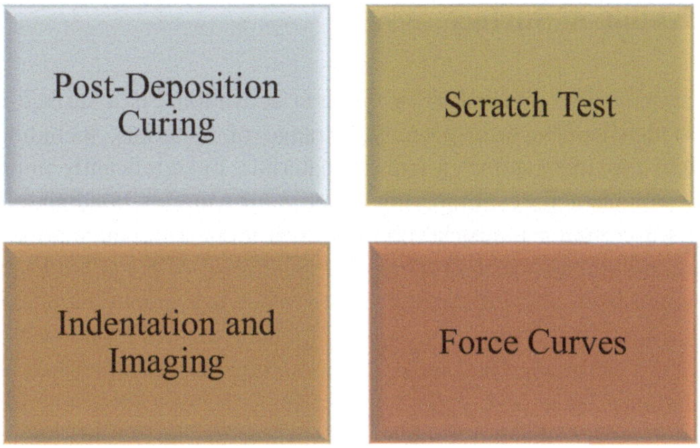

Fig. 7.6 Property of improvement in thin-film adhesion

Long-term dependability depends on the thin film and substrate adhering properly. Delamination is the result of the thin layer separating from the substrate, which causes the product to malfunction. Adhesion is influenced by interfacial treatments, substrate preparation, and deposition technique, among other things [40].

- **Cost and Scalability**

Because high-purity components or specialized equipment are required for certain thin-film deposition processes, the process can be expensive. Furthermore, it can be difficult to scale up the production process for large-scale manufacturing. For academics and engineers, striking a balance between cost-effectiveness, scalability, and performance needs is a major problem.

- **Surface Roughness and Defects**

Thin film's optical, electrical, and mechanical characteristics can be impacted by surface roughness and flaws. Reducing surface roughness and imperfections in films can be achieved by optimizing deposition parameters and post-processing techniques [41].

- **Process Control and Reproducibility**

Thin-film properties must be reliable and consistent for use in industrial applications. For accurate and consistent thin-film deposition, strict procedure regulators and observance of ordinary functioning actions are mandatory.

7.8 Activation of Surface

Plasma treatment, also referred to as "plasma activation," is a surface modification process that can be used on a wide range of materials, including metals, glass, ceramics, textiles, and even natural materials, to significantly improve their surface adhesion capabilities. Preparing surfaces for painting, bonding, gluing, and varnishing is a common industrial process utilization. This outcome is achieved using plasma treatment, which combines the reduction of metal oxides, ultra-fine surface cleaning from organic contaminants, surface topography modification, and deposition of functional chemical groups [42]. Flame treaters usually use a laminar flame, which is a well-mixed fuel and oxidizer mixture before combustion. The terms "premixed laminar flame" and "deflagration" refer to the same thing when discussing the combustion process that is accompanied by a drop in density and pressure as well as an increase in velocity (as opposed to the propagation known as "detonation"). A laminar flame profile has three primary zones Fig. 7.7 that are indicative of distinct reactions [43].

- **Zone of Anticipation**

The "dark zone," as this area is often known, is characterized by a distinctive dark bluish color. This area of a flame is the coolest because, although part of the hydrogen generated is oxidized to water, the burning procedure has not yet stretched

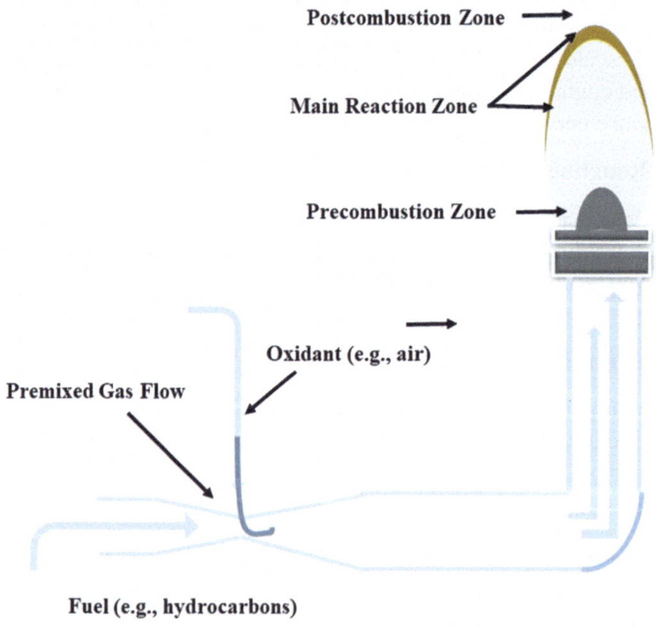

Fig. 7.7 Main zones in a laminar flame

the point of blast, therefore there has been very little net energy released. The sole free radical that is common in this area is the hydrogen atom, which reacts with oxygen and hydrocarbons quickly to prevent the radical pool from forming. This zone is also referred to as the "plummeting zone" for this reason. Since it does not affect the corrosion of the flexible substrate, this area is irrelevant and inefficient for surface-activation [44].

- **Principal Area of Reaction**

The mixed reaction zone, also known as the "luminous zone," is distinguished through the combustion system's greatest temperature, which for combinations based on propane can reach 1900–2000 °C. The concentration of the reactant is negatively impacted by the sharp rise in radical content in this zone. Unlike the previously mentioned decreasing zone, this region is aggressively oxidizing due to the high concentration of radical species. For polyolefin to be flame-treated effectively, such an oxidizing zone is useful. The fuel/air ratio determines the color of this zone: when the combination is gas-lean (because of excited CH radicals), a deep bluish violet radiation is created, and the flame almost completely transparencies as the gas amount decreases; on the other hand, green radiation develops [42].

- **Zone After Combustion**

This is the biggest zone out of the three that are present in a typical laminar flame profile. The reason the temperature stays high here is because of the exothermic oxidation process (partial or entire) of CO into CO_2. While intermediate species like CH_3, C_2H_2, and CH_2O are exclusive to the luminous region, the post-combustion zone can also detect radicals like the Hradical dot, OH radical dot, and Oradical dot. In a laminar flame profile, the concentration of radicals typically makes up around 10−3 of the reactants, whereas the concentration of ion species of which H_3O^+ is the most abundant takes up far less (10–6 of the reactants). Usually, they are situated just past the bright area of the flame [45]. Laminar flame can be understood in terms of convective fluxes of unburned gases from the dark to the light zone and the diffusion of radical species from the high-temperature zone to the pre-heating region, which is the opposite direction of the convective flow [46]. Laminar flame velocity, or the speed at which unburned gases travel throughout the combustion wave in the direction corresponding to the wave surface, is what defines the flame's subsonic wave characteristics in a combustion system [47]. The phenomenon of Main Zones in a Laminar Flame is illustrated in Fig. 7.7.

- **Using Flame to Treat Polyolefin**

All polymers whose general formula is CnH_2n and which are derived from an olefin as an initial monomer are collectively referred to as polyolefin. The two most used polyolefin in wrapping are polyethylene (PE) and polypropylene (PP). Both polymers are known to be intrinsically hydrophobic, while having distinct specialized characteristics. This contributes to their generally low wettability, particularly in aqueous systems. Because of this, surface-activation of polyolefin is typically required prior to the deposition of metals, coatings, paints, inks, and adhesives.

While it hasn't been used as much as corona treatment thus far, flame treatment is a useful method for raising the surface energy of polyolefin. Nonetheless, due to advancements in technical safety concerns, it is getting [48].

- **Surface/Flame Gap**

It is commonly known that one of the most important factors in determining the level of activation attained by the treatment is the distance between the flame and web surface, or the gap between the tips of the bright flame cones and the polyolefin surface. Generally speaking, it has been seen that the treated surface's wettability rapidly decreases as the film moves through the flame. Surface initiation reduces with increasing distance between the flame cone and the film surface, although the treatment's positive effects are still noticeable up to around 20 mm [49].

Since the flame can still be seen on the treated film surface, there are some benefits since the flame temperature and concentration of oxygen radicals are developed in this area (post-combustion) than they are in the dark zone. The characteristic feature of the curve that results from graphing the surface energy standards as a function of the film-to-flame distance is explained by this fact. According to Fig. 7.8 [50], this curve is asymmetric about the highest surface energy value, which is detected at a film-to-flame distance of around 2 mm. This suggests that the flame treatment's beneficial effects are still somewhat noticeable in the post-combustion zone, but that they rapidly disappear in the dark area [51].

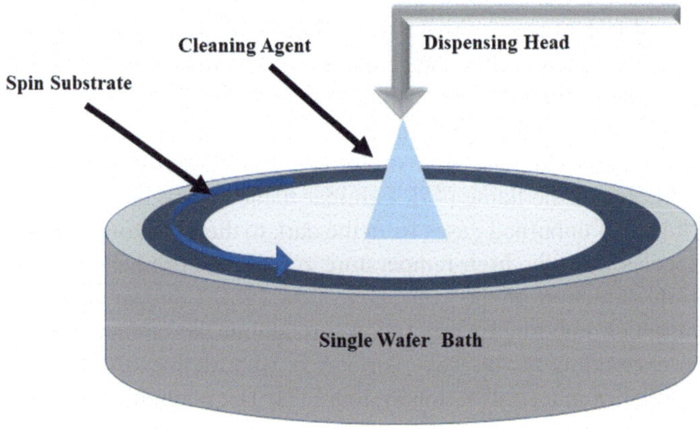

Fig. 7.8 Working mechanism of cleaning of substrate

7.9 Cleaning of Substrate

The glass substrate's surface cleanliness has become more crucial for the manufacturing of hard disks. The deposited films are getting thinner to attain better areal storage densities in modern hard disk drives. This implies that to avoid creating any flaws in the layer, the films must be placed on incredibly smooth and clean surfaces. There has been evidence of a beneficial impact of substrate cleaning through etching procedures on the hard disk's magnetic performance. It was discovered that thin seed layers combined with substrate etching could decrease grain size. Heating the disk to about 250 °C is how production machines clean their substrates. Following this process, there should be a significant decrease in the quantity of organic and inorganic contamination on the disk surface. However, as XPS studies have demonstrated, some contamination is still present on the surface. First, research using an ion source capable of producing ions with energies as high as 350 eV has demonstrated that this technique can be applied to further lessen pollution and improve surface cleaning. For the creation of innovative semiconductor devices at the nanoscale, cleaning silicon (Si) surfaces is a crucial step. Conventional techniques for eliminating organic contaminants and natural or synthetic oxide are frequently paired with elevated temperature desorption phases. However, if, e.g., dopant out-diffusion is not stopped, small feature devices will not work. This work introduces two potential methods for cleaning at low temperatures: A hydrogen DC plasma source integrated into a UHV cluster tool, and an atomic hydrogen source based on dissociative adsorption of hydrogen at a heated tantalum (Ta) surface. The impact of atomic hydrogen on the removal of carbon and oxides is reviewed, along with the current model for native oxide etching using an argon/hydrogen cleaning increases the thin-film coating's adherence to the substrate, which enhances its performance and dependability [52]. By eliminating impurities that may result in variations in density and homogeneity, which could impair characteristics like transmission and reflectance, pre-cleaning also improves the thin film's optical or electrical capabilities. To further enhance adhesion, particularly for oxide films, several pre-cleaning techniques may chemically activate the surface or micro-roughen the substrate. Depending on the substrate material and what needs to be removed, a specialized pre-clean process may be utilized, such as sputter etching and oxide layer removal with an energetic ion source [53]. To produce thin films of superior quality, the deposition apparatus usually needs to be pre-cleaned in situ. It is difficult to monitor how well the cleaning process is working because sampling and testing the surface can contaminate the substrate [53]. New developments in laser fluorescence-based plasma analysis could improve in situ cleaning process monitoring. To sum up, the process of thin-film deposition requires pre-cleaning the substrate to eliminate impurities, increase adhesion, and improve the characteristics of the film that is deposited. The particular cleaning technique employed is determined by the needs of the film and substrate [53]. The working mechanism of cleaning of substrate is shown in Fig. 7.8.

Which techniques are most frequently employed to prepare substrates for thin-film deposition?

- **Wet Chemical Cleaning**

This includes sonicating the substrates in acetone and isopropyl alcohol (IPA) for ten minutes each after soaking them in chromic acid for the entire night. After that, the substrates are baked for two hours at 120 °C in an oven.

- **RF Glow Plate**

This technique eliminates impurities like water molecules and hydrocarbons with very little energy. It has a low chance of causing damage to the substrate surface and covers a significant area [54].

- **Gridded Ion Source**

This method physically sputter etching and removing oxide layers from the substrate surface with high-energy ions. Although active substrate cooling is necessary because of the high-energy ion bombardment, it is the preferable method for oxide etching.

- **Griddles Ion Source**

Griddled ion sources provide moderate to aggressive pre-cleaning along with a wider variety of ion energies than gridded ion sources. They work well with substrates that are sensitive to temperature, such as plastics and polymers [55].

- **Plasma Pre-Treater**

For pre-cleaning, this technique uses high voltage and high-intensity ion bombardment. To avoid damage from the high-energy ions, active substrate cooling might be necessary. Pre-cleaning techniques are selected based on several criteria, including substrate sensitivity, desired cleaning depth, deposition method, and economic concerns. To produce high-quality thin films, in situ pre-cleaning inside the vacuum chamber is typically necessary.

- **The Value of Cleaning**

Impurities must be removed from the substrate surface to prevent them from affecting the film's homogeneity, adhesion, and general quality. The substrate material itself, handling and storage, as well as the surrounding environment, can all be sources of contaminants [56].

The following benefits of cleaning the substrate:

- **Eliminate Severe Contamination**

Eliminate any obvious impurities, such as oils, particles, or other materials that can obstruct the deposition process, as the first stage in the cleaning procedure.

- **Diminish Surface Roughness**

The homogeneity and adherence of the deposited layer may be impacted by surface roughness, which can be lessened with cleaning.

- **Enhance Film Properties**

By eliminating contaminants that could deteriorate the deposited film's optical, electrical, or mechanical qualities, cleaning can help [56].

7.9.1 Techniques for Cleaning

Substrates are cleaned using a variety of techniques in thin-film deposition, such as:

Solvent cleaning is the process of removing impurities using solvents, abrasion, etchants, fluxes, saponifiers, and emulsifiers.

- **Etching**

Particulates and oxide layers are two examples of the particular pollutants that can be eliminated by chemical etching [57].

- **Deposition Method**

The choice of cleaning technique may be influenced by the deposition technique employed. For instance, sputtering systems and evaporation systems cannot coexist with ion source technologies.

- **Depth of Cleaning Required**

The type and quantity of impurities that need to be eliminated will dictate the necessary cleaning technique.

- **Visual Inspection**

The easiest and most popular approach, this one entails examining the substrate surface for any obvious evidence of dust, grime, stains, or cleaning residues. Visual inspection can be used to determine which areas require additional care or cleaning.

- **Triphosphate Adenosine (ATP) Bioluminescence**

The quantity of ATP, a chemical that stores and releases energy in living cells, that is visible on the surface is determined by ATP bioluminescence. Although this approach offers instantaneous feedback and validation of the cleaning procedure, it is susceptible to influence from other chemicals and is unable to differentiate between different kinds of germs [58].

- **Microbiological Sampling**

In this technique, surfaces are sampled using swabs, wipes, or contact plates, and the samples are subsequently sent to a lab for examination. Microbiological sampling is a useful technique for determining the precise kinds and quantities of microorganisms present on surfaces, as well as their susceptibility or resistance to disinfectants.

- **Fluorescent Markers**

These are materials that, when exposed to a specific wavelength, release fluorescent light. They can be applied to surfaces before cleaning. After cleaning, you can use a tool called a fluorometer to scan the surfaces to detect if any fluorescent indicators are still present. If so, the cleaning was either insufficient or unsuccessful [58].

- **Environmental Indicators**

These measurements show how the cleaning and disinfection procedure affects the environment as a whole. Patient happiness, odor, noise, temperature, humidity, and air quality are a few examples [59].

7.10 Conclusion

The substrate material and surface condition have a significant impact on the characteristics of thin films. Good adhesion and homogeneous film formation require the substrate surface to be properly cleaned and activated. Because of their high surface-area-to-volume ratio and thin thickness, thin films are an important class of materials that have special qualities not seen in bulk materials. It is possible to customize the composition of thin films by using the right deposition methods and materials. Film composition is examined using methods such as *X*-ray diffraction. Temperature and rate of deposition have an impact on the surface morphology of thin films, which includes roughness and grain structure. Surface morphology is described using microscopy techniques. By activating the substrate surface with methods like plasma treatment, reactive spots are created that enhance the deposited film's adherence. Thin films can have thicknesses ranging from several micrometers to fractions of a nanometer (monolayer). Controlling the qualities of a film requires accurate thickness measurement. Equally crucial to the thin film's best adhesion is substrate preparation, which calls for a clean and correctly active substrate. The substrate's reactivity is increased by surface-activation methods including plasma treatment, and impurities that can impair adhesion are eliminated by thorough cleaning. Tailoring these qualities requires composition analysis, which is frequently accomplished with the use of methods like *X*-ray diffraction. Since characteristics like roughness and grain structure can have an impact on adhesion and functionality, surface morphology is crucial to the performance of thin films. The effectiveness of thin-film technology in real-world applications is ultimately determined by how these aspects interact.

References

1. Karki, S., et al. 2016. Thin films as an emerging platform for drug delivery. *Asian Journal of Pharmaceutical Sciences* 11 (5): 559–574.

2. Patel, V.F., F. Liu, and M.B. Brown. 2011. Advances in oral transmucosal drug delivery. *Journal of Controlled Release* 153 (2): 106–116.
3. Borges, A.F., et al. 2015. Oral films: Current status and future perspectives: I—Galenical development and quality attributes. *Journal of Controlled Release* 206: 1–19.
4. Morales, J.O., and J.T. McConville. 2011. Manufacture and characterization of mucoadhesive buccal films. *European Journal of Pharmaceutics and Biopharmaceutics* 77 (2): 187–199.
5. Debe, M.K. 2013. Tutorial on the fundamental characteristics and practical properties of nanostructured thin film (NSTF) catalysts. *Journal of the Electrochemical Society* 160 (6): F522.
6. Wang, W., et al. 2014. Device characteristics of CZTSSe thin-film solar cells with 12.6% efficiency. *Advanced Energy Materials* 4 (7): 1301465.
7. Mohammad, S.N. 2002. Electrical characteristics of thin film cubic boron nitride. *Solid-State Electronics* 46 (2): 203–222.
8. Wang, H., S.V. Garimella, and J.Y. Murthy. 2007. Characteristics of an evaporating thin film in a microchannel. *International Journal of Heat and Mass Transfer* 50 (19–20): 3933–3942.
9. Ranjan, R., J.Y. Murthy, S.V. Garimella. 2009. Analysis of the wicking and thin-film evaporation characteristics of microstructures. *Journal of Heat Transfer*
10. Jeong, J.W., et al. 2010. The response characteristics of a gas sensor based on poly-3-hexylithiophene thin-film transistors. *Sensors and Actuators B: Chemical* 146 (1): 40–45.
11. Park, J., et al. 2019. A review of thickness measurements of thick transparent layers using optical interferometry. *International Journal of Precision Engineering and Manufacturing* 20: 463–477.
12. Fingas, M. 2012. How to measure slick thickness (or not). In *Proceeding of the 35th Arctic and Marine Oil spill Program Technical Seminar*
13. Giurlani, W., et al. 2020. Measuring the thickness of metal coatings: A review of the methods. *Coatings* 10 (12): 1211.
14. Buehl, W., et al. 2006 Comparison of three methods of measuring corneal thickness and anterior chamber depth. *American Journal of Ophthalmology* 141 (1): 7–12.e1.
15. Shearer, C.J., et al. 2016. Accurate thickness measurement of graphene. *Nanotechnology* 27 (12): 125704.
16. O'Leary, S.A., B.J. Doyle, and T.M. McGloughlin. 2013. Comparison of methods used to measure the thickness of soft tissues and their influence on the evaluation of tensile stress. *Journal of biomechanics* 46 (11): 1955–1960.
17. Sun, L., et al. 2011. Structure, composition and optical properties of Cu2ZnSnS4 thin films deposited by pulsed laser deposition method. *Solar Energy Materials and Solar Cells* 95 (10): 2907–2913.
18. Ojeda-G-P, A., M. Döbeli, and T. Lippert. 2018. Influence of plume properties on thin film composition in pulsed laser deposition. *Advanced Materials Interfaces* 5 (18): 1701062.
19. Barreca, D., et al. 2001. Composition and microstructure of cobalt oxide thin films obtained from a novel cobalt (II) precursor by chemical vapor deposition. *Chemistry of Materials* 13 (2): 588–593.
20. Aragon, C., V. Madurga, and J. Aguilera. 2002. Application of laser-induced breakdown spectroscopy to the analysis of the composition of thin films produced by pulsed laser deposition. *Applied Surface Science* 197: 217–223.
21. Tam, P.L., and L. Nyborg. 2009. Sputter deposition and XPS analysis of nickel silicide thin films. *Surface and Coatings Technology* 203 (19): 2886–2890.
22. Kim, Y., et al. 2010. Effect of composition and deposition temperature on the characteristics of Ga doped ZnO thin films. *Applied surface science* 257 (1): 109–115.
23. Williams, E.D. 1994. Surface steps and surface morphology: Understanding macroscopic phenomena from atomic observations. *Surface Science* 299: 502–524.
24. Megelski, S., et al. 2002. Micro-and nanostructured surface morphology on electrospun polymer fibers. *Macromolecules* 35 (22): 8456–8466.
25. Li, W., and D. Li. 2006. Influence of surface morphology on corrosion and electronic behavior. *Acta Materialia* 54 (2): 445–452.

26. Javed, F., et al. 2011. Implant surface morphology and primary stability: Is there a connection? *Implant Dentistry* 20 (1): 40–46.
27. Dietrich, W.E., et al. 1992. Erosion thresholds and land surface morphology. *Geology* 20 (8): 675–679.
28. Clark, T., et al. 2000. A new application of UV – ozone treatment in the preparation of substrate-supported, mesoporous thin films. *Chemistry of Materials* 12 (12): 3879–3884.
29. Huang, Z., and H.K. Lee. 2012. Materials-based approaches to minimizing solvent usage in analytical sample preparation. *TrAC Trends in Analytical Chemistry* 39: 228–244.
30. Fan, M., G.F. Andrade, and A.G. Brolo. 2011. A review on the fabrication of substrates for surface enhanced Raman spectroscopy and their applications in analytical chemistry. *Analytica Chimica Acta* 693 (1–2): 7–25.
31. Liu, H., et al. 2009. Preparation of superhydrophobic coatings on zinc as effective corrosion barriers. *ACS Applied Materials Interfaces* 1 (6): 1150–1153.
32. Ahmadi, S., et al. 2014. The role of physical techniques on the preparation of photoanodes for dye sensitized solar cells. *International Journal of Photoenergy* 2014 (1): 198734.
33. Chapman, B. 1974. Thin-film adhesion. *Journal of Vacuum Science and Technology* 11 (1): 106–113.
34. Bull, S. 1992. Techniques for improving thin film adhesion. *Vacuum* 43 (5–7): 517–520.
35. Mittal, K. 1976. Adhesion measurement of thin films. *Active and Passive Electronic Components* 3 (1): 21–42.
36. Li, W., et al. 2004. Significant improvement of adhesion between gold thin films and a polymer. *Applied Surface Science* 233 (1–4): 227–233.
37. Jacobson, S., B. Jönsson, and B. Subdqvist. 1983. The use of fast heavy ions to improve thin film adhesion. *Thin Solid Films* 107 (1): 89–98.
38. Jiang, X., et al. 2014. Improvement of adhesion strength and scratch resistance of fluorocarbon thin films by cryogenic treatment. *Applied Surface science* 288: 44–50.
39. Tamboli, S.H., V. Puri, and R.K. Puri. 2010. Improvement in adhesion and decrease in stress of MgO thin films due to vapour chopping. *Journal of Alloys and Compounds* 503 (1): 224–227.
40. Iwamori, S., et al. 1998. Effect of an interfacial layer on adhesion strength deterioration between a copper thin film and polyimide substrates. *Vacuum* 51 (4): 615–618.
41. Bansal, N., et al. 2019. Growth control of molybdenum thin films with simultaneously improved adhesion and conductivity via sputtering for thin film solar cell application. *Vacuum* 161: 347–352.
42. Colman, R.W. 1984. Surface-mediated defense reactions. The plasma contact activation system. *The Journal of Clinical Investigation* 73 (5): 1249–1253.
43. Nie, L., et al. 2017. Activation of surface lattice oxygen in single-atom Pt/CeO2 for low-temperature CO oxidation. *Science* 358 (6369): 1419–1423.
44. Nagase, H. 1998. Cell surface activation of progelatinase A (proMMP-2) and cell migration. *Cell Research* 8 (3): 179–186.
45. Mills, A.P., Jr. 1979. Thermal activation measurement of positron binding energies at surfaces. *Solid State Communications* 31 (9): 623–626.
46. Han, Y., et al. 2006. Surface activation of thin silicon oxides by wet cleaning and silanization. *Thin Solid Films* 510 (1–2): 175–180.
47. Huynh, V., N.K. Ngo, and T.D. Golden. 2019. Surface activation and pretreatments for biocompatible metals and alloys used in biomedical applications. *International Journal of Biomaterials* 2019 (1): 3806504.
48. Hotchkiss, K.M., et al. 2016. Titanium surface characteristics, including topography and wettability, alter macrophage activation. *Acta Biomaterialia* 31: 425–434.
49. Zhang, Y.-J., et al. 2014. Effects of steam activation on the pore structure and surface chemistry of activated carbon derived from bamboo waste. *Applied Surface Science* 315: 279–286.
50. Sudaryanto, Y., et al. 2006. High surface area activated carbon prepared from cassava peel by chemical activation. *Bioresource Technology* 97 (5): 734–739.
51. El-Hendawy, A.-N.A., et al. 2008. Effects of activation schemes on porous, surface and thermal properties of activated carbons prepared from cotton stalks. *Journal of Analytical and Applied Pyrolysis* 82 (2): 272–278.

52. Zhang, D., et al. 2022. Removal mechanism of blue paint on aluminum alloy substrate during surface cleaning using nanosecond pulsed laser. *Optics & Laser Technology* 149: 107882.
53. Song, J.S., et al. 2004. Wet chemical cleaning process of GaAs substrate for ready-to-use. *Journal of Crystal Growth* 264 (1): 98–103.
54. Magnuson, C.W., et al. 2014. Copper oxide as a "self-cleaning" substrate for graphene growth. *Journal of Materials Research* 29 (3): 403–409.
55. Sonoda, T., et al. 2001. A surface cleaning method for sputter deposition of pure titanium film onto TiNi shape memory alloy substrate. *Vacuum* 60 (1): 197–199.
56. Cras, J.J., et al. 1999. Comparison of chemical cleaning methods of glass in preparation for silanization. *Biosensors and Bioelectronics* 14 (8): 683–688.
57. Sun, X., et al. 2021. Substrate cleaning threshold for various coated Al alloys using a continuous-wave laser. *Photonics*
58. Zhao, Y., et al. 2021. Effect of surface cleaning on interface bonding performance for 316H stainless steel joints manufactured by additive forging. *Materials & Design* 210: 110025.
59. Arnold, N., G. Schrems, and D. Bäuerle. 2004. Ablative thresholds in laser cleaning of substrates from particulates. *Applied Physics A* 79: 729–734.

Chapter 8
Role of Thin Film in Deposition Techniques

Abstract This chapter fully discusses the synthesis of thin films and explains the techniques involved, namely spin coating, CVD, PVD, laser ablation, and atomic layer deposition (ALD). These are very important in the synthesis of thin films and all contribute uniquely toward the formation of high-tech materials with specific features. In the production of films on substrates, it is very commonly practiced to utilize the CVD method that comes from chemical reactions between precursors in the gaseous phase. It is very useful in the production of excellent motion pictures with few defects and outstanding uniformity. CVD is excellent for electronics and optics applications, since it can have numerous modifications-low pressure and plasma enhanced CVD, which enables depositing films at relatively low temperatures, with an enhanced control of film properties. PVD is comprised of a lot of physical deposition methods such as sputtering and thermal evaporation. In the case of PVD, the deposited materials are evaporated in a vacuum before condensation. Vacuum-induced vaporization of solid materials is followed by condensing onto a substrate in PVD. This method is well-known for producing high-performing, long-lasting coatings that improve surfaces' mechanical and chemical characteristics. PVD is widely used in making optical coatings, ornamental finishes, and semiconductors. Applying homogeneous polymer or sol-gel thin films is the main application for the adaptable spin coating technology. Using centrifugal force, a liquid solution is applied to a substrate and quickly spun to distribute the material uniformly. Applications in microelectronics and photonics benefit greatly from spin coating's ability to produce thin films with precise thickness.

8.1 Role of Thin Film in Deposition Techniques

While the metallurgical phase diagram does not impede the growth of thin films, as the nature of deposition procedures used in making thin films is non-equilibrium, this greatly gives thin films an advantage over bulk materials. The development of new thin-film materials to satisfy industries in great need of materials with multiple dynamics and versatility depends on deposition technology. Indeed, almost all the

T. I. Awan et al., *Thin Film Deposition Techniques*,
https://doi.org/10.1007/978-981-96-1364-9_8

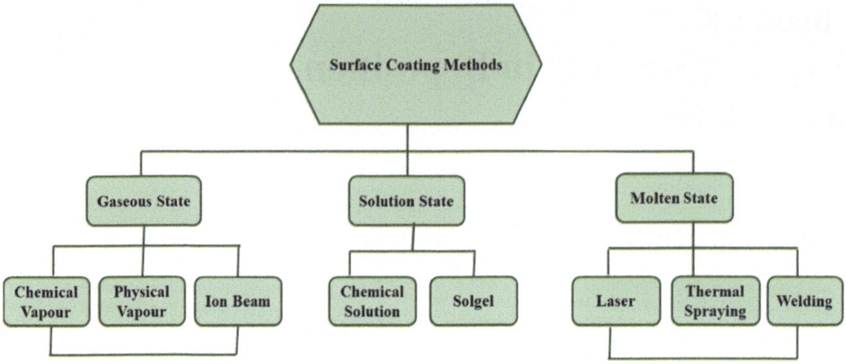

Fig. 8.1 Variants of thin-film deposition techniques

belongings of a thin film are defined by deposition processes, which can also be used to alter the existing qualities [1]. Accordingly, deposition procedures should be carried out with much care, since various deposition methods provide microstructure, surface morphology, tribological, electrical, and biocompatibility features. A single material can, therefore, be applied to several uses and tailored to possess qualities that meet even the most exacting specifications by the application of several deposition processes. It is also possible to realize a hybrid deposition technique in which different methods are combined in such a way that each one contributes to the result of the thin layers. Most deposition techniques follow these three main steps:

1. Synthesis of the deposition species,
2. Conduction from one medium to another [2].

There are numerous deposition techniques, but since the scope of this analysis will fall on thin-film deposition, the process of producing layers thinner than one micron will be given to the two major methods employed in thin-film deposition. Among the popular subsets of deposition techniques are Physical vapor deposition, PVD, and chemical vapor deposition, CVD. The characteristic that sets PVD apart from CVD is vapor [3]. While the vapor in CVD forms a thin film on the substrate by a chemical reaction, in PVD, it consists of atoms and molecules that just condense on it. Figure 8.1 illustrates the types of thin film deposition that are purely in the gaseous state, solution state, and molten or semi-molten state [3, 4].

8.2 Role of Thin Film in PVD Processes

Thin films have evolved over the last century to become an indispensable component of daily living. Its apparent uses are numerous and span a variety of industries, including microelectronics, vehicle components, windows in tall buildings, and the metallic coatings found inside potato chip bags. Thin-film characteristics frequently diverge greatly from nothing more than a more streamlined form of what is known as bulk material [5]. For instance, the optical characteristics of a thin dielectric layer. Layers may be tailored to transmit by changing the thickness of each layer or absorbing light in extremely certain colors or wavelengths, and this differs significantly from how the macroscopic or the big glasses we have become used to [6]. The electrical, optical, and physical characteristics of the selected materials have a significant impact on thin-film applications [7]. Consequently, the methods and procedures employed to create the films have a significant impact on these characteristics. With minor adjustments to the parameters, the same deposition equipment and method that yields a transparent, dielectric film with exceptional optical qualities may also produce opaque, metallic films more suited for electrical conduction than optical uses.

The history of thin-film deposition tools and processes dates back to the 1850s and includes the entirety of the twentieth century [8]. Thin-film technologies were extensively employed in the early 1900s for optical applications, including telescope mirrors. By applying a protective layer to a surface, thin-film deposition modifies, shields, and enhances the surface characteristics of base metals. Many elements are commonly employed in coatings, including Cr, Ti, and Al. Many procedures and techniques are used in the synthesis of films with different thicknesses, compositions, and characteristics [8]. A few of them include magnetron sputtering, ion beam sputtering, physical vapor deposition (PVD), and ion plating. Because of these approaches' wide range of applications, they have become popular and piqued researchers' curiosity. Al-Cr-N, Ti-Al-N, Cr–N, Ti-C-N, and other well-known coatings are used to improve the surface qualities of different substrates employed in daily existence [9]. Physical vapor deposition (PVD) thin-film surface structure is essential as it impacts the thin films' functionality in several applications. Usually, a lot of topographical flaws on the surface reduce the quality of films; in certain situations, they can even lead to their disastrous failure.

Three preparatory processes determine the surface topography of thin films at the microscopic level:

(i) Mechanical substrate pretreatment;
(ii) Substrate ion etching; and
(iii) The deposition process.

The mechanical pretreatment process, which typically consists of polishing and grinding, can leave the substrate material's surface with a variety of abnormalities, such as ridges, scratches, and grooves. Insufficient impacts arise at inclusions that are softer than the ferrous matrix in non-homogenous materials like tool steel, whereas additional shallow growths emerge at harder inclusions. Because of the shear forces, some of the projecting inclusions may tear off during polishing, leaving pits in their wake. Furthermore, the substrate surface may include polishing residue. The topography of the deposited thin film is directly impacted by each of these substrate imperfections [10]. Ion etching is often used as the final stage of substrate surface preparation for cleaning. This stage is carried out to eliminate contaminants from the earlier pretreatment processes, with a focus on eliminating native oxide layers and

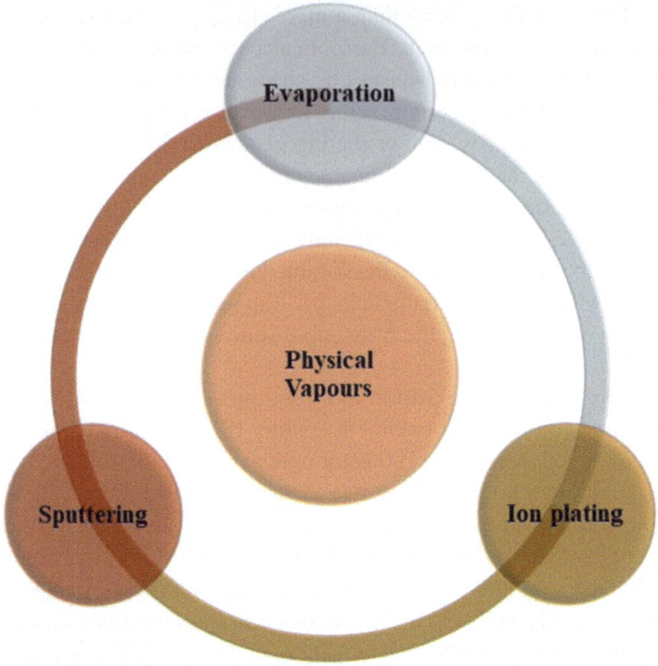

Fig. 8.2 Variants of physical vapor deposition

chemically activating the substrate surface to enhance film adhesion [11]. Ti-Al-N coatings have a low-temperature dependency, which reduces COF and boosts hardness, corrosion, and wear resistance. Because of its superior heat resistance, Al-Cr-N is used to extend the life of tools and boost their cutting capability. In recent times, researchers have focused their attention on the structure, morphology, and content of the Ti-CN coating. Since Ti-C-N offers a solid solution of both TiC and TiN, it has been noted to have the benefits of both TiN and TiC coatings. This synthesis/deposition process creates a mechanical link and, to some extent, metallurgical advantages, such as increased tool/component life and improved component performance. These thin coatings can withstand corrosion, erosion, wear, and oxidation [12]. The major variants of PVD are listed in Fig. 8.2, but the three basic processes of PVD are vacuum or thermal evaporation, ion plating and sputtering.

8.2.1 Thin Film's Function in PVD Processes

Thin-film deposition onto a variety of substrates is done extensively by physical vapor deposition, more popularly known as PVD. The properties and functionality of the final product are almost exclusively dependent on the thin film [13]. Thin films in PVD play the following critical roles:

8.2.2 Surface Alterations

Surface properties of a substrate can be dramatically modified by depositing a thin layer over it such as hardness, wear resistance, corrosion resistance, and aesthetic appeal. This is very important especially for applications like cutting tools, where the thin layer of the ceramic coating enhances wear resistance and the risk of adhesion to the work piece is minimized [14]. The main source of the depositing species in PVD is a solid or liquid rather than a gas, and its vapor pressure is significantly lower than the deposition system's operating pressure. This is how PVD varies from chemical vapor deposition. Nonetheless, in PVD systems, chemical reactions are possible and are observed, almost invariably near the depositing film surface [15]. This might be as simple as adding vacuum system leftover gases to the PVD films that are being deposited. As an alternative, a reactive gas species, like oxygen or nitrogen, for instance, can be purposefully added to the chamber during deposition to appear as an oxide or nitride in the film [16].

- **Optical Characteristics**

  The optical properties of a surface can be altered by modifying thin coatings. Examples include reflectance, transmittance, and color. Applications for examples are window covering, architectural panels, and decorative coatings [17].

- **Electrical Characteristics**

  Such electrical devices as semiconductors and thin-film transistors can be produced using thin films. As all the devices depend on the intrinsic features of the thin film, the intrinsic characteristics are usually considered to be conductivity and dielectric constant [18].

- **Packing**

  Thinner layers of materials are incorporated into food packaging to enhance barrier properties and prolong shelf life. Some examples include aluminum oxide barrier coatings, which can prevent moisture and oxygen from entering the container when applied as a coating to polymer films.

- **Solar Energy**

  Thin-film solar cell is one of the photovoltaic cells produced with thin sheets. The thin film is an integral part of the use of solar energy applications since it both collects light and generates power [19].

8.3 Role of Thin Film in CVD Processes

Deposition of thin film refers to the creation of thin-film coatings and then depositing them onto the substance of a substrate. Coatings can be prepared from virtually any type of material, including metals, oxides, compounds, and others. Moreover, the wide variety of characteristics of the thin-film coatings can be used for adjustment or improvement of specific aspects of the substrate performance [4]. Examples include those that are transparent, those that are extremely strong and scratch-resistant and ones that change how signals are transmitted or the conductivity of electricity itself. A group of thin-film deposition techniques called chemical vapor deposition (CVD) involves heating two chemical precursors to the point of vaporization, while a substrate is inside of a vacuum chamber [20]. When the material reacts with the substrate surface, a thin-film coating with high performance is formed. Coatings made by CVD are suitable for silicon wafers, automobile parts, medical equipment, and many other diverse applications [21]. One of the advantages of CVD techniques is their versatility in coating complex or intricate topographies on a wide range of substrates. Additionally, thin films produced by CVD usually hold their bonds effectively at high-stress conditions. The term "chemical vapor deposition," or "CVD," refers to a broad variety of thin-film deposition techniques that are widely used to produce solid coatings or polymers with superior performance and quality [22]. Essentially, all of the many different CVD processes share one thing in common: A gaseous chemical precursor acts on the substrate, either because of heat or plasma, and creates dense thin films through some chemical reaction. CVD grows dense, pure coatings that are consistent in thickness, both rapidly and in a very adaptable fashion. In thermal CVD, the substrate is heated and a precursor reactant gas is injected into the deposition chamber. The gas can either form an intermediate reactant in the gas phase and be subsequently deposited on the substrate or be directly adsorbed onto the surface of the substrate to be coated. It is carried out at atmospheric pressure or higher temperatures [23].

8.3.1 Used of Vapor Deposition of Chemicals

The electronics, solar panel, and optical industries which include make-believe glasses, optical storage, and sunglasses all rely heavily on the widely used CVD production method. One of its most widely used and cost-effective methods of generating films for packaging, such as potato chip bags, is polymerization by CVD, which is the process of small molecules joining chemically to form a very long chainlike molecule or polymer. Often employed on surfaces to stop weather-related corrosion, CVD coating can be extremely hard, finely grained, and impermeable. Tribology the study of moving parts that includes lubricity and hardness is applied to high-performance automotive and aerospace components utilizing this technique [24].

8.3.2 Equipment for Chemical Vapor Deposition Coating

All CVD coating equipment and procedures are unique to their respective sectors and come in a wide range of variations, yet they all have the same basic design. After being heated or exposed to a plasma, the substrate to be coated is placed inside the deposition chamber. Electromagnetic power sources specifically designed for each application power all CVD reactions. A controlled introduction of one or more reactant precursor gases is made into the deposition chamber [25]. In this case, they can either start reacting chemically with the substrate surface right away or combine to create intermediate gases in the deposition chamber, which react chemically with the substrate surface to produce thin-film coatings [26]. To render the chemical by-products and unreacted atoms or molecules that were taken from the chamber with the exhaust safe for humans and the environment, they must first be treated because they are frequently poisonous, combustible, or harmful to the pumps. Wet scrubbers, vents, and cold traps are typically used to catch them. To preserve the pumps, they are subsequently treated by being run through a water or chemical trap. Particular attention is needed for flammable gases.

8.3.3 CVD Precursors

- Halides: $SiCl_2$, $TiCl_4$, $HSiCl_3$, WF_6
- Hydrides SiH_4, GeH_4, NH_3, $AlH(NMe_3)_3$
- TEOS, or Tetrakis Dimethylamino Titanium, is a metal alkoxide (TDMAT)
- Titanium (NMe_2) Metal Dialkylamides
- Copper Diketonates: Acacia
- Nickel-based metal carbonyls
- Ti $(OiPr)_4$ metal alkoxides
- $AlMe_3$ Ti (CH_2tBu) is an organometallic
- Oxygen.

Transport of the precursors to the substrate occurs by gas diffusion, liquid flow, or a mix of the two once they are inside the deposition chamber. The resident time is the amount of time the atom or molecule must remain on the surface to form a chemical bond [27]. To make place for more atoms, those that don't bond are then forced away. Handling the kinetics and thermodynamics of temperature, pressure, and concentration yields allows for the achievement of the required coating.

8.3.4 Kinds of CVD Procedures

- **The Atmospheric Pressure and Low-Pressure CVD**

The gas is drawn through the deposition chamber using a vacuum pump in LPCVD. Lower pressure can lessen gas-phase interactions and produce a more uniform deposition rate. Pumps are frequently not needed when atmospheric pressure is present, but the deposition rate may be slower. CVD on the hot and cold wall. The chamber is heated for a consistent temperature when using hot-wall CVD. Whereas overheating of the substrate can be an issue, cold-wall CVD allows for faster cooling because only the substrate is heated [28].

- **Through the Use of Atomic Layer Deposition**

In atomic layer deposition (ALD) consecutive layers of various materials can be deposited with a high degree of homogeneity, and exact thickness control. Precursors are progressively added to the chamber, where they undergo self-limiting absorption and surface reactions before being expelled by the gas. After that, the next coating is added to the chamber to create easily controlled, extremely high-quality thin films [29].

- **MOCVD, or Metal–Organic CVD**

Vaporized liquids are frequently employed as metal–organic precursors. One of the most widely used methods for creating III-V semiconductor materials on a chip is this technique, which involves growing multiple intricate layers with different compositions atop one another. Along with solar cells and photodetectors, MOCVD is now widely employed in the manufacturing of optical devices such as light-emitting diodes and lasers. The volatile reaction of by-products and unused precursor species is a characteristic of the CVD process [30]. Several volatile by-products of CVD reactions, such as H_2, Cl_2, HCl, HF, or water vapor, are extremely dangerous. When employing CVD, appropriate safety precautions must be used. Cleaning and venting eliminating unreacted chemicals and by-products is crucial to CVD procedures. CVDs can be categorized according to the energy that fuels the chemical reaction. Energy sources include photons and lasers or thermal (warm) temperatures. The various forms of the CVD thin-film deposition processes are depicted in Fig. 8.3.

8.3.5 Advantages of Thin-Film Deposition by CVD

CVD provides a vast spectrum of available coating materials including metals, alloys, and ceramics. Alloys can be created with the chemical processes defining CVD. For a lot of variant types of batch production runs, it can provide a method that is easily controllable and scalable, offering considerable economies of scale concerning costs. CVD depositions include films of purity over 99.995%. They tend to be fine-grained and allow for very high levels of hardness, especially in diamond-like carbon coatings. Generally, most CVD processes for depositing thin films do not require a vacuum, let alone as high of one as PVD procedures. Most CVD equipment is self-cleaning except for the chemical by-products that are created during reactions at the outflow [31].

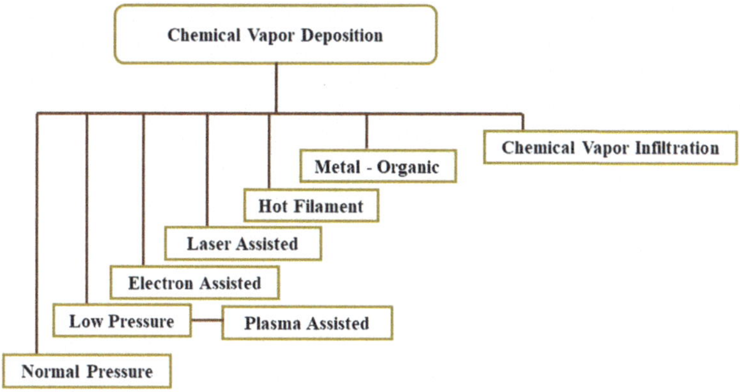

Fig. 8.3 Variants of chemical vapor deposition

8.3.6 Drawbacks of Thin-Film Deposition by CVD

Numerous by-products of CVD pose a risk since they can be extremely poisonous, explosive, or corrosive. Careful thought must be made to manage them in a way that is safe for both the environment and people. This can be costly depending on the particular antecedents.

The sort of substrate that can be coated without being damaged can vary depending on the heat-driven nature of thermal CVD techniques. The heat used in the process might cause stresses and failures between films with varying heat expansion coefficients. Certain precursor gases can be expensive, especially those that are metal–organic compounds that are frequently employed in the production of films [32].

8.4 Role of Thin Film in Vacuum-Based Evaporation

Next-generation TF PVs face serious challenges in competitiveness with silicon-based PV in efficiency, cost of manufacture, and volume of production, and, in the longer term, in containing only abundant non-toxic constituents exhibiting environmental stability without significant degradation. However, TF did not continuously have such a hard time in the PV market. In particular, amorphous Si (a-Si), cadmium-telluride (CdTe), and copper-indium-(Gallium)-Sulfide/Selenide (CIS/CIGS) realized important gains in the market between 2003 and 2009 due to the increased interest in those technologies because of the lack of polysilicon material. However, since 2009, the price of polysilicon has sharply declined as a result of a significant expansion in manufacturing capacity, and PV based on crystalline silicon (c-Si) has regained over 85% of the market [33]. Furthermore, overproduction and production scale-up factors caused a sharp drop in the cost of c-Si-based PV modules. This resulted in a significant market shakeout and the current bankruptcy of several TF (as well as significant c-Si) PV firms. Investors are extremely wary of TF techniques and associated start-ups as a result, and it is now harder for such technologies to break into the market. Vacuum-based evaporation thin-film deposition is important for many sectors, such as packaging, electronics, and optics. To create a thin-film coating, a source material is heated until it evaporates and condenses onto a substrate surface. The following are some important thin-film properties for vacuum-based evaporation [34]:

- **Better Qualities**

To improve the basic material's characteristics, thin-film coatings can adapt it to a range of uses. For instance, to enhance the reflectivity, transmission, and protective qualities of lenses, many thin-film layers are evaporated in the optical field. Thin coatings improve a material's ability to transmit electricity and withstand corrosion in electronics [35].

- **Precise Control**

Thin-film deposition may be precisely controlled in terms of both composition and thickness in terms of vacuum-based evaporation. The evaporation rate, pressure, and substrate temperature may all be precisely adjusted to produce homogeneous, excellent thin films with the required properties.

- **Wide Range of Materials**

Vacuum evaporation thin-film deposition applies to a broad class of materials, such as oxides, nitrides, and metals. The material selection is determined by the intended thin-film qualities as well as the particular application.

- **Scalability**

Systems for vacuum-based evaporation can be expanded to handle large substrates and high throughput. Because of this, the procedure may be used to produce goods on an industrial scale, including coating packaging foils or eyeglass lenses [36].

- **Flexibility**

Thin-film deposition is capable of being applied to build structures ranging from single-layered simple coatings to multi-layered complex structures. Because it is an adaptable technique, advanced materials can be synthesized in large volumes, demonstrating specific properties.

In a nutshell, vacuum-based evaporation thin-film deposition is a critical process that can now be used to manufacture materials of high performance with enhanced properties, with high precision and scalability. This procedure will be indispensable in the majority of sectors because it can deposit nearly all types of materials in a vacuum.

Large part of homogeneity in the deposition process of a thin film is ensured by the vacuum system. The vacuum environment provides several significant contributions toward uniformity in the thin film.

- **Reduction in Collisions**

Evaporated particles have much bigger mean free path within a vacuum, which allows them to move onto the substrate without hitting other gas molecules. Less scattering and knowing that the particles actually follow the desired course ensures deposition is more uniform over the substrate surface.

- **Regulation of the Rate of Deposition**

With the vacuum system, the rate at which the deposited material can be deposited with precise accuracy is controlled. A constant vacuum pressure gives the operators ample opportunity to control the evaporation parameters such as the distance of the source and its temperature, which affects the thickness of the film and the homogeneity, by holding at a constant vacuum pressure [37].

- **Purity of the Deposited Material**

A high vacuum environment minimizes impurities that may compromise the homogeneity of the film. The material evaporated must be pure since impurities can cause the thin film to have flaws and in homogeneities. The vacuum system improves the quality of the final film by effectively isolating the deposition process from outside impurities.

- **Thermal Stability**

The vacuum system also assists in the process to keep the temperature stable. Whenever the film thickness changes when the temperature increases and decreases, the vacuum system conditions pave the way for uniform film growth.

Geometry and Source Configuration

To enhance uniformity, the design of the vacuum system can be optimized, incorporating the geometry of the substrate and the evaporation source. To further enhance film uniformity, e.g., the distance between the source and the substrate can be altered to generate a more uniform dispersion of the particles that are evaporated. With the minimization of particle collision, regulation of deposition rates, guaranteed material purity, retention of thermal stability, and improvement of source geometry, it is in the vacuum system that makes the production of homogeneous thin films come to pass. All these factors combine to make the outcome better and of quality when the thin films are deposited [38].

8.5 Role of Thin Film in Spin Coating

Introduction

Applying homogeneous thin films on flat substrates is done by the spin coating process. Typically, a small puddle of fluid resin is placed in the center of a substrate, and the substrate is then rapidly rotated. The resin will gradually spread to the substrate's edge and peel off due to centrifugal force, leaving a thin layer of resin on the top. The type of resin used and the settings used for the spin process will determine the final film thickness as well as other characteristics (such as surface tension, percentage of particles, drying rate, and viscosity) [39]. The qualities of coated films are defined in part by variables including acceleration, fume exhaust, and final rotating speed. Spin coaters, or simply spinners, are machines used for spin coating. Until the required thickness of the film is reached, rotation is maintained while the fluid spins off the substrate's edges. Typically, the applied solvent evaporates at the same time as it becomes volatile. Therefore, the thinner the coating, the faster the angular speed of spinning [40]. Repeatability is one of the key elements of spin coating. Significant differences in the coated film can arise from subtle changes in the parameters defining the spin process. Although there hasn't been much research on the subject, the spin coating approach provides various advantages, including inexpensive cost, easy control and handling of chemicals and substrates, faster thin-film creation, and less need for high annealing temperatures. In this study, a thorough explanation of the theory and practice of spin coating will be examined, along with the design and production of reasonably priced spin coating apparatus [41].

8.5.1 Description of the Thin-Film Deposition by Spin Coating Process

- **A Dispensing Phase**

A standard spin technique involves depositing the resin fluid onto the substrate surface during the dispensing step. There are two popular dispensing techniques: dynamic and static. Simply placing a tiny fluid puddle in the middle of the substrate is known as static dispensing.

Depending on the size of the substrate to be coated and the fluid's viscosity, this can range from 1 to 10 cc, large substrates or higher viscosities usually demand a larger puddle to provide complete substrate coverage during the high-speed spin stage. The technique known as "dynamic dispensing" involves dispensing while the substrate is rotating slowly [39].

- **A Phase of Substrate Acceleration**

This period is characterized by a strong mass flow of fluid ejected by the rotational motion from the wafer surface. In this period, spiral vortices may also appear temporarily because of the initial depth of fluid on the wafer surface. This is due to an inertia effect of the top of the fluid layer which tries to twist the wafer. Any indication of fluid thickness variations eventually disappears when the fluid gets sufficiently thin to fully co-rotate with the wafer. When the target speed of the wafer is achieved and the fluid is sufficiently thin, then the rotating accelerations are precisely balanced by the viscous shear drag.

Spin coating is a well-known technique in the laboratory for the rapid preparation of membranes. Normally, this method is used to prepare a thin membrane by spinning

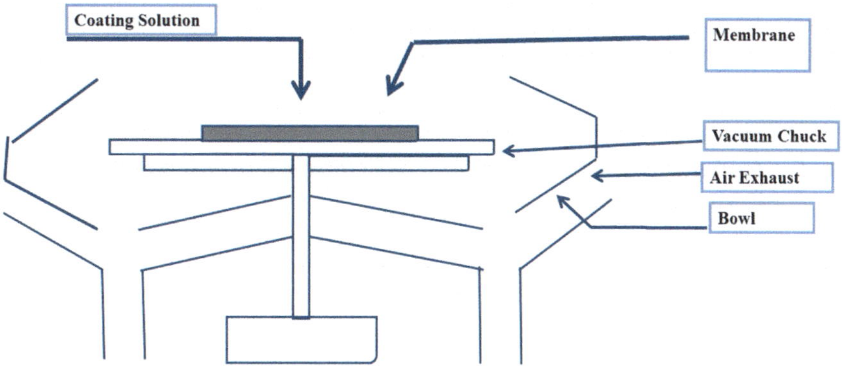

Fig. 8.4 Thin-film deposition by spin coating

out the solution of precursors over the edge of the support. The coating solution is to be thrown onto the support surface, and under centrifugal force, the solution spreads in a thin film; the solvent evaporates rapidly to yield a uniform coating on the surface of the substrate [42], as shown in Fig. 8.4.

8.6 Role of Thin Film in Laser Ablation

Precise cutting, drilling, surface cleaning, compositional analysis, and thin-layer deposition are some of the laser processing applications that appear to originate from this very basic concept of laser ablation. The latter application had been straightforward since a plate, slide, or wafer may be placed in front of the plasma plume and may function as a collector for the hot ablated material which eventually condenses into a thin film. PLD means pulsed laser deposition, and that is what this deposition technique is called [43]. Smith and Turner tried thin-film deposition for the first time in 1965; success in the stoichiometric transfer of a compound with complex molecular structure, which was extremely hard to obtain using other deposition techniques, marked the real breakthrough in the field. It is proper to say that the congruent ablation in this case was achieved. The greater delivered energy of laser beams, through the reduction in pulse time, greatly expanded the range of materials that could be ablated. Some of the PLD variations arose from the need to deposit more complicated materials, or elements, which degrade at high temperatures. Instead of using a single laser beam as in classical PLD, two laser beams can be used simultaneously to ablate two targets placed on a carousel system. This will generate a plasma mixture that grows thin films of different compositions across the surface. This form of PLD is referred to in the literature as combinatorial pulsed laser deposition, or CPLD. Another variation is the matrix-assisted pulsed laser evaporation, developed from the need to protect compounds that have long and fragile molecular chains. MAPLE utilizes a frozen mixture that contains the active material to be deposited and a buffer matrix that preponderantly absorbs the laser beam [43].

8.6.1 Techniques for Deposition and Compositional Analysis Using Pulsed Laser Ablation

- **Deposition of Pulsed Lasers (PLD)**

In this context, the collector is called the "substrate," and the material to be vaporized by the laser beam is called the "target." They have to be plan-parallel and placed inside a deposition chamber that must be sealed under vacuum conditions. A high-intensity laser, serving as an energy source to ablate the target material for depositing the thin film, is placed outside the deposition compartment. The deposition process does not contain contamination or impurity because it is the photons that cause the target to vaporize. On the other hand, despite the experimental assembly of PLD being truly simple, the laser material interaction that forms the basis is an extremely complicated physical event comprising a sequence of distinct processes [44]. The following are these processes,

- Coupling of the optical energy to the target material.
- Melting of the surface.
- Vaporization of a thin upper layer of the molten surface in the form of a plume.
- Photon absorption by the vaporized species, which eventually will limit the laser fluence reaching the target surface.
- The propagation of the plume along the direction normal to the target.
- Returning to stage A after about a few nanoseconds from the end of the pulse with a re-solidified surface.

8.6.2 Factors that Cause PLD Deposition

The following are some of the variables that affect the laser ablation process:

(i) Target material properties,
(ii) Laser beam parameters,
(iii) Deposition conditions, and
(iv) By rotating and translating the target about the laser beam, uniform ablation of the target is achieved.

Typically, a few centimeters separate the target and substrate. By rotating the substrate holder, e.g., or moving the substrate in the direction of the plasma, the layer homogeneity can be increased. A critical factor influencing the microstructure, morphology, and crystallinity of the deposited films is the substrate temperature [33].

8.6.3 The Use of Lasers in Laser Ablation

Ruby lasers were among the earliest forms of pulsed lasers to be utilized for solid material ablation. These days, excimer or lasers are used in laser ablation systems. The use of solid-state systems is justified by their low cost, low maintenance requirements, minimal laser beam divergence, and ease of integration into minimal marketable ablation structures. Furthermore, the ray outline at the base frequency typically resembles a super-Gaussian or Gaussian distribution. Halogen gas-filled laser hollows, as opposed to solid-state quartzes, are the foundation of excimer lasers. Managing problems requires more attention to detail, and laser equipment is quite large. But excimer lasers are far more powerful and usually offer superior definition [45].

There are many benefits of using PLD for thin-film growth such as,

- It is possible to ablate approximately any material or amalgamation of materials due to the flexibility of wavelength and power density.
- The system does not include the laser. Consequently, a significant amount of flexibility in the ablation geometry is achievable.
- A fine control over the growth rate is made possible by the employment of a pulsed laser beam.
- The majority of ablated materials or combinations of materials can result in the congruent transfer of the composition.
- Modulation of evaporated species' kinetic energy to regulate a film's development characteristics and growth modes.

Performing PLD has drawbacks as well. A portion of them are technical, while others are inherent to the process of ablation and the electromagnetic collaboration that occurs among photons and materials.

- Certain plume species have high kinetic energy, which leads to sputtering as well as developing film and substrate surface imperfections [46].
- An angular energy distribution and inhomogeneous vigor dissemination in the laser plume result from an inhomogeneous energy distribution in the laser ray outline.
- Compared to heavier elements, light elements such as oxygen or lithium have differing growth rates and angular disseminations in a spiral. As a result, an additional source such as sufficient background gas or a modified target configuration is needed to complement these components to achieve the intended film composition [47].

8.7 Role of Thin Film in Atomic Layer Deposition

The ALD technique is grounded on the irregular pulsation of originator gases and vapors against the substrate's external surface, monitored by the precursors' surface reaction or chemisorption. In between the precursor pulses, an inert gas is used to cleanse the reactor. The procedure moves forward in several phases when the experimental circumstances are properly adjusted. In these circumstances, each deposition cycle sees a steady rise and consistent increase in thickness. The self-limiting growth process makes it easier for conformal thin films to grow over wide surfaces at an appropriate thickness. Different multilayer structures evolve easily as well. The ALD process is appealing to microelectronics for the production of integrated circuits of the next generation because of these benefits [48]. The technique known as atomic layer epitaxy (ALE) was first developed for the production of amorphous Al_2O_3 insulator films and polycrystalline luminous ZnS: MN films for electroluminescent flat-panel displays. Amorphous oxide films and polycrystalline II–VI compounds were the subjects of the early ALE literature. Epitaxial development of III–V, and II–VI compounds has drawn a lot of attention since 1985. However, the complex surface chemistry in this field prevented any significant breakthrough from being made. Since the middle of the 1990s, silicon-based microelectronics have been the source of a fast-growing interest in ALD. This growth results from integrated circuits (IC) ever-decreasing device size and rising aspect ratios. Furthermore, the necessary film thickness has frequently dropped to a few nanometers or less. Recently, ALD has been discussed in several studies that concentrate on various topics, such as nanotechnology, catalysts, and electronic and optoelectronic materials. The ALD method's fundamentals are discussed in this overview, with a focus on the significance of surface and precursor chemistry. Selected ALD procedures are given as examples, based on recent literature, and designed for films used in microelectronics [49]. Nitride barriers to prevent diffusion, oxide coatings for dielectrics, and metals for metallization are a few examples. Binary or more reactants, also known as precursors, are injected alternately into a reactor at a specific pressure and temperature to deposit a material on the surface of a desired substrate that is also put in one compartment linked to the reactor. We refer to this as ALD or amorphous layer deposition. In contrast to CVD and comparable techniques, which permit the simultaneous injection of precursors, ALD pulses the precursors one after the other [50].

8.7.1 Mechanisms of Surface Reactions

In a typical ALD process, a substrate is sequentially and non-overlapping exposed to two reactants, A and B. Unlike chemical vapor deposition (CVD), where thin-film formation occurs in a steady-state manner, each reactant in ALD interacts with the surface in a self-limited manner, meaning that the reactant molecules can only react with a finite number of reactive sites on the surface. When all of those locations are occupied, the reactor's expansion stops. Only when the remaining reactant molecules have been flushed out of the reactor is reactant B injected. One method of depositing a thin layer is to alternately expose A and B. In the ALD process, it is necessary to specify the dosage multiplied by the length of time a surface is exposed to a precursor as well as the purge multiplied by the intervals between doses that the precursor must exit the chamber. In a binary ALD process, the dose-purge-dose-purge sequence constitutes an ALD cycle. Furthermore, rather than using the concept of growth rate, ALD processes are described in terms of their growth per cycle [51].

8.7.2 *Thermal Atomic Layer Deposition (ALD)*

Temperatures between ambient temperature (– 20 °C) and 350 °C are needed for ligand exchange or surface reactions similar to combustion in thermal ALD. Surface reactions cause it to happen, allowing for precise thickness control regardless of reactor design and substrate geometry (subject to aspect ratio) of the most well-known thermal ALD instances, the synthesis of Al_2O_3 from trimethylaluminum (TMA) and water is one. TMA dissociative chemisorbs on the substrate surface during the TMA exposure, and any leftover TMA is pushed out of the chamber. TMA's dissociative chemisorption leaves an $AlCH_3$-covered surface behind [52].

8.7.3 *Atomic Layer Deposition (ALD) Plasma*

In contrast to thermal ALD, a greater variety of precursors can be utilized in PA-ALD, allowing for the deposition of a greater variety of materials. Additionally, the great reactivity of the plasma classes permits lowering the deposition temperature without sacrificing the excellence of the film [53].

8.7.4 *Geographic Atomic Layer Deposition (ALD)*

A purging phase in progressive ALD isolates the distinct originator and co-reactant dosages from one another in intervals. On the other hand, these gases are given at distinct sites in latitudinal ALD (s-ALD), resulting in their three-dimensional departure. Precursor and co-reactant are continually supplied in atmospheric pressure s-ALD, and a gas curtain keeps them apart to avoid gas-phase reactions. Usually, nitrogen injection and exhaust places make up such a gas curtain [54].

8.7.5 *Photo-Assisted Atomic Layer Deposition (ALD)*

UV light is used in this ALD variant to quicken substrate surface reactions. As in plasma-assisted ALD, the reaction temperature can therefore be lowered. Although the activation is less than with plasma-assisted ALD, it is frequently simpler to regulate by varying the illumination's wavelength, intensity, and time [55].

8.7.6 Metal Atomic Layer Deposition (ALD)

The mandate for copper using an interrelated material and the comparative simplicity of the thermal deposit of copper has drawn a lot of attention to copper metal ALD. Among the first-row transition metals, copper is the most readily reduced metal and has a positive standard electrochemical potential [28]. As a result, many ALD procedures have been created, some of which employ hydrogen gas as the co-reactant. Higher temperatures may cause deposited copper to aggregate, hence continuous films with minimal surface roughness are best achieved by performing copper metal ALD at $\leq$ 100 °C [56].

- **Advantages**

  ALD offers a highly regulated technique for creating a film with an atomically precise thickness. Furthermore, the development of numerous coated arrangements is unpretentious. The apparatus's compassion and precision variety are highly advantageous for entities occupied in the grounds of nanotechnology and microelectronics to yield small however operational semiconductors. Thermochemical preferred catalysts and relatively low temperatures are commonly used in ALD. While functioning with delicate surfaces of substrates, e.g., organic and biological samples, the lesser temperature is beneficial. If the decomposition rate of certain thermally unstable precursors is quite modest, they can still be used [57].

- **Disadvantages**

  Substrates must be extremely pure, which means that enormous expenditures will follow. Although this expense might not seem significant in comparison with the equipment required, it could take many trials to identify the ideal circumstances for the intended product. It could be necessary to eliminate extra precursors from the finished product once the layer has been created and the procedure is finished. Certain finished items have fewer than 1% of contaminants in them [58].

8.8 Conclusion

Advances in material science and technology depend on the functions of Chemical Vapor Deposition (CVD), Physical Vapor Deposition (PVD), Spin Coating, Laser Ablation, and Atomic Layer Deposition (ALD) in the creation of thin films. The range of thin-film technologies is increased by the distinct benefits and particular uses of each technology. CVD is well-known for its capacity to use chemical processes involving gaseous precursors to make uniform, high-quality films. This technique is suitable for the application in the semiconductor and solar industries because it allows for control of the thickness and composition of the thin film. PVD yields a hard coating which improves the surface properties through the evaporation of the materials through physical processes. The spin coating is very suitable for the uniform layers of polymers or sol–gel materials, where the thickness is controlled through centrifugal action. This method is widely used in microelectronics and optics where uniformity is critical. It introduces yet another layer of flexibility and accuracy in the deposition of intricate materials that can be obtained by laser ablation wherein concentrated laser beams are used to remove material from a target. This approach is useful in yielding certain structural and compositional characteristics of thin films. A distinguishing characteristic of ALD is that it can deposit extremely thin films with atomic accuracy, which proves useful in applications where conformal coatings on intricate geometries are required. The self-limiting nature of the ALD process guarantees high consistency and reproducibility, which is very important for nanotechnology and advanced semiconductor fabrication. In a nutshell, the various deposition techniques of spin coating, laser ablation, CVD, PVD, and ALD all have unique advantages that come in handy in being able to fulfill the various technical requirements and serve the purpose of thin-film development. The realization of the desired nature and characteristics of film performance is highly dependent on the choice of proper deposition techniques, thereby setting an emphasis on such techniques in the long run of new materials' and applications' development in many spheres.

References

1. Arunkumar, P., S.K. Kuanr, and K.S. Babu. 2015. Thin film: Deposition, growth aspects, and characterization. *Thin Film Structures in Energy Applications* 1–49.
2. Sequeda, F. 1986. Thin film deposition techniques in microelectronics. *JOM Journal of the Minerals Metals and Materials Society* 38 (2): 55–65.
3. Mavukkandy, M.O., et al. 2020. Thin film deposition techniques for polymeric membranes—A review. *Journal of membrane science* 610: 118258.
4. Morosanu, C.E., *Thin films by chemical vapour deposition*. Vol. 7. 2016: Elsevier.
5. Lugscheider, E., et al. 1996. Comparison of the structure of PVD-thin films deposited with different deposition energies. *Surface and Coatings Technology* 86: 177–183.
6. Panjan, P., et al. 2020. Review of growth defects in thin films prepared by PVD techniques. *Coatings* 10 (5): 447.
7. Hopwood, J.A. 2000. The role of ionized physical vapor deposition in integrated circuit fabrication. In *Thin Films*, 1–7. Elsevier.
8. Baptista, A., et al. 2018. Sputtering physical vapour deposition (PVD) coatings: A critical review on process improvement and market trend demands. *Coatings* 8 (11): 402.
9. Hubler, G., and J. Sprague. 1996. Energetic particles in PVD technology: Particle-surface interaction processes and energy-particle relationships in thin film deposition. *Surface and Coatings Technology* 81 (1): 29–35.
10. Moarrefzadeh, A., and M. Branch. 2012. Simulation and modeling of physical vapor deposition (PVD) process. *WSEAS Trans. Appl. Theor. Mech* 7: 106–111.
11. Shahidi, S., B. Moazzenchi, and M. Ghoranneviss. 2015. A review-application of physical vapor deposition (PVD) and related methods in the textile industry. *The European Physical Journal Applied Physics* 71 (3): 31302.
12. Karamiş, M., and H. Sert. 1998. The role of PVD TiN coating in wear behaviour of aluminium extrusion die. *Wear* 217 (1): 46–55.
13. Gulbiński, W. 2010. Physical vapor deposition of thin film coatings. In *Surface Properties and Engineering of Complex Intermetallics*, 83–92. World Scientific.

14. Liang, F., et al. 2022. A review of thin film electrolytes fabricated by physical vapor deposition for solid oxide fuel cells. *International Journal of Hydrogen Energy* 47 (87): 36926–36952.
15. Velmurugan, R., and B. Subramanian. 2021. *Physicochemical Approaches for Thin Film Energy Storage Devices through PVD Techniques*, in *Management and Applications of Energy Storage Devices*. IntechOpen.
16. Harsha, K.S. 2005. *Principles of Vapor Deposition of Thin Films*. Elsevier.
17. Abreu, C., et al. 2015. Tribological characterization of TiO_2/Au decorative thin films obtained by PVD magnetron sputtering technology. *Wear* 330: 419–428.
18. Ichou, H., et al. 2024. Exploring the advancements in physical vapor deposition coating: A review. *Journal of Bio-and Tribo-Corrosion* 10 (1): 3.
19. Usui, H. 2011. Preparation of polymer thin films by physical vapor deposition. In Functional Polymer Films: 2 Volume Set, 287–318.
20. Rashid, H.U., et al. 2015. Catalyst role in chemical vapor deposition (CVD) process: A review. *Reviews on Advanced Materials Science* 40 (3): 235–248.
21. Pedersen, H., S.T. Barry, and J. Sundqvist. 2021. Green CVD—Toward a sustainable philosophy for thin film deposition by chemical vapor deposition. Journal of Vacuum Science & Technology A 39 (5).
22. Kleijn, C., et al. 2007. Multi-scale modeling of chemical vapor deposition processes for thin film technology. *Journal of Crystal Growth* 303 (1): 362–380.
23. Crowell, J.E. 2003. Chemical methods of thin film deposition: Chemical vapor deposition, atomic layer deposition, and related technologies. *Journal of Vacuum Science & Technology A: Vacuum, Surfaces, and Films* 21 (5): S88–S95.
24. Mathur, S., et al. 2004. Chemical vapor deposition of $MgAl_2O_4$ thin films using different Mg−Al alkoxides: Role of precursor chemistry. *Chemistry of Materials* 16 (7): 1304–1312.
25. Kern, W., and V.S. Ban. 1978. Chemical vapor deposition of inorganic thin films. *Thin Film Processes* 257–331.
26. Ashfold, M.N., et al. 1994. Thin film diamond by chemical vapour deposition methods. *Chemical Society Reviews* 23 (1): 21–30.
27. Ehsan, M.A., et al. 2022. Recent advances in processing and applications of Heterobimetallic oxide thin films by aerosol-assisted chemical vapor deposition. *The Chemical Record* 22 (7): e202100278.
28. Bruno, L., C. Pijolat, and R. Lalauze. 1994. Tin dioxide thin-film gas sensor prepared by chemical vapour deposition: Influence of grain size and thickness on the electrical properties. *Sensors and Actuators B: Chemical* 18 (1–3): 195–199.
29. An, W.-J., E. Thimsen, and P. Biswas. 2010. Aerosol-chemical vapor deposition method for synthesis of nanostructured metal oxide thin films with controlled morphology. *The Journal of Physical Chemistry Letters* 1 (1): 249–253.
30. Fau-Canillac, F., and F. Maury. 1994. Control of the uniformity of thickness of Ni thin films deposited by low pressure chemical vapor deposition. *Surface and Coatings Technology* 64 (1): 21–27.
31. Wang, M., et al. 2017. CVD polymers for devices and device fabrication. *Advanced Materials* 29 (11): 1604606.
32. Yudasaka, M., et al. 1995. Effect of Ni on graphite thin-film formation from organic materials by chemical vapor deposition. *Journal of Vacuum Science & Technology A: Vacuum, Surfaces, and Films* 13 (4): 2142–2145.
33. Petit, R.R., et al. 2021. Atomic layer deposition on polymer thin films: On the role of precursor infiltration and reactivity. *ACS Applied Materials & Interfaces* 13 (38): 46151–46163.
34. Taylor, D. 2015. Vacuum-thermal-evaporation: The route for roll-to-roll production of large-area organic electronic circuits. *Semiconductor Science and Technology* 30 (5): 054002.
35. Abzieher, T., et al. 2021. From Groundwork to Efficient Solar Cells: On the Importance of the Substrate Material in Co-Evaporated Perovskite Solar Cells. *Advanced Functional Materials* 31 (42): 2104482.
36. Eberspacher, C., et al. 2001. Thin-film CIS alloy PV materials fabricated using non-vacuum, particles-based techniques. *Thin Solid Films* 387 (1–2): 18–22.

37. Memarian, N., et al. 2017. Deposition of nanostructured CdS thin films by thermal evaporation method: Effect of substrate temperature. *Materials* 10 (7): 773.
38. Lee, D., and K. Yong. 2013. Non-vacuum deposition of CIGS absorber films for low-cost thin film solar cells. *Korean Journal of Chemical Engineering* 30: 1347–1358.
39. Tyona, M. 2013. A comprehensive study of spin coating as a thin film deposition technique and spin coating equipment. *Advances in Materials Research* 2 (4): 181.
40. Kotsuki, K., et al. 204. The importance of spinning speed in fabrication of spin-coated organic thin film transistors: Film morphology and field effect mobility. *Applied Physics Letters* 104 (23).
41. Acuña, D., et al. 2016. Growth and properties of lead iodide thin films by spin coating. *Bulletin of Materials Science* 39: 1453–1460.
42. Chaudhary, K.T. 2021. Thin film deposition: Solution based approach. In *Thin Films*. IntechOpen.
43. Tyona, M. 2013. A theoritical study on spin coating technique. *Advances in Materials Research* 2 (4): 195.
44. Nisticò, R., D. Scalarone, and G. Magnacca. 2017. Sol-gel chemistry, templating and spin-coating deposition: A combined approach to control in a simple way the porosity of inorganic thin films/coatings. *Microporous and Mesoporous Materials* 248: 18–29.
45. Sneh, O., et al. 2002. Thin film atomic layer deposition equipment for semiconductor processing. *Thin Solid Films* 402 (1–2): 248–261.
46. Yasmeen, S., et al. 2023. Atomic layer deposition beyond thin film deposition technology. *Advanced Materials Technologies* 8 (20): 2200876.
47. Jin, Y., H. Yu, and X. Liang. 2021. Understanding the roles of atomic layer deposition in improving the electrochemical performance of lithium-ion batteries. *Applied Physics Reviews* 8 (3).
48. Zaera, F. 2012. The surface chemistry of atomic layer depositions of solid thin films. *The Journal of Physical Chemistry Letters* 3 (10): 1301–1309.
49. Mondloch, J.E., et al. 2013. Vapor-phase metalation by atomic layer deposition in a metal–organic framework. *Journal of the American Chemical Society* 135 (28): 10294–10297.
50. Kim, K., et al. 2014. Selective metal deposition at graphene line defects by atomic layer deposition. *Nature Communications* 5 (1): 4781.
51. Knisley, T.J., L.C. Kalutarage, and C.H. Winter. 2013. Precursors and chemistry for the atomic layer deposition of metallic first row transition metal films. *Coordination Chemistry Reviews* 257 (23–24): 3222–3231.
52. Dasgupta, N.P., et al. 2015. Atomic layer deposition of metal sulfide materials. *Accounts of Chemical Research* 48 (2): 341–348.
53. Ramos, K.B., M.J. Saly, and Y.J. Chabal. 2013. Precursor design and reaction mechanisms for the atomic layer deposition of metal films. *Coordination Chemistry Reviews* 257 (23–24): 3271–3281.
54. Lu, J., J.W. Elam, and P.C. Stair. 2013. Synthesis and stabilization of supported metal catalysts by atomic layer deposition. *Accounts of Chemical Research* 46 (8): 1806–1815.
55. Hagen, D., M.E. Pemble, and M. Karppinen. 2019. Atomic layer deposition of metals: Precursors and film growth. *Applied Physics Reviews* 6(4).
56. Hämäläinen, J., M. Ritala, and M. Leskelä. 2014. Atomic layer deposition of noble metals and their oxides. *Chemistry of Materials* 26 (1): 786–801.
57. Griffiths, M.B., et al. 2016. Atomic layer deposition of gold metal. *Chemistry of Materials* 28 (1): 44–46.
58. Lim, B.S., A. Rahtu, and R.G. Gordon. 2003. Atomic layer deposition of transition metals. *Nature Materials* 2 (11): 749–754.

Chapter 9
Emerging Trends and Issues Related to Thin Films

Abstract In this chapter, emerging trends and related issues of thin films have been described in detail. With the emerging scope of technology, thin-film growth production is increasing greatly. The recent innovations and improvements influencing the direction of thin-film deposition processes are explored in the chapter on emerging trends and associated concerns in thin-film technology. The growth in acceptance of leading-edge depositions like CVD and PVD is driven by the industry need for materials with improved performance characteristics. For optimized step coverage, adhesion, and film characteristics on challenging substrates, this chapter addresses hybrid approaches that merge the best features of PVD and CVD. New material discussed in the chapter is sophisticated materials, namely nanomaterials and metamaterials. These are fast growing and start being more popular in the thin-film application in several industries, such as electronics, optical, and renewable energy fields. It addresses the aspect of sustainability in thin-film production, considering scalability, cost, and environment-oriented matters. Additionally, the chapter takes into account new technologies, such as machine learning and artificial intelligence, which could be applied to improve deposition procedures and predictive maintenance. Therefore, the chapter exhaustively addresses changing trends and future projections regarding the field of thin films. It advocates for the role of thin films in raising technological applications and solving present-day problems in material science.

9.1 Introduction

A thin layer whose thickness is between a few nanometers and several micrometers is called a thin film. The most ordinary example of a thin film in our house is the mirror. It is a thin reflective coating of metal on the back of a glass to form a reflecting surface. An important part of forming thin layers is nucleation. Nucleation is crucial in making thin layers. Nucleation has been defined as the process whereby a substance or mixture takes on a new thermodynamic phase or structure by self-assembly or self-organization [1]. We can style nucleation by describing the surface process of adsorption, desorption, and surface diffusion. They are deposited on the surface

T. I. Awan et al., *Thin Film Deposition Techniques*,
https://doi.org/10.1007/978-981-96-1364-9_9

of a substrate by different techniques of deposition, such as thermal evaporation technique, physical vapor deposition, chemical vapor deposition, electrodeposition, sol–gel method, sputtering, and laser beam-assisted deposition or electron beam deposition. Molecular beam epitaxy, atomic layer deposition, and the Langmuir–Blodgett method allow to deposit of a single layer of atoms or molecules. In physical vapor deposition (PVD), chemical reactions are not involved [2]. It uses mechanical, electromechanical, or thermodynamic means to form a thin film of solids. In PVD, material source is heated to high temperature either by electron beam or thermal methods in vacuum. Material is vapor transported to the target in a vacuum. The film quality in the case of PVD is not as good as sputtered film (the process of sputtering involves the bombardment of an ion beam on a pre-coated target which makes the beam even denser). Thin films are non-conformal and direct in the case of the PVD process. Quartz balance is a device that can be used to measure the thickness of the film formed by the PVD process [3].

One layer of a conformal thin film is deposited at a time using a gaseous precursor in atomic layer deposition and its sister technology, molecular layer deposition. To achieve complete layer saturation before starting the next layer, the procedure is divided into two half-reactions, run sequentially, and repeated for each layer. As a result, the substrate undergoes a chemical reaction that forms the required composition when the first reactant is deposited, followed by the second reactant. The method can be used at low temperatures, although it is slower than chemical vapor deposition because of the stepwise. Atomic layer deposition on polymeric substrates can transform into sequential infiltration synthesis, in which reactants permeate the polymer and engage with functional groups [4].

Thin film is not a relatively new technology rather it has been used for thousands of years. The earliest historical application of thin films can be traced back to the ancient Assyrians, who utilized tin glazing on brickwork and pottery for decorative purposes. Archeological evidence, including glazed brick panels recovered from sites, such as Nimrud dating back to approximately 890 B.C., attests to this early usage. Over time, various forms of glazes have been developed to address pottery porosity and for aesthetic embellishment. The significance of thin films in optics was recognized in the seventeenth century, notably by scientists like Hooke and Newton. Newton's seminal work in 1675 involved the observation and analysis of colors in thin films of transparent materials, particularly air or water confined between differently curved glass surfaces [5]. His findings, such as the generation of concentric colored rings known as Newton's rings, contributed to the understanding of interference phenomena, despite initial skepticism regarding the wave nature of light. These early investigations not only elucidated the fundamental properties of thin films but also laid the groundwork for their diverse applications in contemporary industrial settings. Thin films exhibit distinctive optical characteristics, alongside mechanical, electrical, and magnetic properties, underpinning a range of modern technologies including antireflection coatings, optical filters, surface acoustic wave devices, electronic components (both discrete and integrated), magnetic data storage systems, and superconductors [5].

Since then, this technology has been widely applied in different squares of life. Not only they are thin films extremely adaptable in terms of their applications, but they also come in a wide variety of forms, everything from optically clear coatings to hard titanium and all in between. Thin films may be mechanical, optical, electronic, magnetic, thermal, or chemical. Mechanical thin films are usually made of materials that are resistant to wear, corrosion, and abrasion. Thin mechanical films can be used in coating applications to offer extra strength against abrasion and friction to almost any surface, including metals, plastics, glass, and so on. Additionally, they provide excellent anticorrosive protection, which makes them useful in a variety of fields and applications, including transportation logistics and aerospace engineering [4]. Typically, materials are coated with optical thin films to give them their proper optical properties. They are among the developments in the field of solar power that allow the manufacturing of flexible, lightweight, and sustainable solar panels. These particular coatings can change color, enhance performance, or boost reflectivity depending on the film's durability and the underlying layer mix. These special thin films offer the potential to assist any product that is exposed to sunshine since they protect against UV radiation and sun-induced fading [6].

In electrical or electronic applications, thin films especially those made of aluminum, copper, and alloy offer greater versatility than their bulkier counterparts. Heat transfer is made possible by these thin films' superior insulation compared to thick film components. The thin coating minimizes power loss and boosts sensor sensitivity when applied to circuitry. They are therefore very compatible with a wide range of surfaces, including semiconductors, insulators, and integrated circuits. When using magnetic qualities in engineering and industrial applications, magnetic thin films offer an alternative to conventional materials. They are extremely thin, frequently having a width of less than one nanometer. They nevertheless retain all of the characteristics of ordinary magnets, namely attraction and repulsion. They also have several benefits, like remarkable durability and resistance to outside interference. These characteristics make them perfect for lengthy life in consumer products or operations with difficult equipment. Various kinds of magnetic thin films find application in automation systems, sensing apparatus, memory disks, and other data storage devices [7].

In several industries, thermal thin films also referred to as insulating thin films help with electrical resistance, heat dissipation, and insulation. The polymers used to create these films have high-performance thermal properties, and they also contain unique additives that strengthen the films' resilience to thermal cycling and other harsh environments. For more flexibility and quicker application periods, thermal thin films can also be applied without the use of textiles or adhesives. In comparison with other materials, thermal thin films produce insulating layers that provide superior airtight sealing while using less electricity. Their immense value for industrial applications stems from this [8].

9.2 Chemical Thin Films

Chemical thin films are ranked among the most interesting and complex materials in modern industry. From corrosion-resistant coatings to electrical conductivity and optically active materials, these thin films form a wide range of qualities. Due to chemical thin films' protection, even simple products can take advantage of added durability. Combining different parts makes it possible to achieve customized compositions unique to the situation. With more research unfolding every day, chemical thin films provide an unlimited number of creative possibilities for engineers and manufacturers alike [9]. Thin films have very small thicknesses (maybe in nanometers), and they are widely applicable to different places. They are formed by techniques, such as CVD, PVD, sputtering, thermal evaporation, pulsed laser plasma deposition, and electrohydrodynamics. From mobile screens to processors of mobile and other integrated circuits and from solar panels to aerospace technology, they are used everywhere. In short, thin film has become an essential technology in the modern era and there are still emerging new trends in thin-film technology [10].

9.3 Emerging Innovations and Trends

Over the past century, thin films have garnered considerable attention from scientists and engineers, becoming a focal point of extensive research and development endeavors. This timeframe has witnessed the emergence of materials science and engineering as a pivotal discipline driving advancements across various fields of applied science and engineering. Within this context, the investigation of thin-film materials has served as a unifying thread in the evolution of the field. This area encompasses a spectrum of materials, including films affixed to comparatively thicker substrates, multilayered structures, patterned films on substrates, and freestanding films. The propulsion toward contemporary engineering practices can be attributed to notable strides in the synthesis, processing, and nanoscale characterization techniques pertinent to these materials. At the outset of the twenty-first century, the United States National Academy of Engineering conducted a project to identify the most significant engineering achievements of the preceding century. This initiative underscored thin-film materials as a cornerstone of advancements in materials technology during the latter half of the twentieth century, impacting diverse domains, such as electronics, computing, healthcare technologies, and optical sciences. The refinement and proliferation of numerous methodologies and techniques for thin-film growth and manipulation have facilitated their widespread adoption and utilization across industries. The early industrial integration of thin-film technology has played a pivotal role in catalyzing further enhancements and innovations in deposition methodologies [11].

Creating highly flexible electronic devices necessitates the use of materials that exhibit exceptional ductility. One suitable approach for achieving this is the deposition of thin metal films onto flexible substrates to form highly malleable interconnects. Thin-film systems typically face a fundamental challenge with ductility, even when comprised of face-centered cubic (FCC) materials that exhibit ductile behavior on a macroscopic scale. Despite this, they can only endure a limited elongation, typically just a few percent, before experiencing failure. An effective strategy for attaining significant elongation in thin films involves depositing them onto a substrate, particularly a flexible one. This method helps to distribute strain across a larger area and mitigate localized necking. Research has demonstrated that film failure occurs when localized deformation leads to detachment from the substrate. In such systems, the presence of a singular defect does not result in catastrophic failure since the substrate carries the load. Additionally, the compound formed by the film–substrate combination better represents potential applications compared to standalone films and offers the advantage of easier handling. In recent times, there has been a growing interest in materials possessing high stretchability and low modulus due to their potential applications in various fields, such as soft robotics, wearable electronics, prosthetics, and artificial intelligence. These soft materials offer numerous advantages including lightweight, affordability, remarkable flexibility, and stretchability [11]. Achieving such properties often involves combining multiple strong or weak interactions, such as hydrogen bonds, ionic bonds, and crosslinking networks. While a low modulus facilitates safe and efficient shape changes, it also tends to result in easy fracture and reduced durability. However, the phenomenon of strain hardening has been identified as an effective strategy for enhancing materials with superior mechanical properties and safeguarding them against damage caused by extensive deformations. Without ductile electronics, such objectives would remain a nightmare. Amorphous silicon (a-Si) has been studied since the 1960s. Because pure a-Si lacks long-range ordering, it has an extremely high carrier recombination rate and dangling bond density, which limits its semiconductor behavior. Under specific manufacturing settings, amorphous silicon was claimed to display semiconductor capabilities between the late 1960s and mid-1970s. Although it was not recognized at first, the presence of hydrogen during the thin film's development resulted in hydrogenated a-Si or a-Si:H, where the hydrogen broke dangling bonds. Bandgaps greater than c-Si (about 1.7 eV) resulted from this decrease in the density of gap states. The first solar cells with an efficiency of up to 2.4% were reported by Carlson and Wronski in 1976, which launched this solar technology and subsequently a burgeoning business. The most common device form was a p-i-n structure, where the produced carriers could survive long enough to be collected because of the "intrinsic or i-layer." Additionally, composition primarily the quantity of hydrogen added to the layer could change the a-Si:H bandgap. This resulted in first to three-junction topologies that might be used with tandem devices and then absorb more of the sun spectrum flowing in, resulting in increased productivity [12]. Over the first ten years, the original 2–5% research efficiencies in single-junction devices more than doubled. The modules greater than 7–8% and verified cell efficiencies above 10% by the 1990s are obtained.

Nanostructured thin films represent a highly active area of research, particularly notable for their applications in fields, such as photovoltaics, photocatalysis, and sensor technologies. The thin films are going to be highly engineered in thickness, crystallinity, porosity, and also optical properties. They may be synthesized on any suitable substrate under a vast methodology array including the sol–gel process, chemical solution deposition, electrochemical etching, and more conventionally through chemical vapor deposition (CVD) and physical vapor deposition (PVD). Commonly, characterization of these films involves an array of surface analysis techniques that include *X*-ray diffraction (XRD), scanning electron microscopy (SEM), transmission electron microscopy (TEM), atomic force microscopy (AFM), ellipsometry, electrochemistry, small-angle *X*-ray scattering (SAXS), reflectance spectroscopy, scanning tunneling microscopy (STM), *X*-ray photoelectron spectroscopy (XPS), secondary ion mass spectrometry (SIMS), electron spectroscopy for chemical analysis (ESCA), *X*-ray topography, and diffusion-ordered spectroscopy nuclear magnetic resonance (DOSY-NMR) [13]. This review accounts for the synthesis and characterization of nanostructured thin films, especially nanocrystalline titanium dioxide and modified silicon, with multiform applications in solar cells, water treatment, water splitting, self-cleaning textiles, sensors, optoelectronic devices, and lab-on-chip systems.

Thin films have emerged as a surface modification technique for permanent implant materials, which tries to improve biodegradation resistance, mechanical strength, and biocompatibility with several other characteristics. Ideally, an implant should provide the characteristics of resistance against degradation, promotion of osseointegration, prevention of bacterial adhesion, and lowering of prosthetic infections. The efficacy of implants in clinical settings hinges upon the interaction between cells and material surfaces, a pivotal aspect. Recent advancements in biomaterials have led to the introduction of various coatings on implant materials, addressing these concerns. These coatings enhance the biocompatibility, mechanical properties, and bioactivity of the biomaterial.

Layered crystalline materials, known as 2D materials, hold immense promise in the realm of nanotechnology, particularly for next-generation electronics. Given their atomic thinness, preserving the original properties of these materials during fabrication is crucial for realizing devices based on them. Atomic layer deposition (ALD) emerges as an ideal method for precisely adding materials at an atomic scale to nanomaterials. However, the effectiveness of ALD on 2D materials is heavily influenced by the surface characteristics of these materials due to the surface-sensitive nature of ALD reactions. While the majority of materials exist in three-dimensional (3D) forms, we commonly understand the concept of two-dimensional (2D) shapes. For instance, a sheet of paper appears to possess an ideal 2D shape due to its minimal thickness, although other 2D shapes may not appear as ideal. Similarly, in the case of 2D materials, their thickness can approach the scale of atomic layers, while the other two lateral dimensions do not, akin to a sheet of paper. Specifically, 2D materials are layered substances where robust in-plane crystalline atomic layers are connected by weak out-of-plane van der Waals forces, allowing individual atomic layers to be readily separated into single or multiple layers. Thin films are being deposited on such 2D materials nowadays [14].

9.4 Ductile Electronics

Ductile electronics are those that can bear mechanical deformation while keeping functionality. Being the opposite of stiff traditional electronics, ductile electronics bend, stretch, or twist and hence find applications in fields where flexibility and durability are required. In material science, "ductile" means that the material can be deformed plastically to a large degree before rupture, which in turn again may become an advantageous feature, for example, in flexible or wearable technologies [12].

Therefore, the importance of ductile electronics is in their transformative abilities, which will cut across many industries, from flexible displays to wearable health-monitoring systems, smart textiles, or environmentally responsive sensors. These flexible devices thus provide improvements over traditional electronics in enhanced comfort for wearable devices from increased resilience against mechanical stresses and the ability to be contoured around non-traditional shapes and surfaces [15].

9.4.1 Materials Used in Ductile Electronics

Ductile materials are quite instrumental in developing flexible, wearable, and stretchable electronic devices. Here is an overview of some key materials of ductile electronics:

Conductive polymers, like PEDOT, can be put into the region. Poly (3, 4-ethylenedioxythiophene) polystyrene sulfonate is an organic materials that conduct electricity. The combination of electrical conductivity and mechanical flexibility makes these materials highly valued. Unlike traditional conductive materials, conductive polymers are light and processable into thin films, hence perfect for many modern applications that require both conductivity and pliability. Extensive usage is in flexible displays, sensors, and organic solar cells. PEDOT: Among others, it is outstanding in its excellent film-forming ability, stability, and high conductivity, mostly considered one of the key materials for flexible electronics [16].

Among metal nanowires, the use of CNTs (carbon nanotubes) and AgNWs (Silver Nanowires) in flexible electronic materials has been realized due to their excellent electrical and mechanical capabilities. AgNWs have higher conductivity and better optic transparency; hence, they can be applied potentially in both transparent conductive films and flexible touchscreens. On the other hand, carbon nanotubes have demonstrated outstanding mechanical strength and electrical properties that have found their applications in the making of flexible transistors and sensors. The nanomaterials of the present invention will lead to the progress of components for

Green electronics whose functionality is intact upon bending or stretching, a central factor to be taken into consideration while formulating the next range of flexible devices.

Graphene is a single atomic layer of carbon atoms in a hexagonal lattice. It has exceptionally high electrical conductivity, mechanical strength, and flexibility. Unique attributes of graphene help it capture colossal shares in flexible battery sensors and transistors. The graphene can be formed into a very conductive film where flexibility is needed especially for developing those elements that need to form durable stretchy parts. Other 2D materials that have been suggested for applications in flexible electronics are the molybdenum disulfide and MoS_2. MoS_2 itself has a few unique attributes to its electronic and optical properties that make it very useful in applications, such as flexible semiconductors and optoelectronic devices [12].

These can be related directly to flexible electronics, such as silicon and polyurethane. The materials involved in this category are designed to bear stretch, bend, or twist without any loss in structural integrity or functionality. Of these, silicon is particularly preferred because of its excellent properties in elasticity and biocompatibility, which can meet all medically related requirements, such as in the case of wearable health-monitoring devices or electronic skins. Because it is both strong and flexible, polyurethane finds use in flexible connectors and protective coatings for electronic components. It is possible to create foldable electronics that bend and take on different curvatures and movements in all types of shapes, basically with such substrates, and this becomes highly desirable in wearable and portable devices [15].

Elastomeric composites borrow stretchability properties from elastomers and conductivity from metals or carbon-based materials. Their design details the process of stretching these materials with recovery in shape while sustaining electrical conductance, involving very much the point being taken into consideration in flexible electronics applications. Conductive elastomers usually combine metal particles or carbon nanomaterials and polymer matrices, finding major applications in flexible circuits and wearable devices, including soft robotics. These materials lie at the core of developing flexible high-performance electronic systems because they bear mechanical deformation without electrical compromise.

The name organic semiconductor quite literally means carbon-based material with semiconductor properties. Some examples of organic semiconductors include pentacene and P_3HT (poly (3-hexylthiophene)). These materials are flexible, with low processing temperatures compared to traditional inorganic semiconductor materials. Organic semiconductors can be used for organic field-effect transistors (OFET), organic light-emitting diode (OLED), and organic photovoltaic cell (OPV) applications. Because of their light weight and ability to result in films of thin and flexible nature, they are most suitable for flexible and lightweight electronic devices [17].

9.4.2 Applications of Ductile Electronics

Ductile electronics will become a completely new paradigm of electronic device design for flexibility, durability, and adaptability that goes beyond state-of-the-art handling in rigid electronics. Thus, such kinds of cutting-edge materials will be very important for industries in the future by continuing to enhance user experience, functionality, and potential behind electronic devices with innovative applications. From wearable technology to flexible displays, ductile electronics will redefine what your interaction is going to be with consumer electronics in daily life and let in a view into a future where devices can blend seamlessly into surroundings and activities.

One of the high-growth avenues where the application of ductile electronics can penetrate exists within wearable health-monitoring systems, coupled together by vital sign trackers, such as heart rate, temperature, and blood pressure, among others. These devices get fixed to clothes or are incorporated into several wearable accessories by flexible sensors and several conductive materials that become conformable to body shape, ensuring comfort and continuous contact with the skin to give genuine data for medical diagnosis and healthcare management [18].

Ductile electronics extend beyond conventional functionalities, thus opening a pathway to smart textiles. These fabrics have integrated sensors, actuators, and communication modules into their structure. Temperature control, moisture detection, or interaction with clothes is done through fabrics. It is the application of ductile materials that allows SMART textiles to have enough flexibility and mechanical resistance to sustain applications in domains related to sports performance monitoring, military equipment, or fashion clothing with integral electronic functionality.

Ductile electronics now make possible flexible displays of technologies, such as organic light-emitting diodes and electronic paper. Thin, flexible layers of light-emitting organic material, OLEDs are subjected to an electric current to produce illumination, which provides vivid colors with high contrasts. The e-paper displays in use with e-readers are not much different from real ink on paper. They have the benefit of low power consumption and readability in any light in handy use. Such ductile materials do not damage the electronic parts inside; these displays can bend, roll, and fold. This is where the road is going: foldable smartphones, rollable TVs, and portable e-signage [18]. Ductile electronics powering flexible displays have several advantages in many industries. For advertising, flexible screens are arched and lively display systems that capture more eyes toward them and engage users. In consumer electronics, the strong point of lightness is already among the major driving forces of device portability and user experience for flexible displays. Films on Kapton HN exhibited distinct cracking behavior. This substrate's surface had rolling scratches that were inherited from substrate manufacture, as well as dirt particles that adhered to the surface. The statistical distribution of these inherent faults had an impact on the film-breaking behavior at least as strong as composition and heat treatment did, which significantly changed the cracking behavior [19].

9.5 Applications in Various Fields

Extend into automotive interiors, whereby mounting on curved surfaces and dashboards will provide advanced navigation and entertainment systems.

Ductile electronics contribute to the development of flexible environmental sensors for the sensing of a variety of surroundings about air quality, humidity, and pollutant contaminants. They are greatly enforced in wearables or places where one cannot shoehorn traditional rigid sensors. That is because the flexibility of the ductile material in conformance to irregular surfaces makes it resistant to environmental stresses, which enables environmental sensors to collect data reliably for applications of environmental monitoring and research.

These flexible biomedical sensors are developed based on ductile electronics for use in healthcare and medical diagnostics applications. Such biochemical sensors could be inlaid either in wearable devices or directly on the skin and hence continuously observe certain physiological parameters like glucose levels, oxygen saturation, or muscle activity. The flexibility and biocompatibility of ductile materials ensure comfortable and continuous monitoring, avoiding skin irritation or discomfort hence, finding an application in long-term health [20].

Ductile electronics will therefore play a key role in making the development of flexible solar cells possible, which can be bonded onto many surfaces, thereby turning sunlight into electrical energy for use with portable supplies by a great deal of wearable electronics and remote sensors. Flexible solar cells, compared with their traditional rigid photovoltaic panels, will bring an added weight, flexibility, and facile integration advantage by being based on ductile materials; therefore, they would probably open up new areas for the application of solar energy harvesting in urban and off-grid environments. Exploiting ductile materials to make mechanical vibrations or movements is get converted to electrical energy. These generators therefore can become fitted inside a shoe, a bridge, or even any industrial machinery to collect the ambient mechanical energy and store it for the supply of power to small electronic devices or sensory equipment. Compared to this application, the ductile materials allow the piezoelectric generators to have a longer time of cyclic mechanical stress and deformation with continuous energy harvesting, hence reducing maintenance time. The numerous applications harvested energy has in these MEMS devices include powering wireless sensors that monitor the status of structural health in smart buildings or bridges, and powering wearable technologies toward the enhancement of energy efficiency [21].

9.6 Thin-Film Nanostructures

Nanostructured thin films represent a remarkable advancement in material science, leveraging nanotechnology to imbue thin layers of materials with extraordinary properties and functionalities not achievable in conventional films. These nanostructured films exhibit features, such as nanoparticles, nanorods, nanotubes, nanowires, or nanoporous networks, endowing them with enhanced strength, hardness, corrosion resistance, thermal stability, optical clarity, electronic conductivity, energy storage capabilities, and sensor sensitivity [22].

9.6.1 Nanorods

In the realm of nanotechnology, nanorods represent a specific morphology among nanoscale entities, characterized by dimensions typically falling within the range of 1–100 nm. These structures can be synthesized from either metals or semiconducting materials. Typically, nanorods exhibit standard aspect ratios, ranging from 3 to 5, denoting the ratio of their length to width. Nanotubes represent a distinctive category of materials wherein their properties are influenced not solely by their composition but also by their geometry. Factors, such as diameter, number of walls, length, chirality, van der Waals forces, and overall quality, collectively determine the characteristics and functionality of nanotubes [23].

9.6.2 Nanowire

A nanowire denotes a nanostructure characterized by a diameter on the order of a nanometer ($10-9$ nm). Typically, nanowires possess thicknesses or diameters of tens of nanometers or less, with lengths varying. At this scale, quantum mechanical effects play a significant role, leading to the occasional reference to nanowires as "quantum wires." Various types of nanowires have been developed, including metallic (e.g., Ni, Pt, and Au), semiconducting (e.g., Si, InP, and GaN), and insulating (e.g., SiO_2 and TiO_2) variants [24].

9.6.3 Nanoporous Organic Polymer Networks

Nanoporous organic polymer networks represent a category of materials composed exclusively of lighter elements from the periodic table. These materials hold promise for various applications, such as storage, separation, and catalysis. In this overview, we discuss the different classes of Nanoporous polymer networks, covalent organic frameworks, hyper-cross-linked polymers, conjugated microporous polymers, and polymers of intrinsic micro porosity. This has allowed a widening set of synthetic pathways which enable to prepare a very large variety of polymer networks from crystalline to amorphous types. Furthermore, these materials are also modular, and hence, inclusion of a wide variety of functional groups is also quite easy [25].

9.7 Fabrication Techniques

The physical properties of nanowires are governed by parameters, which include their morphology, diameter-dependent bandgap, and carrier density of the material state. Since nanowires tend to display aspect ratios (length-to-width ratio) in the order of 1000 or more, nanowires are generally considered to be 1-D materials. One of the most fundamental techniques for making these nanostructured thin films is physical vapor deposition—sputtering technology. In PVD, energetic particles like ions or electrons bombard a target material causing atoms and/or molecules to egress from the target surface. These ejected species then condense on a substrate, forming precise films.

All this enables target material to be engineered, plasma parameters manipulated, or self-assembled monolayers introduced such that tailoring growth of nanostructured thin films with desirable characteristics is feasible. Techniques, such as grain boundary engineering, could be utilized for creating nano-twinned ultra-fine grain copper films, well known for their remarkable thermal stability and corrosion resistance. Similarly, dual-phase nanostructuring yields crystalline metal and amorphous metallic glass nanocomposites with superior strength and toughness [26].

9.8 Advancements and Challenges in Nanostructured Coatings

Thin film- and coating-based techniques are pivotal in numerous emerging nanotechnologies. Some nanostructured coatings exhibit properties that defy current explanations or models, while exciting theoretical concepts remain untestable until the requisite coating structures are engineered. Consequently, this field is dynamic, challenging, and expanding. Non-vacuum techniques encompass electrolytic and chemically deposited layers, polymer coatings, and dip or spin coating dense layers of preformed nanoparticles, potentially aided by chemical modifications inducing self-assembly into patterns. Nanostructured films containing precisely engineered arrays of nanoholes can also be generated within otherwise thin continuous layers post-deposition, either via standard lithographic etching or by depositing onto a controlled planar array of nonmetallic nanoparticles, subsequently removed by selective chemical etching or mechanically. Notably, routine vacuum deposition techniques, such as sputtering, can also yield nanostructures depending on deposition conditions and materials employed. Consequently, film material properties may significantly deviate from bulk values, even with a small volume fraction of nanovoids and tightly packed nanograins or columns. Often, a hierarchy or multiple types of nanostructures coexist in thin films, influencing the properties achieved by deliberately engineered nanofeatures. For instance, the electromagnetic properties surrounding 20–40 nm grains may affect the overall response to the presence of nanoholes, with occasional fine voids around 0.5 nm wide. Similarly, isolated nanoparticles and nanocolumns may possess influential substructures. To accurately model the influence of larger nanofeatures, any produced models must consider coexisting smaller nanostructures or other effects causing deviations from bulk values [27].

9.9 Evolution and Applications of Nanostructured Thin Films

Until the mid-1980s, the presence of columnar nanostructures, often accompanied by voids, in vacuum-deposited thin films was predominantly viewed as a hindrance despite its scientific interest and prevalence. The surge in adoption of ion-assisted deposition techniques around 1990 stemmed from the imperative to eliminate these columnar voids, ensuring stability and durability in precision optics. Concurrently, energetic deposition methods, such as cathodic arc and unbalanced magnetron sputter deposition, were discovered to yield densely packed arrays of nanograins and nanocolumns, nearly reaching bulk density. This realization underscored the mechanical advantages offered by some nanostructured films, including enhanced hardness and scratch resistance. A recent breakthrough involves the utilization of nanocomposite films like TiN nanoscale inclusions within Si_3N_4 matrices (referred to as nc-TiSiN), achieving super- and ultra-hardness comparable to diamond. This burgeoning field of nanostructured films, akin to diamond films, holds significant promise for optics, where resilient surfaces are paramount. While some nanostructured films require protective layers to enhance their robustness, these protective coatings also influence the resulting optical characteristics. However, the majority of these films exhibit considerable strength and resilience [28].

Unlike the 1980s, nanostructured thin films are now acknowledged as a rich source of novel and practical opportunities across various applications, notably in optics, mechanics, chemical and biochemical sensing, and catalysis. Given their diverse functionalities, modeling plays a crucial role, yet current design tools have limited applicability compared to classical thin-film systems. Our interest in columnar thin films stemmed not only from their potential for innovative applications but also because the novel structures they presented offered an excellent platform for developing and critically assessing effective medium models. These models are instrumental in describing the effective optical response, or relative permittivities, of the overall structure [29].

9.10 Re-evaluating Effective Medium Concepts in Metal-Containing Nanostructures

Recent advancements in metal-containing nanostructures prompt a re-examination of the concept and scope of effective mediums. Controversies surrounding the possibility of effective mediums with apparent negative refractive indices underscore this need. Disagreements have arisen as physicists, both supporting and opposing, have treated these nanostructured materials as effective mediums in situations where a homogeneous representation of the composite material is debatable.

In operational terms, if an incident plane waves onto an inhomogeneous medium and produces a plane-reflected wave in the far-field zone, the medium can be represented by effective refractive indexes. Similarly, if the medium transmits light, a plane-transmitted wave should also emerge. Conceptually, the effective medium representation elucidates how electromagnetic energy appears to traverse a composite medium rather than its actual traversal. Consequently, uniform reflected and transmitted plane waves, observed at distances much larger than the wavelength λ_0 in vacuum, appear to emanate from a homogeneous material. Intensities and phases of these waves can then be utilized with standard thin-film models to determine the effective optical constants. Within the near-field zone and inside the composite material, energy flows exhibit inhomogeneity but on a scale significantly smaller than the vacuum wavelength λ_0.

9.11 Diverse Applications of Nanostructured Thin Films in Modern Technology

The potential applications of nanostructured thin films span various industries. In optics, they find use in antireflective coatings, filters, polarizers, and holographic materials. In electronics, they contribute to the fabrication of transistors, diodes, capacitors, and memory devices. Energy storage systems benefit from their inclusion in batteries, supercapacitors, and fuel cells. Furthermore, in sensor technology, they plays a crucial role in biosensors, gas sensors, and pressure sensors.

Antireflection coatings serve to diminish the reflection of one or multiple wavelengths, commonly applied on lens surfaces to minimize light loss. A straightforward approach involves designing a coating to mitigate reflection at the interface between two materials by introducing an additional material for light interaction. This strategy reduces the overall reflection coefficient by enabling light reflection from two interfaces, each characterized by a smaller disparity in refraction indices compared to the original interface. Thin-film polarizers (TFP) serve to differentiate the polarization directions of light (s- and p-polarization). When positioned beneath the Brewster angle, specific to the material, the TFP achieves complete elimination of the p-polarized component [30].

Thin-film memory, created through a government-funded initiative by Sperry Rand, emerged as a rapid alternative to magnetic-core memory. Rather than assembling individual ferrite cores on wires, this innovation involved depositing 4 μm thick dots of perm alloy, an iron-nickel alloy, onto small glass plates via vacuum evaporation methods and a mask. Subsequently, drive and sense lines were incorporated using printed circuit wiring over the alloy dots. While enabling remarkably swift access times of approximately 670 ns, this technology proved prohibitively expensive to manufacture.

Nanoelectronics is a consequence of nanostructured thin films. Nanoelectronics involves the utilization of nanotechnology in the development of electronic components, encompassing a wide array of devices and materials. What distinguishes these components is their incredibly small size, necessitating extensive exploration of interatomic interactions and quantum mechanical properties. Among the candidates are hybrid molecular/semiconductor electronics, one-dimensional nanotubes/nanowires, such as silicon nanowires or carbon nanotubes, and advanced molecular electronics [31].

Critical to nanoelectronics are devices with dimensions typically ranging between 1 and 100 nm. Notably, recent advancements in silicon MOSFET (metal–oxide–semiconductor field-effect transistor) technology have reached this scale, exemplified by nodes, such as 22 nm CMOS (complementary MOS) and subsequent generations like 14, 10, and 7 nm FinFET (fin field-effect transistor). Considered by some as disruptive technology, nanoelectronics presents a departure from traditional transistors due to the significant differences observed in present candidates.

As we continue to integrate PVD sputtering technology with nanotechnology, the future of nanostructured thin films appears exceedingly promising. These materials hold the potential to revolutionize industries by offering novel materials with unprecedented properties and functionalities, paving the way for innovation and advancement across various sectors.

9.12 Thin-Film Biocompatible

Biomaterials serve as substitutes for damaged biological tissues, aiming to restore structural integrity through a careful balance of design, material characteristics, and physiological compatibility. Over the next few decades, research in biomaterials is expected to continue expanding, with a focus on enhancing properties and performance. Key requirements include biocompatibility, inertness to bodily environments (such as pH 7.4 and 37 °C temperature), nontoxicity, and non-carcinogenicity. These materials must endure in the body without adverse reactions or degradation, necessitating a combination of favorable material and biological properties tailored to specific applications. The interaction of biomaterials within the human body environment is a crucial consideration. Established biomaterial categories include metals/alloys, ceramics, and polymers, each with unique advantages and limitations in various applications [32].

9.12.1 Key Considerations in the Design of Biomaterials

The design of biomaterials is tailored to their specific intended functions, with a focus on ensuring prolonged, safe, and effective performance without inducing adverse reactions. Key criteria for optimal biomaterials include excellent biocompatibility, enhanced resistance to corrosion and wear, and improved osseointegration capabilities. Biocompatibility refers to a material's ability to elicit an appropriate response from the host under specific conditions. It entails ensuring that the biomaterial interacts with tissues without being rejected and, in some cases, undergoes controlled degradation within the body. An ideal biomaterial should possess properties that align with those of damaged tissues, facilitating strong bonding at the biomaterial–tissue interface. Moreover, it should not impede normal cellular functions and must be free from toxicity. Additionally, the material should facilitate protein adsorption, cell adhesion, and growth within its environment. The success of biomaterials hinges on their compatibility with the human body's response to implants, which serves as a measure of biocompatibility.

Materials intended for implantation must be devoid of allergenic and toxic components and resistant to corrosion. Therefore, alloys are carefully crafted with compositions that are inert in bodily fluids. Furthermore, surface modifications are implemented to enhance the material's immunity within the body environment, thereby mitigating corrosion and enhancing biocompatibility. Osseointegration, the vital connection between bone and implant, facilitates healing and the generation of new bone tissue, crucial for bio-implants. When fibrous tissue develops around the implant, it hampers bonding with nearby tissues, leading to micromotion and potential implant loosening. Hence, achieving an optimal surface on implants is imperative for secure attachment to adjacent bone. Surface chemistry and topography play significant roles in fostering favorable osseointegration [33].

Surface properties of biomaterials, encompassing both chemical and physical characteristics, dictate the biological response of biomedical devices. Customizing these properties through modifications offers a practical approach to enhancing tissue–material interfaces and mitigating adverse reactions. The material's surface significantly influences its biological response within the body environment, often necessitating different surface properties than those of the bulk material. For instance, bone implants require enhanced bone formability for optimal biological integration, while blood-contacting devices like heart valves prioritize blood compatibility. Surface modification techniques aim to retain excellent bulk properties while improving necessary surface properties, such as resistance to corrosion, wear, and biocompatibility.

The response of biomaterials' biocompatibility hinges on surface properties. While significant strides have been made in biomedical research, achieving the ideal biomaterial that fulfills all requirements remains a challenge. Surface modification of implants allows alteration of composition, structure, and morphology, typically categorized into methods for improved bone bonding and accelerated bone healing. Enhanced bone-bonding methods alter surface topography to mechanically interlock the implant with bone, increasing surface area and energy to enhance protein absorption, cell proliferation, and osseointegration. Coating implant surfaces with inorganic bone materials enhances bone-forming cells' activity and biochemical interlocking with bone [34].

Efficient coating techniques, altering morphology, structure, and composition, are employed for surface modification, including thin-film coating methods. Various coatings, such as transition nitride coatings, thin-film metallic glasses (TFMGs), diamond-like carbon (DLC), and bioceramics, are utilized to improve biocompatibility, corrosion and wear resistance, and bioactivity in biomedical applications. This review focuses on discussing these biocompatible coatings and their mechanical, corrosion, and biological properties relevant to biomedical applications.

Scientists are drawing attention to nitride coatings due to their advantageous properties, including outstanding resistance to corrosion, hardness, ductility, and biocompatibility. Binary or ternary nitrides of transition metals exhibit promising mechanical properties, corrosion resistance, as well as tribological and biocompatibility attributes. Coatings, such as TiN, TiAlN, ZrN, TaN, NbN, and VN, are considered suitable for implants as protective layers, enhancing their lifespan by minimizing wear and corrosion.

9.12.2 Assessing Biocompatibility and Biological Performance of Nitride Coatings in Biomaterials

Biomaterials are closely integrated with living tissue, making it crucial for implanted materials to avoid inducing adverse reactions in the human body. Thus, the surfaces of bio-implants must be nonallergenic and devoid of toxic elements. Biocompatibility of hard coatings is assessed through various biological studies, including cytotoxicity for tissue compatibility, hemocompatibility and hemolysis for blood compatibility, and antibacterial activity through in vitro cell line studies, and in vivo animal studies. For instance, Antonio et al. evaluated the in vivo antibacterial activity of titanium nitride (TiN) coated onto titanium by fixing coated and uncoated acrylic devices in the molar-premolar region of volunteers' jaws, revealing a significant reduction in bacterial presence on TiN surfaces after 24 h.

Various cell lines were used to study the biocompatibility and cytotoxicity of nitride coatings, demonstrating good viability and proliferation compared to controls, along with excellent adhesion and preservation of cell morphology. Brunello et al. compared fibroblast proliferation and bacterial biofilm inactivation between TiN and ZrN coatings, showing effective bactericidal activity against multiple strains. Li et al. demonstrated the biocompatibility of tantalum nitride (TaN) through spindle-shaped bone marrow stem cell morphology and enhanced cell adhesion and spreading.

The biological activity of orthopedic implants relies on osteogenesis differentiation, demonstrated through alkaline phosphatase (ALP) activity and calcium nodules indicative of mineralization. TaN coatings exhibited increased ALP activity and calcium nodules, indicating enhanced in vitro osteogenic activity.

Blood compatibility (hemocompatibility) of implants is crucial, with nitride coatings showing reduced platelet activation, indicating their hemocompatible nature [35].

9.12.3 Enhancing Biocompatibility and Functionality of Biomaterials

Surface properties of biomaterials play a significant role in bone formation enhancement or inhibition, with optimization necessary to mitigate adverse effects. Coating surfaces with TiN and thick TiO_2 can reduce metal ion diffusion, potentially improving in vivo applications. In vivo, biocompatibility studies on TiN-coated specimens implanted in various animal models confirmed nontoxicity and bone formation without adverse reactions.

Recent research has prioritized enhancing the biological properties of implant devices by developing a bioactive interface using bioceramics between tissue and metal implants. Bioceramics serve a unique role as biomedical materials, being both bioactive or bioresorbable and nontoxic. They play a crucial role in bone healing and uncemented implant fixation through their composition and coating methods. Utilizing bioceramics as overlays can improve the biocompatibility of metal implants. Acting as resorbable lattices, bioceramics provide a temporary framework that the body degrades and replaces, facilitating tissue regeneration. Their chemical and thermal stability, coupled with good mechanical properties like strength and wear resistance, position them as promising materials for medical implants [36].

9.13 Advanced Applications of Thin-Film Coatings in Medicine

Pharmaceutical scientists are employing thin films as a versatile drug delivery method across various biological systems. This coating technology has yielded several advantages:

- Enhanced drug release (improved onset and efficacy).
- Improved drug stability.
- Controlled release (particularly useful for delayed release).
- Minimized side effects.

The application of thin films on the surfaces of surgical tools, such as blades, has demonstrated improved cutting properties while lowering the likelihood of post-surgical complications. Similar to the advantages seen with implants and drug delivery systems, thin-film coatings on surgical instruments are anticipated to offer the following benefits:

- Enhanced biocompatibility.
- Increased durability and resistance to wear.
- Decreased friction.
- Improved resistance to corrosion.

These benefits have been validated and quantified through the examination of the optical properties of glass and other surfaces using scanning electron microscopy and atomic force microscopy. These methods allow for the calculation of surface roughness, friction, and other indicators of success. As technological progress marches forward, the medical industry stands poised for a wave of new applications. These advancements hold immense potential to enhance patient outcomes, extend longevity, and elevate quality of life within the healthcare sector. With the rise of nanotechnology, the utilization and refinement of thin-film coatings will become increasingly vital due to the incredibly small surfaces and minuscule thickness of devices [37].

9.14 Thin-Film Deposition in 2D Materials

Materials that are only one or two atoms in thickness but larger in all the other dimensions are known as two-dimensional materials. The size of the 2D materials with which scientists work is usually of the order of several square micrometers; they cannot be seen with the naked eye but can be seen using a type of microscope perhaps used in your high school science classes. Scientists are working with a combination of lab-synthesized materials, like CeSil made of cerium, silicon, and iodine along with naturally occurring materials like graphene, a type of ultra-strong carbon first isolated at Columbia in 2004. They typically start working with these materials in three dimensions and then reduce them to two to run tests and see what physical properties like superconductivity or magnetism might emerge when the materials get atom flat. Although the quality of these is currently lacking, scientists are trying to find new ways of creating 2D materials directly from scratch, bypassing the peeling down from 3D.

2D materials, like graphene and TMDs, become imperative to transcend the limitations in power and performance estimates of silicon technologies. It opens up new nanoscale or atomic-scale ways of designing and operational logics of transistors where monolithic integration in 3D will be feasible, along with atomic thin channel transistors realized [38].

9.15 PVD and CVD Fabrication Methods

Physical vapor deposition is a technique by which thin films get deposited onto a substrate from source material vaporized in form. This forms the base for other techniques, such as thermal evaporation and sputtering. As indicated, the case of thermal evaporation involves heating the source material in the solid lumps or pellet form until it gets vaporized; thereafter, the vapor formed is taken onto the substrate and undergoes condensation hence resulting in a thin film. In the case of sputtering, ions are generally used to bombard the target material with such high-energy particles. The impact from such a collision ejects atoms from the target onto the substrate. Different methods of PVD are assigned to high-purity film preparation with good adhesion, which again is related directly to various advanced applications. Although very well suited for coating 2D materials, PVD methods could give ultra-pure films with very good adhesion. However, homogeneous depositions onto 2D materials might be rather challenging due to their intrinsic thinness and defect formation during deposition. Although all the odds are really against it, PVD remains one of the widely used techniques to coat 2D materials, principally for applications that involve high film quality with precisely controlled thicknesses. Successful applications of PVD in 2D materials include electronic devices and sensors where the quality and uniformity of thin films shine through [39].

CVD (chemical vapor deposition), as mentioned earlier, is a thin-film deposition process conducted through the chemical reactions of precursor gases with heated substrates. Thermal CVD and plasma-enhanced CVD are two fundamental variants, both offering alternative mechanisms for deposition. In thermal CVD, the energy driving the chemical reactions leading to the formation of a film comes from high temperatures, while in plasma-enhanced CVD, the plasma provides a route for lowering the requirements of temperature while giving way for higher reaction rates. Such flexibility makes CVD a potential tool for the production of high-quality and homogeneous films over large areas, especially 2D materials. CVD will provide controlled growth of thin films on 2D materials and ensure the uniformity of thickness for the film, along with perfect control over thickness. Tuning conditions of the reaction, especially temperature and pressure under which gases flow, will enable one to conduct a fine adjustment of films' properties. This certainly will be connected with optimizing the performance of 2D materials for various application fields. For instance, the produced quality and uniformity of CVD-grown graphene and TMDs have been proved to be one of the critical materials that would be applied in both electronic and optoelectronic devices.

It is shown that atomic layer deposition works quite effectively for the conformal coating of 2D materials and improves their functional properties. This technique ensures the deposition of thin films uniformly even on complex structures with a high aspect ratio. This will be very useful in the development of advanced electronic, optical, and catalytic devices. While high performance and durability are delivered in a lot of state-of-the-art applications by 2D materials-based devices, this technique has become Dudley-friendly, allowing the deposition of pinhole-free ultrathin films.

On the other hand, high-quality dielectric layers for transistors or protective coatings for energy storage devices have been realized using ALD [40].

9.16 Applications

How thin films are to help in enhancing efficiency and performance for one of the electronic device building blocks, transistors, is very critical. Thin films on 2D materials, such as graphene and molybdenum disulfide, exhibit high electrical properties, including elevated electron mobility in transistors, which ensures low power consumption. That is to say, it can write film corresponding to the film; besides, the thickness of the film can be fine-tuned with atomic precision in trying to realize increased speed and energy efficiency for devices dependent on transistors. This can thus lead to a realization of a smaller electronic device that boasts better performance features.

High-frequency Devices and Sensors: Thin-film architectures based on 2D materials will remain front-runners in the implementation of high-frequency devices and sensitive sensors. For instance, several "weird" properties in 2D materials can be combined; for example, their high conductivity and mechanical flexibility could accomplish many applications. For example, few-layer-thick graphene film will help to realize high-frequency transistors working in the terahertz range, hence providing ultra-high rates of data transfer and processing. Thin film increases the surface area for its interaction with target molecules; therefore, the sensors become much more sensitive and selective. Such materials can function in a great variety of different fields: from environment monitoring to medical diagnostics [41].

Optoelectronic thin-film technologies have been focused on enhancing the efficiency and sensitivity of photodetectors and light-emitting devices. Any 2D material thin film, like TMD, will naturally show rather nice features of light absorption and emission, thus presenting compatible materials for dealing with these phenomena. For example, MoS_2 thin films have applications in a photodetector requiring high responsivity and fast response time for imaging and communication technologies. More performances can be improved in their application by taking thin films of such 2D materials at the forefront, like LEDs with efficient charge transport and light emission to eventually give brighter, more energy-efficient lighting.

Thin-film technologies of 2D materials could be assimilated and cross-fertilized into a huge base of traditional and new optoelectronic applications. On the other hand, integration might realize hybrid devices possessing the finest properties of traditional materials but combined with 2D materials. For instance, the effectiveness of silicon-based photonic devices can be enhanced by incorporating thin films based on 2D materials. Such synergy just opens the door to a plethora of new innovative applications in displays, solar cells, and optoelectronic systems and makes technology evolve.

Their importance is underlined for thin films in the realization of energy storage systems, mainly concerning batteries and supercapacitors. Thin-film electrodes developed on 2D materials, such as graphene or TMDs, provide a large surface area endowed with very nice electrical conductivity, impressing the notions of effective charge storage and transfer. Their energy density and cycle performance are good compared to bulk forms, which turn out to be powerful and long-lived when it comes to energy storage. It can be said to be an advantage concerning supercapacitors having fast charge–discharge cycles and hence is bound to enhance performance for eventual applications in fields, such as electric vehicles or portable electronics [42].

Thin films on 2D materials extremely improve catalytic efficiency and stability for technologies of energy conversion, such as fuel cells. On the other hand, unique surface properties combined with atomic-scale control given by thin-film deposition techniques can further be developed into highly active and durable catalysts. For example, 2D material-based thin-film fuel cell electrodes would exhibit highly enhanced catalytic activity in terms of oxygen reduction reaction and hydrogen evolution. It thereby attains highly efficient and reliable fuel cells, hence clean energy, just by its way of producing sustainable technologies. This large active surface area, thin films, and improved reaction kinetics are instrumental for advancing catalytic processes in energy conversion.

These applications open up new horizons for a whole range of thin-film technologies on 2D atomically plane materials for various industry sectors, ranging from electronics and optoelectronics to energy storage and energy conversion. Research and development in this highly specific area will certainly provide new ways and technological innovations regarding problem-solving ahead [42].

9.17 Problems and Solutions in Thin-Film Deposition

Advanced industries directly related to semiconductor manufacturing process, optics, and coatings involve one process of thin-film deposition. The process here entails the deposition of a thin material of chosen film onto the substrate to obtain a specific property or function. Despite its applications and technological importance, thin-film deposition has problems that may affect the quality and performance of the products.

The main problems in the deposition of thin films are getting the right film thickness on the substrate, getting adhesion between the film and the substrate, and avoiding stressing that leads to cracking or peeling. Other complications are associated with such problems as contamination, defects, and surface roughness that compromise the desired properties of the film. Other issues include the composition control of the film and prevention from oxidation or corrosion.

An understanding of the deposition process, as well as the relevant factors that may be of consequence in these processes, can lead one to certain approaches to solve these types of problems. Such solutions may be in the form of refinements to parameters for example, to adhesion, effected by the use of atomic layer deposition with special undercoats and possibly subsequent post-deposition treatments to “even out” such irregularities.

These methods will be invaluable to surmount the challenges involved in the deposition of films and come up with high-quality, reliable films that would serve optimally for different applications.

Following are some of the problems explained with their solutions [43].

1. **Uneven Thickness**: Imagine you are trying to smear frosting onto a cake, but some places end up too large and some too small. This also occurs under conditions of thin-film deposition where it becomes quite difficult to obtain the same thickness over its whole surface. Now, think of rotating the cake as you are slathering it with frosting. In deposition, a substrate is rotated or otherwise moved around to provide even coverage of the material. Again, advanced techniques particularly atomic layer deposition can help in laying down that perfect, even layer.
2. **Poor Adhesion**: Sometimes, it does not stick that well to the thin film; it is kind of like peeling a sticker off too easily. It may lift or peel from the substrate. Just imagine prepping the wall before painting by cleaning and priming the wall. Well, cleaning the substrate properly and using special undercoats in thin-film deposition can help the film to hold on to it so that it stays in its place.
3. **Stress and Cracking**: Temperature changes or the deposition process itself can stress thin films, whereby they may crack or warp. The cracking of the film due to stress is avoided and the film remains intact by careful control of the temperature in deposition, using layers for stress relief, or gentle heat treatment afterward.
4. **Contamination**: Consider the dust settling on wet paint and ruining the finish. Many types of contaminants, be it dust or some other impurities, may sneak into the thin film and affect its quality and, hence its performance. That can be kept at bay by keeping the environment of deposition super clean, using high-purity materials, and regularly maintaining the equipment. Monitoring can also catch anything that might have happened early on.
5. **Defects and Pinholes**: These small defects or pinholes in the films are very similar to small punctures in a balloon, which may lead to failures in performance. Fine-tuning deposition parameters, quality materials, and filtration systems for removing particulates from the material can all help to realize defect-free films just like baking the perfect cake, which calls for the right ingredients and conditions.
6. **Composition Control**: It could be cookie batter or a thin film, but it is about getting the right mix in the right proportions. Reliance on material composition variations leads to bad performance. Right composition is ensured through regular calibration of deposition sources and process-monitoring techniques in real time. The right mix is maintained by multisource deposition systems.
7. **Surface roughness**: It can be compared to a rough road, quite like a bumpy road, in that such a surface may not be ideal for high-precision applications. As regards the roughness, excess roughness will lower the performance of the film, that is, smoothening the substrate before deposition and controlling the deposition rate, with techniques like chemical–mechanical polishing (CMP) afterward, which can lead to a high-quality, smooth surface.

8. **Oxidation and Corrosion**: That is similar to what happens to metals when they rust: thin films oxidize or corrode, hence losing their power with time. Technology that extends the life of thin films may be increased by protective coatings and deposition, storage in inert atmospheres, and careful material choice to avoid oxidations and corrosion.

9.18 Environmental Consideration

9.18.1 Waste Management and Recycling

Thin-film production involves associated processes that generate wastes in the form of excess materials, solvents, or by-products. Understanding these kinds of wastes is basic to developing a waste management strategy. At the very basic level, waste can be reduced by attaining an optimum usage of materials by fine-tuning the production processes and engaging in cleaner productions. For example, correct depositions can reduce the raw material required in quantity and hence reduce generated waste. Other than saving resources, such strategies are very useful in minimizing the ecological burden of the manufacturing process.

The entire disposal of thin-film products is, therefore, a challenging exercise given the hazardous materials present in them that call for proper handling and disposal. Advanced recycling techniques provide for raw material recovery, retrieval of valuable metals, such as indium and gallium derived from thin-film solar cells, and disposal of toxic materials in an environmentally safe manner. Advanced recycling technologies, in such a way, will be able to separate and redeem those materials without virgin raw materials and minimize environmental contaminations. Setting up robust programs for recycling and enhancing waste infrastructure, therefore, remains essential for attaining sustainable disposal practices.

The various ways by which the environment could be safeguarded from thin-film production and disposal impacts include closed-loop recycling, whereby materials are returned to the production process and reused to reduce waste. In addition, the development of biodegradable or easily recyclable thin-film materials will be of importance in the reduction of harm to the environment. This shall be achieved through reducing, reutilizing, and recycling policies and practices. For example, when designing a product by considering end-of-life disposal, it makes it easy to recycle and reduces the overall impact on the environment [44].

9.18.2 Legislative and Industry Standards

Of the national and international environmental regulations controlling hazardous substances, managing emissions, and ensuring proper waste disposal, only a few seem to relate to thin-film production and use. These limits set what is regarded as an acceptable level of toxic materials, emissions during the production process, and how wastes are disposed of. Accordingly, there exists a need to act within such regulations to minimize impacts caused on the environment and health of members of the public by electricity. These manufacturers keep them updated regarding these regulations, thereby avoiding legal penalties and fostering sustainability in their industry.

Industry standards and certifications developed, including ISO 14001, will set an appropriate environment for environmentally friendly thin-film manufacturing. Such standards guide in making an efficient environmental management system, thus helping companies reduce their impacts on the environment. The protection of the environment is further enhanced through the adoption of voluntary actions and best practice principles by some industrial leaders. This may prevent the implantation of energy-efficient technologies that reduce emissions and offer possibilities for renewable raw materials usage in production. Standards and practices, such as these, give an added advantage to a company in terms of better environmental performance, reputation, and competitiveness [45].

9.18.3 Compliance and Enforcement

These mechanisms encompass the surveillance of compliance and enforcement in light of conformance to set environmental regulations and standards. This is provided for by governmental and nongovernmental institutions through examinations, audits, and assessments against requirement standards. Noncompliance invites penalties in fines, legal suits, or loss of reputation, while incentives take the form of fiscal advantages, grants, or certifications on good environmental performance. All this should be complied with, and measures of enforcement are put in place by the governments. These shall ensure that the companies remain within the boundaries set by governments, concerning the environment, and that operators keep on improving the practice [46].

9.18.4 Future Trends and Innovations

Research and development in the area of thin films have risen to insist increasingly on eco-friendly materials that ought to be based on less toxic constituents and renewable sources to reduce natural environmental impact. Material science innovations resulted in biodegradable thin films and materials that require less energy for formulation. Case studies into these innovative materials showcase improved environmental performance that can replace traditionally used materials in several industries.

The new green production technologies will reduce the ecological footprint of thin-film manufacturing. These include, among others, low-temperature deposition techniques, solvent-free processes, and energy-efficient manufacturing technologies. All such kinds of innovations bring down energy consumption, reducing harmful emissions and decreasing the quantity of hazardous materials used. Not only is the environment favored by adopting green production technologies, but cost savings and efficiency gains are also brought to manufacturers.

Thus, sustainable applications of thin films toward environmental sustainability have increased. Energy-efficient coatings and thin-film solar cells are among the kinds of sustainable applications this kind of renewable energy technologies would find application in. Lifecycle assessment (LCA) approaches correspond to the tools that drive out the environmental performance assessment of thin-film products from production to end-of-life. This holistic approach serves to identify improvements while taking care that the idea of sustainability has already started at the beginning of a product's life from raw material extraction up to the end-of-life stage. Case studies of particular projects or companies that have individually succeeded in minimizing their impacts on the environment are very instructive. They bring new strategies and technologies engendering positive environmental benefits. For example, a company will have extensively introduced some programs of recycling that reduce waste and salvaged valuable raw materials from them, or environmentally friendly technologies of production that reduce emissions and consumption of energy. Based on such analyzed examples, some best practices can be more widely applied in this industry [47].

9.18.5 Lessons Learned from Environmental Mishaps

Learning from past environmental mishaps is very fundamental in avoiding such problems in the future. Any incident where thin-film production or use resulted in an adverse impact on the environment is extremely worth studying. For instance, hazardous materials could have been disposed of inappropriately thus contaminating the environment. The lessons learned from such incidences that is, what took place and why would enable companies to focus their attention on the development of preventive measures and upgrading of practice. Such lessons will thus translate into continual improvement in environmental performance in the thin-film industry [48].

9.19 Adhesion-Related Problems

Indeed, any issue or concern related to the deposition of thin films is related to adhesion; it essentially counts in performance and reliability and very importantly has long-term properties for applications that range from advanced electrònics to protective coatings. These are many of the factors, that can be listed, that may influence thin-film adhesion to substrates: surface contamination, surface roughness or thermal expansion mismatch, chemical incompatibility, deposition parameters, and many others. Therefore, all these issues must be tackled in the applications of thin films [49].

9.19.1 Surface Contamination and Its Impact on Adhesion

Surface contamination is one major cause of adhesion loss. Consequently, substrate surfaces could host residues of many contaminants in the form of dust, oils, or moisture and build a block to proper contact for thin films. Contaminated films will therefore bond only very weakly until the films peel or delaminate under stress. All this delineates that cleaning is essential before deposition since the substrate surface has to be clean if good adhesion is to be realized. This may be achieved using proper solvents that would allow to dissolve or elute diversity in organic contaminants, gentle shake-off of particulates by ultrasonic cleaning followed by high-frequency sound waves, or plasma cleaning to etch away surface contaminant materials and, in some cases, alter molecularly the surface material. Noticeably, while very effective, plasma cleaning is an especially good technique for raising the surface energy of a substrate to promote adhesion to materials [50].

9.19.2 Surface Roughness and Its Influence on Adhesion

Surface roughness remains yet another critical factor influencing adhesion. The roughness in the substrate surface influences mechanical interlocking with the film. Too-smooth surfaces may just not have enough interlocking sites; adhesion can thus be weak. On the other hand, very rough surfaces can lead to or correct concentrated stresses responsible for the cracks or delamination of the film. Other than this, it is in optimizing surface roughness that the key to improving these properties lies. For example, the extremely smooth surfaces can be roughened slightly by controlled etching. The very rough surfaces can be smoothened by techniques, for instance, polishing. For example, thin films of primer or undercoat adhesion-promoting layers can be applied to act at the interface between thin films and substrates to enhance adhesion [51].

9.19.3 Thermal Expansion Mismatch and Its Effects on Adhesion

The mismatch of thermal expansion between thin films and substrates can create enormous stresses, particularly on temperature change. A point to note is that against a thermal variation, the coefficient of behavior of different materials involves different behaviors of expansion and contraction. The film, whose coefficients are going to exhibit high differences concerning the substrate ones, will normally result in cracking or delamination caused by the induced stress under working conditions; such is the case of thermal cycling. In such a case, the selection of material whose coefficient of thermal expansion and contraction may be tried to be close. These buffer layers resemble some sort of intermediate zone fighting between the two parties, hence absorbing the stress and saving the thin film [52].

9.19.4 Chemical Compatibility and Adhesion

Another possible reason for adhesive failure due to incompatibility between film and substrate is known as chemical incompatibility. At the interface, there could be a great variety of chemical reactions, generally forming weak bonds or brittle compounds of low adhesion strength. For instance, the interface may already be weakened by the formation of oxides or other compounds brought about while depositing it. Added aids in the introduction and improvement of chemical compatibility in the form of adhesion-promoting intermediate layers realize an improved adhesion, nevertheless. Logically, these should adhere well to both substrate and thin film and guarantee very good and smooth interfaces. Other than that, various treatments like plasma oxidation alter surface chemistry to result in excellent bonding properties and thus improved adhesion [53].

9.19.5 Impact of Deposition Parameters on Adhesion Quality

Regarding adhesion, deposition parameters include temperature, pressure, and deposition rate, which were quite very crucial in determining the quality of adhesion. Unsuitable deposition parameters lock a low density of film, weak bonding, and unreleased residual stresses within the thin film whenever it is used. For instance, too high temperatures in a deposition may introduce an extremely high degree of thermal stress, while low-temperature depositions result in incomplete bonding. A too high rate of deposit may be variant with poor-quality films. Very low rates can just be slow and impractical. Realistically, too high rate of deposit may come off as a variant with films of poor quality, and the very low ones are only slow and unpractical. Therefore, careful experimentation with processing conditions was made to optimize these parameters. It is the parameters that determine good balance such that the thin film would have been deposited uniformly in a very strong adhesion and reduces internal stress to a minimum. Bad adhesion is manifested by delamination, cracking, general poor performance, and durability diminution in the thin film [54].

9.20 Delamination

Delamination can be defined as a situation where the film peels off the substrate and creates voids or bubbles that can render the entire film useless. This becomes a problem relating to such devices as the electronic, wherein it causes circuit failure through delamination or optical applications that cause distortions. Therefore, delamination results in cracking due to either intrinsic stresses or extrinsic mechanical forces, which afterward propagate into the thin film and most likely will cause either structural application failure or reduced efficiency in an electronic device. Adhesion failure decreases not only performance but also increases maintenance costs; thus, it lowers the reliability and lifetime of the end product. Different solutions can be applied to minimize these effects. Of all the general methods for adhesion promotion, probably surface treatment is the most efficient [55].

9.21 Surface Preparation Techniques for Enhanced Adhesion

Mechanical cleaning methods involve abrasive blasting and polishing. In these methods, impurities are physically removed and the surface is roughened, increasing mechanical interlock. The chemical cleaning, introduced by acids, alkalis, or even solvents, cleans and etches the surface, leaving it prepared for finer adhesion. This is realized by activation and cleansing of the surface by an ionized gas plasma treatment, which increases the surface energy, makes it very reactive, and hence has a positive effect on adhesion. Another utilizes intermediate adhesion layers that glue the thin-film material to the base surface and are chosen to bond well with both. Since they have been chosen to bond, in sequence, to both the substratum and to the thin film, their interface will not just transmit but even increase stress. Among many others, an adhesion layer is applied using materials, such as metals, chromium, titanium, compounds like silicon dioxide, silicon nitride, and so on. This interface layer may also act as a barrier against any undesired chemical reaction. Deposition conditions are optimized to be appropriate for adhesion. This involves temperature optimization to ensure low residual stresses, a depositing rate that ensures compactness of the film with high compactness density and thickness uniformity, and to some extent, optimization of the pressure and atmosphere in the deposition chamber upon realizing a better quality film. It is only through well-controlled experiments and methods that an effective enhancement of the adhesion level of thin films will be established [56].

The increased buffer layers may additionally prevent thermal expansion mismatches and also chemical incompatibilities. This hence makes the several layers the zone that is intermediate and absorptive of differences in thermal expansion during stress taking place. Assurance that weak or brittle compounds do not form at the interface accompanies the use of buffer layers to accommodate the differences. This also applies to the annealing and laser treatment in the post-deposition treatments for best adhesion. Annealing is the process whereby a deposited film gets heated up to relieve stresses that have commenced building up from within the film and to make the bonding in the film optimal. Annealing gives rise to stronger bonding, where atoms are rearranged from the thin film and the substrate toward a more stable structure. Laser treatment is the process aimed at applying focused laser pulses on a film or substrate in such a way that the delivered energy or local modification raises adhesion at some points of the film or substrate, hence increasing force in general [57].

9.22 Issues of Uniformity

It is, therefore, one of the critical parameters that arguably has immense effects on the performance and the reliability of uniformity in thin-film deposition for the resulting films in electronics, optics, and protective applications. Nonuniform thin films may change electrical, optical, and mechanical properties until finally bringing about the failure of a device or causing it to operate erratically in respective applications. Based on this fact, attaining uniform thin films involves careful control of the deposition process along with consideration of a whole host of factors that impact the thickness and composition of the deposited film [58].

9.22.1 Challenges for Measuring and Achieving Uniformity

One evident challenge lies in how exactly to measure and realize a uniform thin film with the deposition technique itself. Each deposition technique has some inherent advantages vis-à-vis uniformity and limitations, found in the cases of PVD, CVD, and ALD. For example, although there is material distribution in both the sputtering and evaporation PVD processes, due to the directional flux of material in these techniques, coupled with the possible nonuniformity across the substrate from geometry effects in the deposition chamber, this is not expected to produce such uniform films. For instance, in the PVD processes of sputtering and evaporation, this can be because the fluxes of materials are directional, and together with the geometry of the deposition chamber, the variation across the substrate is induced. Equally, in the case of CVD, a nonuniform gas flow and temperature gradient in a reactor may result in nonuniformity of the film. While ALD offers excellent control of the film thickness at the atomic level, this technique can nevertheless suffer from uniformity problems in case its precursor distribution is not well managed [59].

9.22.2 Factors Affecting Film Uniformity

The other two major factors for film uniformity would be substrate size and shape. Large or convoluted geometry substrates bring more difficulties in achieving uniform coverage. The distance variations from the source material in PVD or the precursor access variations in CVD may cause very large thickness variations. Often, this has been a problem in many applications where demanding film properties in most cases, for the whole substrate are required, like in semiconductor wafers or optical coatings. Substrate rotation, tilting, or even special fixture designs reduce the problems by presenting different areas of the substrate to the deposition flux or the gases. A further important parameter influencing uniformity is the deposition rate. At high deposition rates, nonuniform films are obtained with limited possibilities for atoms or molecules to diffuse at the surface; columnar growth of other nonideal microstructures occurs. On the other hand, extremely low deposition rates prove impractical for large-scale production and add other problems like contamination or simply a long exposure to reactive gases. The deposition rate is thus optimized, balancing throughput against uniformity [60]. Other techniques, such as pulsed deposition, have to be used in the pursuit of improved uniformity. This allows more time between successive pulses for atoms/molecules to diffuse and settle on the substrate. The other critical parameter relating to temperature control during deposition allows the realization of uniformity in the films that would otherwise not be possible. There can be temperature gradients across the substrate, and this can result in changes in film growth rates and properties. The uniform temperature reactor concept underlies the majority of these processes, like CVD, to provide homogeneous film deposition. This can often be quite challenging due to the interaction between the heating elements, gas flows, and reactor design. Multizone heating or sophisticated temperature control techniques, like substrate cooling, can reduce these swings in temperature and provide for more homogeneous film growth.

The other two critical parameters, turning out to be rather important for the uniformity of the deposited film, include the surface roughness and cleanliness of the substrate. A rough or contaminated surface may result in nonuniform nucleation and growth of the thin film, thus ending with a nonuniform thickness and poor-quality film. Proper substrate preparation involves cleaning and conditioning the surface to be smooth and clean for deposition. Other techniques that can help prepare the substrate are chemical polishing, plasma cleaning, and other thermal treatments to improve the uniformity of films [61].

9.23 Dynamics of Gas Flow

The second critical factor in such a chamber is the dynamics of the gas flows involved in these processes, especially the CVD and ALD. One easily notices that some kind of nonuniform gas flow could just be behind the uneven distribution of precursors leading to variations in thickness. Uniform delivery of gas through the system will be enhanced by flow and computational fluid dynamics simulations, raising uniformity much above for the optimized design in reactors. Other than that, stable pressure and the reproduced deposition rate of flow are going to avoid fluctuations that affect the thickness of the film [62].

9.23.1 Shadowing and Masking

Shadowing and masking effects from features on a substrate or deposition setup in general block the material flux during PVD, leading to nonuniform depositions. A very complex geometry of substrates may combine with sources that have high collimation; then, the so-called shadowing effect may turn out to be very significant. Some general ways of improving this issue are by source-to-substrate distance variation, multiple sources, or just rotating the substrate.

Other techniques that can improve the uniformity of film include aspects of post-deposit treatments involving annealing or chemical mechanical polishing (CMP), which iron out variations in thickness to better the quality of the thin film. Annealing induces atomic inter-diffusion, which relieves most of the stresses inside a film and thereby enables it to have a more homogeneous structure. CMP mechanically removes too much material so that the surface of the film is flat and homogeneous [63].

9.23.2 Contamination and Defects

Afterward, contamination and defects directly influence thin-film performance, reliability, and lifetime for applications like electronics, optics, and protective coatings. These different flaws can initiate various processes, finally changing the properties of the film and generally impairing the functionality of the devices. In such cases, high-quality thin films can only be attained where causes, effects, and mitigation strategies of contamination and defects are known [63].

9.23.3 Sources of Contamination

Probably among the most likely sources of contamination during thin-film depositions is the deposition chamber itself. Most probably, particulates, residual gases, and chemical residues might find entry into the deposition chamber and hence into the deposition process. Their adsorption at the substrate surface would result in defects like pinholes or inclusions. First of all, a high vacuum coupled with ultra-high purity gases and materials needs to be created to reduce this problem. The deposition chamber and its various constituents can be cleaned periodically, which may help clean off any contamination. Other than this, in situ monitoring techniques and cleaning techniques, such as plasma cleaning, can be availed for further reductions in contamination [64].

The next very critical step is substrate handling and preparation, in which chances of contamination can arise. Moreover, the defects in deposited film can also increase due to contamination, which can be transferred from fingerprints, dust, or residue of other process steps onto the substrate surface. Good handling techniques, gloves, and cleanroom conditions can reduce the contamination of substrates. Some examples of substrate cleaning methods, some of which give assurance of the surface removal of contaminants, include cleaning with solvents, ultrasonic cleaning, and plasma cleaning. If the environment in which substrate preparation and handling are done is clean, contamination and defects will be drastically reduced.

The other critical factor is that the purity of starting material is determinant for thin films' quality. Impurities present in the source materials may get incorporated into the films after deposition, which induces unwanted defects like doping and grain boundaries. These defects lead to nonhomogeneous film properties. Such chances of impurity incorporation can be reduced by using high-purity source materials, characterizing the source materials properly before deposition, and other sophisticated techniques. In that case, both methods require the high purity of precursor gases and delivery systems. Calibration, combined with periodical purification techniques and regular clean-up of the gas delivery line, can ensure a high purity of the materials [65].

These, therefore, turn out to be drivers of defects within thin films: process parameters, such as temperature, pressure, and deposition rate control. This could cause inappropriate process conditions, which turn out film stresses, crystallographic defects, and nonuniform growth of the films. For instance, a very high deposition temperature will lead to aggressive diffusion and grain growth, while a low temperature causes incomplete reactions with poor adhesion of the film. In other words, process parameters should be optimally designed or optimized by proper experiments and controls to reduce the number of defects as small as possible. Advanced techniques include the development of in situ monitoring and feedback control systems that can maintain the deposition conditions optimized to reduce incidences of defects [66].

9.23.4 Voids and Pinholes

The second major problem issues associated with thin-film depositions are voids and pinholes. Probably, inadequacies in surface preparation, rapid deposition rate, and feeble material supply may be the causes of defects of this nature. Voids and pinholes may form around these weak points of the film, which should bring down the mechanical integrity, electrical properties, and barrier performance of the films. These kinds of defects can be reduced by ensuring adequate substrate cleaning and optimizing deposition rates with proper material fluxes. Moreover, multilayer deposition and annealing will fill up the voids and make the film homogeneous [67].

9.23.5 Grain Boundaries and Crystallographic Defects

Most of the thin-film materials especially those deposited by PVD and CVD techniques have grain boundaries and Manhattan-like crystallographic defects. These can drastically lower the electrical, optical, and mechanical properties of the film. However, for example, grain boundaries can act as scattering centers that reduce charge carrier conductivity and mobility in a film. Deposition conditions should be controlled to minimize grain boundaries and crystallographic defects; such conditions include temperature, pressure, and deposition rate. This can be complemented by post-deposition treatments that include annealing to enhance crystallinity and reduce the density of defects in thin films [68].

9.23.6 Stress and Strain Management

The second kind of defect emerging from thin films includes the stresses and strains, which may lead finally to cracking, delamination, and warping. It will, in turn, further be divisible into those stresses caused by mismatches between thermal expansion coefficients of the film and substrate, those related to intrinsic deposition stress, or otherwise from external mechanical forces. Careful material selection in which the thermal properties will match, the optimization of deposition parameters for low intrinsic stress, and the use of buffer layers that absorb the stress all go to control the stress and strain. Annealing post-deposition is also vital in the relaxation of the developed stress and enhancing mechanical integrity in the film [69].

9.23.7 Impact of Contamination and Defects on Thin Films

The contamination and defects introduced into the thin films have effects on both the functional properties and reliability and lifetime effects. For instance, in devices, contamination may trigger shortcuts or leakage currents which eventually lead to failure. Otherwise, they may introduce scattering, absorption, or loss into the transparency for optical applications. High stringency quality controls implemented during the deposition process ensure high reliability and performance of the product. Among them, one can mention deposition environment control, material and film characterization, and final product testing [70].

9.24 Characterization Techniques for Defect Analysis

State-of-the-art characterization techniques should be fronted in identifying and analyzing contamination and defects in thin films. Of the methods that give full information concerning morphology, structure, and composition for any kind of film, the major ones are the SEM, TEM, AFM, and XRD methods. With these techniques, it will be possible to track back the sources of contamination and defects in the deposition process and hence provide ground for its improvement. In contrast, both in situ monitoring techniques, quartz crystal microbalance, and optical emission spectroscopy, are correlated with deposition conditions that offer real-time feedback on the conditions under which deposition is being conducted and thus immediate measures to minimize defects [71].

9.25 Conclusion

Thin-film trends and issues are an ultra-broad range of challenges and developments outlining their applicability in the fields of electronics, optics, and protective coating uses. Among the main trends are further developments concerning deposition techniques, for instance, methods of PVD, CVD, or ALD. These can increase the control over the properties of the film while introducing uniformity and partial defect reduction problems. One critical issue is that of adhesion, possibly influenced by surface contamination, roughness, and relative mismatches in coefficients of thermal expansion between the films and substrates. Techniques like plasma cleaning and applying intermediate adhesion layers can improve the adhesion strength. Indeed, the area of uniformity has been very important concerning deposition parameters, substrate geometry, and gas flow dynamics. Deposition control and improvement in substrate handling are necessary so that deposited films are uniform both in composition and thickness. This variety of processes has created defects: voids, pinholes, grain boundaries, and stresses. It is expected to provide a drastic change in the mechanical, electrical, and optical properties of the thin films. The problems can be worked out by various strategies like optimizing process parameters, multilayer deposition, and mitigation by post-deposition treatments, such as annealing. On the other hand, very high reliability and performance can also be realized under rigorous control of the deposition ambiance against contamination and state-of-the-art characterization using SEM, TEM, and in situ monitoring. Coupled with sophisticated material development and nanotechnology in the future, new deposition methods will bring challenges steep enough that today's applications of thin films in flexible electronics, Nanophotonics, and biomedical devices do increase. While further research finds new ways of innovating in materials science and deposition technologies, it will be important to meet trends and problems arising to exploit the possibilities and potential of thin films for further technological development.

References

1. Ellahi, R. 2020. Recent Trends in Coatings and Thin Film: Modeling and Application, 777. MDPI.
2. Karki, S., et al. 2016. Thin films as an emerging platform for drug delivery. *Asian Journal of Pharmaceutical Sciences* 11 (5): 559–574.
3. Patil, A., et al. 2008. Issue and challenges facing rechargeable thin film lithium batteries. *Materials Research Bulletin* 43 (8–9): 1913–1942.
4. Cantarella, G., et al. 2020. Review of recent trends in flexible metal oxide thin-film transistors for analog applications. *Flexible and Printed Electronics* 5 (3): 033001.
5. Adeyinka, A.M., et al. 2023. A review of current trends in thin film solar cell technologies. *International Journal of Energy and Power Engineering* 17 (1): 1–10.
6. Anderson, A.-L., et al. 2016. Thin films for advanced glazing applications. *Buildings* 6 (3): 37.
7. Wang, Q., et al. 2016. Progress in emerging solution-processed thin film solar cells–part I: Polymer solar cells. *Renewable and Sustainable Energy Reviews* 56: 347–361.
8. Kumar, P., et al. 2018. Hybrid porous thin films: Opportunities and challenges for sensing applications. *Biosensors and Bioelectronics* 104: 120–137.
9. Morosanu, C.E. 2016. *Thin Films by Chemical Vapour Deposition*, vol. 7. Elsevier: Amsterdam.
10. Mane, R., and C. Lokhande. 2000. Chemical deposition method for metal chalcogenide thin films. *Materials Chemistry and Physics* 65 (1): 1–31.
11. Malik, K., and A. Jasińska-Biliczak. 2018. Innovations and other processes as identifiers of contemporary trends in the sustainable development of SMEs: The case of emerging regional economies. *Sustainability* 10 (5): 1361.
12. Lohmiller, J., N. Woo, and R. Spolenak. 2010. Microstructure–property relationship in highly ductile Au–Cu thin films for flexible electronics. *Materials Science and Engineering: A* 527 (29–30): 7731–7740.
13. Ojanen, V. 2014. Maintenance innovations-Types, patterns and emerging trends. In *2014 IEEE International Conference on Management of Innovation and Technology*. IEEE.
14. Holland, S.W., and S.Y. Nof. 1999. Emerging trends and industry needs. In *Handbook of Industrial Robotics*, 31–40.
15. Xu, W., T. Lu, and F. Wang. 2010. Effects of interfacial properties on the ductility of polymer-supported metal films for flexible electronics. *International Journal of Solids and Structures* 47 (14–15): 1830–1837.
16. Cordill, M.J., O. Glushko, and B. Putz. 2016. Electro-mechanical testing of conductive materials used in flexible electronics. *Frontiers in Materials* 3: 11.
17. Kim, D.H., et al. 2010. Stretchable, curvilinear electronics based on inorganic materials. *Advanced Materials* 22 (19): 2108–2124.
18. Lupinacci, A., et al. 2013. A study of solder alloy ductility for cryogenic applications. In *2013 IEEE International Symposium on Advanced Packaging Materials*. IEEE.
19. Susilawati, S., et al. 2024. A systematic review of the trends thin film characteristics research as electronic device (2015–2024). *Jurnal Penelitian Pendidikan IPA* 10 (6): 313–322.
20. Bakis, C., et al. 2001. Self-monitoring, pseudo-ductile, hybrid FRP reinforcement rods for concrete applications. *Composites Science and Technology* 61 (6): 815–823.
21. Ji, X., et al. 2022. Revealing the ductile-to-brittle transition mechanism in polycrystalline body-centered tetragonal tin (Sn) for cryogenic electronics. *Journal of Alloys and Compounds* 903: 163948.
22. Arya, S.K., et al. 2012. Recent advances in ZnO nanostructures and thin films for biosensor applications. *Analytica chimica acta* 737: 1–21.
23. Numanoğlu, H.M., B. Akgöz, and Ö. Civalek. 2018. On dynamic analysis of nanorods. *International Journal of Engineering Science* 130: 33–50.
24. Tian, B., T.J. Kempa, and C.M. Lieber. 2009. Single nanowire photovoltaics. *Chemical Society Reviews* 38 (1): 16–24.
25. Dawson, R., A.I. Cooper, and D.J. Adams. 2012. Nanoporous organic polymer networks. *Progress in Polymer Science* 37 (4): 530–563.

26. Nayak, R., et al. 2012. Recent advances in nanofibre fabrication techniques. *Textile Research Journal* 82 (2): 129–147.
27. Farooq, S.A., et al. 2022. Nanostructured coatings: Review on processing techniques, corrosion behaviour and tribological performance. *Nanomaterials* 12 (8): 1323.
28. Benelmekki, M., and A. Erbe. 2019. Nanostructured thin films–background, preparation and relation to the technological revolution of the 21st century. In *Frontiers of Nanoscience*, 1–34. Elsevier.
29. Walgraef, D. 2003. Nanostructure evolution during thin film deposition. *Physica E: Low-Dimensional Systems and Nanostructures* 18 (4): 393–401.
30. Marichy, C., M. Bechelany, and N. Pinna. 2012. Atomic layer deposition of nanostructured materials for energy and environmental applications. *Advanced Materials* 24 (8): 1017–1032.
31. Ogugua, S.N., O.M. Ntwaeaborwa, and H.C. Swart. 2020. Latest development on pulsed laser deposited thin films for advanced luminescence applications. *Coatings* 10 (11): 1078.
32. Rajan, S.T., B. Subramanian, and A. Arockiarajan. 2022. A comprehensive review on biocompatible thin films for biomedical application. *Ceramics International* 48 (4): 4377–4400.
33. Mirzaali, M.J., et al. 2022. Additive manufacturing of biomaterials—Design principles and their implementation. *Materials* 15 (15): 5457.
34. Wang, L., et al. 2022. Key considerations on the development of biodegradable biomaterials for clinical translation of medical devices: With cartilage repair products as an example. *Bioactive Materials* 9: 332–342.
35. Ion, R., et al. 2014. Long-term corrosion performances and cytocompatibility of nitrided Ti and Ti–6Al–4V alloy in severe functional conditions. *Materials and Corrosion* 65 (6): 593–604.
36. John, A., et al. 2015. Physico-chemical modification as a versatile strategy for the biocompatibility enhancement of biomaterials. *RSC Advances* 5 (49): 39232–39244.
37. Vendra, V.K., L. Wu, and S. Krishnan. 2010. Polymer thin films for biomedical applications. *Nanomaterials for the Life Sciences* 5.
38. Kim, H.G., and H.-B.-R. Lee. 2017. Atomic layer deposition on 2D materials. *Chemistry of Materials* 29 (9): 3809–3826.
39. Bhardwaj, N., and S.C. Kundu. 2010. Electrospinning: A fascinating fiber fabrication technique. *Biotechnology Advances* 28 (3): 325–347.
40. Barbulovic-Nad, I., et al. 2006. Bio-microarray fabrication techniques—A review. *Critical Reviews in Biotechnology* 26 (4): 237–259.
41. Wood, M. 2007. Colloidal lithography and current fabrication techniques producing in-plane nanotopography for biological applications. *Journal of the Royal Society Interface* 4 (12): 1–17.
42. Pawar, A.B., and I. Kretzschmar. 2010. Fabrication, assembly, and application of patchy particles. *Macromolecular Rapid Communications* 31 (2): 150–168.
43. Schwartz, R.W. 1997. Chemical solution deposition of perovskite thin films. *Chemistry of Materials* 9 (11): 2325–2340.
44. Wilson, D.C., C. Velis, and C. Cheeseman. 2006. Role of informal sector recycling in waste management in developing countries. *Habitat international* 30 (4): 797–808.
45. Song, D., et al. 2022. Review on legislative system of photovoltaic industry development in China. *Energies* 15 (1): 306.
46. Castro, L., and C. Scartascini. 2015. Tax compliance and enforcement in the pampas evidence from a field experiment. *Journal of Economic Behavior & Organization* 116: 65–82.
47. Thakur, R., S.H. Hsu, and G. Fontenot. 2012. Innovation in healthcare: Issues and future trends. *Journal of Business Research* 65 (4): 562–569.
48. Weber, R., D.W. Aha, and I. Becerra-Fernandez. 2001. Intelligent lessons learned systems. *Expert Systems with Applications* 20 (1): 17–34.
49. Parker, M., et al. 2004. Colorectal surgery: The risk and burden of adhesion-related complications. *Colorectal Disease* 6 (6): 506–511.
50. Ames, D.P., and S.J. Chelli. 2004. Surface contamination effects on film adhesion on metals and organic polymers. *Surface and Coatings Technology* 187 (2–3): 199–207.
51. Croll, S. 2020. Surface roughness profile and its effect on coating adhesion and corrosion protection: A review. *Progress in Organic Coatings* 148: 105847.

52. Göstemeyer, G., et al. 2012. Effect of thermal expansion mismatch on the Y-TZP/veneer interfacial adhesion determined by strain energy release rate. *Journal of Prosthodontic Research* 56 (2): 93–101.
53. Stöckelhuber, K.W., et al. 2010. Contribution of physico-chemical properties of interfaces on dispersibility, adhesion and flocculation of filler particles in rubber. *Polymer* 51 (9): 1954–1963.
54. Montero-Ocampo, C., E. Ramírez-Ceja, and J. Hidalgo-Badillo. 2015. Effect of codeposition parameters on the hardness and adhesion of TiVN coatings. *Ceramics International* 41 (9): 11013–11023.
55. Gioia, G., and M. Ortiz. 1997. Delamination of compressed thin films. *Advances in Applied Mechanics* 33: 119–192.
56. Frascio, M., et al. 2021. Appraisal of surface preparation in adhesive bonding of additive manufactured substrates. *International Journal of Adhesion and Adhesives* 106: 102802.
57. Palmieri, F.L., et al. 2016. Laser ablation surface preparation for adhesive bonding of carbon fiber reinforced epoxy composites. *International Journal of Adhesion and Adhesives* 68: 95–101.
58. Kenward, P., and B. Wichmann. 1991. Approved uniformity issues. *ACM SIGAda Ada Letters* 11 (1): 114–120.
59. Velez, N.L., J.K. Drennen, and C.A. Anderson. 2022. Challenges, opportunities and recent advances in near infrared spectroscopy applications for monitoring blend uniformity in the continuous manufacturing of solid oral dosage forms. *International Journal of Pharmaceutics* 615: 121462.
60. Agarwal, S. 2021. Major factors affecting the characteristics of starch based biopolymer films. *European Polymer Journal* 160: 110788.
61. García-Muñoz, S., and D.S. Gierer. 2010. Coating uniformity assessment for colored immediate release tablets using multivariate image analysis. *International Journal of Pharmaceutics* 395 (1–2): 104–113.
62. Zhang, H., et al. 2017. Large eddy simulation of turbulent flow using the parallel computational fluid dynamics code GASFLOW-MPI. *Nuclear Engineering and Technology* 49 (6): 1310–1317.
63. Liu, C., et al. 2012. Theoretical design of shadowing masks for uniform coatings on spherical substrates in planetary rotation systems. *Optics Express* 20 (21): 23790–23797.
64. Rodushkin, I., E. Engström, and D.C. Baxter. 2010. Sources of contamination and remedial strategies in the multi-elemental trace analysis laboratory. *Analytical and Bioanalytical Chemistry* 396: 365–377.
65. Lye, D.J. 2009. Rooftop runoff as a source of contamination: A review. *Science of the Total Environment* 407 (21): 5429–5434.
66. Li, P., et al. 2021. Sources and consequences of groundwater contamination. *Archives of Environmental Contamination and Toxicology* 80: 1–10.
67. Jin, X., et al. 2020. Root cause analysis of pinhole defects on painted galvanized steel panel. *Engineering Failure Analysis* 115: 104598.
68. Patala, S. 2019. Understanding grain boundaries–The role of crystallography, structural descriptors and machine learning. *Computational Materials Science* 162: 281–294.
69. Fung, I.W., et al. 2022. A Stress-Strain Model for resilience engineering for construction safety and risk management. *International Journal of Construction Management* 22 (12): 2308–2324.
70. Panjan, P., et al. 2020. Review of growth defects in thin films prepared by PVD techniques. *Coatings* 10 (5): 447.
71. Doshvarpassand, S., C. Wu, and X. Wang. 2019. An overview of corrosion defect characterization using active infrared thermography. *Infrared Physics & Technology* 96: 366–389.

Chapter 10
Thin-Film Applications in Different Fields

Abstract This chapter describes the application of thin films in various fields. We look at the applications of thin film in sensors, actuators, and solar cells and the many types of sensors in thin films. We have also discussed the use of thin films in optoelectronics and biocompatible applications. This chapter also discusses the use of thin-film biosensors and their roles in tissue engineering, cardiovascular disease, and drug delivery systems. The following section also discusses the use of thin films in electronics and advanced electronics. This article discussed the future possibilities and challenges associated with thin film and the deposition of green thin films. Overall, thin-film incorporation into various other developments has been discussed.

10.1 Application of Thin Film

Thin-film technology has traditionally been employed in a wide variety of applications, beginning with aesthetic reasons expanding to photographic objectives later on, and expanding to a nearly limitless variety of possibilities with the introduction of sophisticated deposition techniques, aided by the explosive growth of vacuum apparatus and power generation. In general, thin films have been utilized to boost bulky substances' performance by spreading an extra layer with the necessary physical and chemical properties. The subsequent chapter provides a quick explanation of the most technologically significant thin-film applications [1]. Thin-film studies have either initiated or expanded numerous new fields of inquiry in solid-state physics and chemistry, centered around characteristics that are peculiar to the film's thickness, geometry, and composition. Substances are processed into thin films, allowing for simple incorporation into a variety of electronics. Thin films are extremely thermally resilient and somewhat rigid, despite being sensitive. Organic materials, on the other hand, offer adequate thermal insulation and toughness, although they are soft. Thin-film mechanical characteristics can be assessed by tensile testing of unsupported films and the microbeam suspension deflection technique, although nanoindentation is the easiest approach. Optical experiments are an effective approach to examining the features of semiconductor materials [2].

T. I. Awan et al., *Thin Film Deposition Techniques*,
https://doi.org/10.1007/978-981-96-1364-9_10

10.1.1 Piezoelectric Films for Devices as Sensors, Actuators, and Harnessing Energy

Piezoelectric microelectromechanical systems (MEMS) enable high-sensitivity detectors and large relocation, low-voltage actuators. The latest developments in the deposition of perovskite thin films, in particular, indicate the development of MEMS devices susceptible to significant shifts at voltage levels appropriate with analogous metal-oxide semiconductors. Furthermore, if the electronics are installed in extremely loud situations, they can be employed for generating electricity. All of these uses rely on being capable to achieve high piezoelectric coefficients and maintain them during the micro-manufacturing procedure. The development of exceptionally fast piezoelectric sensitivities in films throws up a broad range of potential uses. Several applications will be given, spanning low-voltage radio frequency MEMS switches and resonators, actuators for millimeter-scale robotics, droplet ejectors, power scavengers for unguarded sensors, and medical ultrasound actuators [3].

10.1.2 Thin-Film Actuator

Actuators that are capable of being activated in a variety of methods are necessary for developing microsystems. Magnetostriction, the reverse of the piezoelectric impact, shape remembering, and the impact of the bimetallic blade will all be employed as actuation mechanisms. Magnetic fields, electric fields, and temperature fluctuations are the factors that are activated in these conditions. Thin-film actuators are ideal for microsystem innovation as they can be easily miniaturized downward to the μm region and manufactured using microsystem fabrication procedures. These substances share an aptitude to convert electrical energy into mechanical energy via the aforementioned effects. Magnetostrictive and piezoelectric materials are particularly interesting since they facilitate the conversion of energy in both vertical and horizontal directions. These opposing consequences are therefore able to be utilized [4].

10.1.3 Bistable Thin-Film Shape Memory Actuator

Bistable shape-retaining actuators, characterized for application in tactile graphic displays, were prepared by applying microsystem technology processes. The actuators were prepared as sputter-deposited buckled metallic thin-film carriers with structured Ti–Ni–Cu and Ti-Ni-Hf shape memory alloys at the leading edge and bottom, respectively. They were deposited on wavy-structured substrates whose longitudinal dimensions ranged from 2.2 to 3.5 mm wide and 1–3 mm long. The ranges of the current and voltage applied to the actuators for control varied between 0.2 and 0.8 V and between 0.2 and 0.8 A, respectively. Maximum force achievable was 2.2 mN with a corresponding displacement of 0.7 mm. A dedicated test configuration was designed to further enhance the performance. A dedicated test configuration was designed to further increase the reliability. The bistable actuators were cast onto a planar surface with dimensions at the sides between 6 and 8 mm wide and 3 and 6 mm long. The actuators were reconfigured with actuation voltages from 0.6 to 1.6 V and currents from 0.6 to 1.8 A. Hence, a force of 16 mN and a displacement of 1.2 mm were obtained [5].

10.1.4 Transparent Thin-Film Acoustic Actuators and Sensors Made of Carbon Nanotubes

Transparent thin-film acoustic transducers that function as both microphones and speaker systems are produced. The sensor is made of a thin film of piezoelectric poly (vinylidene fluoride) (PVDF) wrapped with transparent carbon nanotube (CNT) conductors. The conductors are created using CNT acid treatment and layer-by-layer nanoassembly on the PVDF substrate surface. The created thin*film detectors have good auditory sensitivity across a wide band of frequencies and are translucent thin, adaptable, and cheap. Transducers can transform windows, computer screens, touch boards, and posters into undetectable amplifiers and headsets with microphones significantly improving room ambiance [6].

10.1.5 A New Actuating System Based on Ferromagnetic SMA Thin Films

A new actuation mechanism that takes advantage of the interaction of inherent magnetically and shape retrieval forces functioning on a ferromagnetic shape memory alloy (SMA) microactuator in a magnetic field. This mechanism is characterized by high actuation and low biasing forces throughout each activation orientation, which leads to a long stroke. Because no additional biasing devices are required, ultra-compact designs are possible. An optical microscanner measuring 7 mm × 3 mm × 5 mm demonstrates huge scanning angles of up to 120° in off-resonance mode. The scanning performance maintains frequency independence below a critical frequency defined by thermal transfer durations [7].

10.1.6 A Silicone Elastomer with Exceptional Durability and Self-healing Properties Used for Thin-Film Thermal Actuators

The positive aspects of thin-film actuators include their simplicity and instantaneous reflexes. However, once injured, it presents significant issues about athletic consistency. A self-healing thermal actuator was created through the combination of a healing oneself silicone polymer and a polyimide sheet. The silicone elastomer with exceptional regeneration and durability was created using a triple dynamic reversible bond consisting of a hydrogen bond, an oxime bond, and a metal conjunction bond. Polyurea groups are supplied with a substantial amount of connecting points to facilitate supramolecular hydrogen bonding, and nitrogen atoms on the oxime group form complex with various metal ions (e.g., Cu^{2+}, Zn^{2+}, Fe^{3+}), greatly improving mechanical strength (up to 4818 kPa). Silicone elastomer demonstrated quick and effective self-healing (99.9% within 10 h in the room). Furthermore, the actuator can be chemically ambitious, allowing for an approximate bend of 59.9° at 100 °C. The actuator demonstrates remarkable stability while cycling over 100 cycles and can maintain the initial twisting angle following mending. The demonstration of healing oneself capability on bendable actuators establishes a preliminary foundation for future creation of robotic limbs [8].

10.1.7 Miniaturized Extendable Thin-Film Sensor for Accessible, Multipurpose Respiratory Surveillance

Respiratory signals are incredibly essential physiological signs that can often indicate or foretell the severity and intensity of numerous disorders. However, developing an implantable pulmonary tracking device with ease, outstanding longevity, and high accuracy remains an important problem. We created an easy-to-fabricate, portable, badge reel-like retractable self-powered sensor (RSPS) with excellent clarity, compassion, and sturdiness for ongoing surveillance of critical indications, such as respiration rate, apnea, and pulmonary ventilation. A small rotating thin-film turboelectric nanogenerator (RTF-TENG) was created by combining three sets of interdigital electrode architectures with phase differences and flexible printed circuit board (FPCB) fabrication technologies. The RSPS uses discrete sensing technology with a resolution of 0.13 mm, sensitivity of 7 P mm^{-1}, and solidity of over 1 million straining cycles with minimal hysteresis and good anti-environmental interference capabilities [9].

10.1.8 The Long-Lasting Durability of Thin-Film Electrodes for Neurostimulation

Implantable electrodes that can accurately assess brain function and give an electrical stimulus to a target region are currently being used to treat a variety of neurological and neuropsychiatric conditions. Retractable thin-film electrodes have received increased interest in recent years due to their ability to reduce infiltration and injury during implantation. Previous research has focused on improving the electrode's electrical and mechanical properties; nevertheless, chronic stability must be confirmed before electrodes can be translated from laboratory to clinical use. Neurostimulation electrodes, which continuously inject impose, have yet to successfully demonstrate continuous operation in vivo for ten years or longer, which is the acknowledged clinical viability standard. Stability for an extended period can only be obtained by shifting the focus to understanding the reasons why such devices fail [10].

10.1.9 Shape Memory Alloy Actuators

Thin films of shape memory alloys (SMAs), such as nickel titanium (NiTi), can be placed on substrates to form microactuators. When subjected to heat, the thin-film transforms and creates movement. SMA actuators are utilized for:

- Extremely invasive surgical instruments that can traverse throughout the body.
- Stents and catheters can alter morphology to fit the anatomy.
- Microfluidic blender and regulators.

- **Piezoelectric Actuators**

Microelectromechanical (MEMS) actuators are made from thin films of piezoelectric materials, such as lead zirconate titanate (PZT) or zinc oxide. These actuators may produce exact movements and forces when exposed to an electrical signal, allowing purposes, such as:

- Microfluidic valves and pumping devices for lab-on-a-chip systems adjustment stages for surgical machines and binoculars. Vibration-based tactile interfaces for medical instruction.

10.2 Application of Thin Film in Photovoltaic

Thin-film photovoltaic is a solar cell technique that uses very thin films of semiconductor materials to turn sunlight into power. In contrast to standard solar cells, which are often produced from crystalline silicon wafers, thin-film solar cells are built up of remarkably thin layers of active substances, frequently only a few micrometers broad. This provides increased adaptation and versatility in deployments.

- **Reducing Greenhouse Gas Emissions**

Thin-film PV systems typically release much less greenhouse gases during operation than fossil fuel power facilities. They emit approximately 20 g of CO_2/kWh (g CO_2/kWh), whereas fossil fuel plants discharge between 500 and 1000 g CO_2/kWh. This considerable decrease in toxins helps to mitigate global warming and improve the cleanliness of the air [11].

- **Low Priority Pollutants**

The development and execution of thin-film solar cells produce only 2–4% of the top contaminants discharged by fossil fuel facilities. The consequent decrease in dangerous pollutants reduces the total ecological impact and health hazards connected to energy generation.

- **Efficient Material Use**

Thin-film methods often use fewer resources than classic crystalline silicon solar cells, resulting in reduced consumption of resources and waste generation. Thin-film PV is lightweight, making it easy to set up and incorporate into a variety of programs, notably building-integrated photovoltaic (BIPV) and flexible solar panels.

- **Potential for Recycling**

While several thin-film technologies, such as cadmium telluride (CdTe), require toxic chemicals, the packaging methods utilized during production can successfully encapsulate these toxins. The ability to recycle thin-film PV materials, especially expensive elements, such as tellurium, can reduce environmental implications at the end of their lives. Encouraging recycling programs can help minimize trash and reclaim important resources.

- **Thin-Film Photovoltaic**

Indoor photovoltaics have gained popularity due to their immense potential for delivering energy for devices that use little power underneath indoor lighting in our daily lives. Third-generation thin-film solar cells, which include solar cells with dye sensitization, perovskite solar cells, and organic solar cells, have advanced rapidly in terms of their material formulation and performance for photovoltaic systems [12].

- **Thin-Film Shear Modulus**

Thin-film solar cells (TFSC) have matured and are now poised to contribute significantly to global electricity generation. They may outperform c-Si solar modules in terms of vast area coverage, reduced production costs, and a variety of purposes. The variables that restrict TFSC module effectiveness in comparison with champion cell functionality are explored, as is the significance of enhanced throughput and yield. Several studies suggest that manufacturing TFSC on a large scale (> 100 MW) with parallel lines of cloned machinery, high material utilization, and spray-on encapsulation can reduce prices to < $1/W. Small investors and large organizations are showing considerable new economic fascination with TFSC, which validates the thin-film technique. Unique properties of TFSC, which provide it a benefit over c-Si in two marketplaces, are small rural solar house systems and integrated photovoltaic installations. TFSC outperforms c-Si in annual energy production (kWhr/kW), has outdoor endurance comparable to c-Si, and is employed in MW-scale installations around the world. Advocates for TFSC should highlight their advantages over c-Si, such as less money spent, higher kWhr/kW output, more substantial battery storage current, desirable appearance as flexible materials, stability over the long term comparable to c-Si, and multiple deposition routes with room for creativity and progressive enhancement. Regardless of current efficiency levels, there is a large market for TFSC if expenses are decreased [13].

10.3 Applications of Thin Films in Optoelectronics

- **Antireflective (AR) Coatings**

AR coatings are intended to reduce reflection and improve light transmission across optical surfaces, such as lenses and displays. These coatings are made from thin films using procedures, such as:

- **Multilayered Structures**

Thin films can be created by layering materials with varying refractive indices to interfere with light waves and reduce reflections at specified wavelengths. This is especially crucial in systems, such as camera lenses, spectacles, and rooftop solar panels where maximum light transmission is required.

- **Longevity and Abrasion Resistance**

Hard coatings, which are frequently produced from thin films, offer scratch resistance and durability to optical instruments. These coatings are critical for preserving the integrity and lifespan of surfaces subjected to external wear and tear.

- **Optical Filters**

Thin films are also widely utilized in the production of optical filters, which absorb or reflect specific wavelengths of light. Key applications include.

- **Bandpass Filters**

Allow some wavelengths to pass while blocking others, which renders them useful in spectral analysis, the field of photography, and telecommunications. Thin-film technology allows for exact management of filter properties including bandwidth and center wavelength.

- **Reflective Filters**

Used in devices, such as mirrors and beam splitters, reflective filters use thin-film coatings to create high reflectivity at certain wavelengths. This is especially relevant in laser applications and optical devices.

- **Color Filters**

Thin films in display technologies can produce brilliant color filters that improve the visual experience by deliberately filtering out undesirable wavelengths.

- **Thin Films used in Contact Lenses**

Smart contact lenses have a variety of thin-film elements and are made using various thin films. They were created to identify and treat disorders including intraocular pressure, and ophthalmology employing tear fluid. This research examines how extremely thin films are treated to acquire the functionality essential to their use in intelligent contact lenses. The various kinds of thin films employed, along with the many uses, are reviewed, beginning with the historical advances that led to the discovery of smart contact lenses and ending with cutting-edge technology related to thin films [14].

10.4 Application of Thin Film in Buffer and Window Layers

Thin films can also act as buffers and display layers in solar cells. These layers increase luminescence and decrease electron–hole recombination distortions, which improves general cell functionality. Transparent conductive oxides (TCOs) are extensively utilized as glass coatings that permit sunlight through while transmitting electricity [15].

10.4.1 Application of Thin Film in Transistor and Electronics

Although liquid crystal displays (LCDs) have been around for more than fifty years, they have traditionally been regarded as niche market items due to their mediocre efficiency: narrow viewable angles, lengthy response times, and a shortage of large-area displays. The circumstances have altered considerably since the introduction of active matrix (AM) LCDs, namely amorphous silicon (a-Si: H) thin-film transistor (TFT)-driven LCDs. Panels made with TFT LCDs will be an important impetus for the progress of TFT technology in the not-so-distant future. Versatile electronics, integrated circuits, sensors, detectors, as well light-emitting diodes, and other new applications will benefit from TFTs. Industry R&D efforts have aimed to improve the

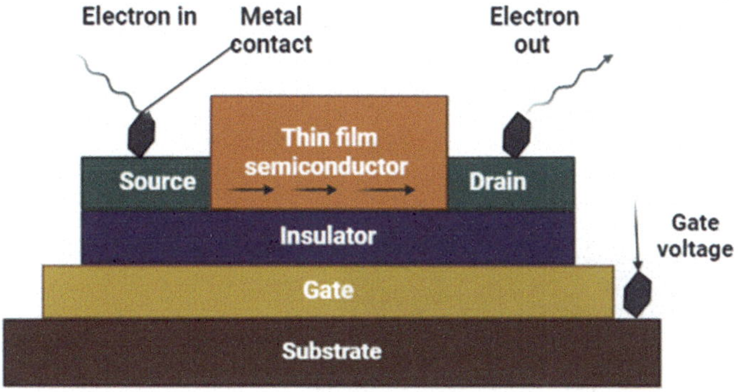

Fig. 10.1 Working mechanism of thin-film transistor

expense, production, and circulation of large-area screens [16]. Figure 10.1 describes the general mechanism of thin-film transistor.

10.5 Thin-Film Solar Cell

TFSCs are appealing devices that have the potential to drastically reduce the amount of active material required. Some semiconductors can be used in thin-film form to produce powerful solar cells on an extensive basis, despite their high cost and scarcity. The functionality of thin-film solar cells continues to rise due to a greater awareness of their distinctive structural, chemical, and optoelectronic attributes. These features can be used for passivation as well as stimulation, photon scattering/recycling, surface electric field generation, and assessed bandgaps. A deeper awareness of the computerized role of connections in the layered structure of devices, as well as an examined consideration of photons and excited carriers, is projected to reduce the gap between attained and hypothetical efficiency gains. For specialized applications, a-Si: H-based thin-film solar cells can still be made on a megawatt scale. Despite having restricted productivity, inadequate effectiveness, stability issues, and expenses equivalent to crystalline Si cells, the cell technology benefits from a straightforward and standardized fabrication method for integrated modules on rigid yet malleable substrates. Despite improvements in productivity and security, the multifunction cell will still fall behind the more robust, successful, and cost-effective c-Si cell [17]. Figure 10.2 generally describes the mechanism of thin-film solar cell.

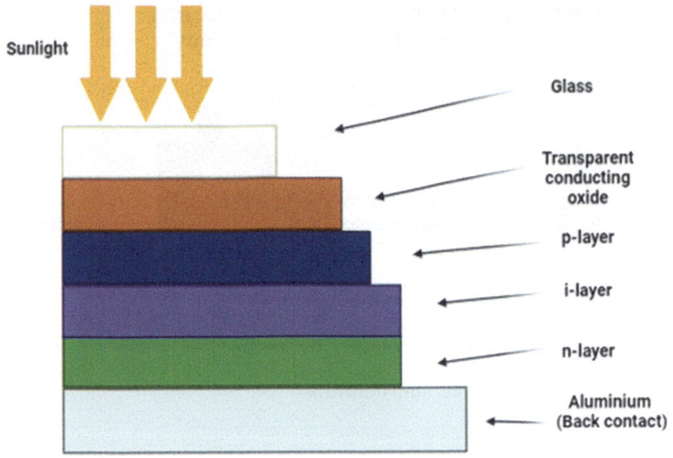

Fig. 10.2 Working mechanism of thin-film solar cell

10.6 Thin-Film LCD

A thin-film-transistor liquid–crystal display (TFT LCD) is a form of liquid–crystal panel that employs thin-film transistor technology to increase image characteristics like as addressability and clarity. A TFT LCD is an active matrix LCD, as opposed to passive-matrix LCDs or straightforward, direct-driven LCDs with a few sections. TFT LCDs are found in television sets, computer displays, cellular phones, video gaming systems, digital personal assistants, GPS systems, displays, and gauges in some vehicles and medium- to high-end motorbikes [18].

10.7 Thin-Film Transistor as Biosensors

Thin-film transistor (TFT) technology might have an important role in the rapidly increasing sector of chemical and biological sensors, enhancing efficiency and achieving semiconductor technologies for biosensor development. A TFT is a MOSFET made completely of thin-film components on an insulating substrate. TFTs, like enhancement-mode MOSFETs, typically work on the same principle, with very little current flowing between the source and drain because Vgs (the gate voltage concerning the source) is low due to the active layer's high resistance. When the Vgs is high, a charge is induced near the oxide–semiconductor interface, establishing a channel between the source and drain. As a result, the TFT functions as a switch under the control of the Vgs. TFTs can be used as circuitry elements for molding and reading the signal of prospective label-free electrical localization techniques, or as potentiometric sensors for identifying various bimolecular collaboration, such as electrochemical impedance spectroscopy. Being capable to identify bimolecular

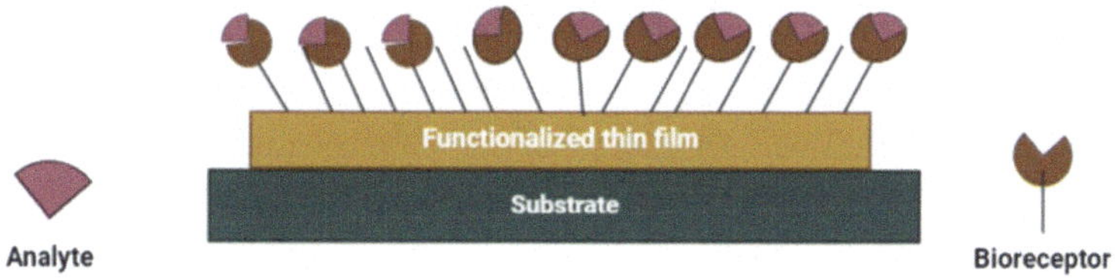

Fig. 10.3 Schematic representation of thin film as a biosensor

interactions is critical for pharmaceuticals, medicinal, and biotechnology applications. Most often used methods for discovering such links rely on optical approaches, specifically fluorescence identification of labeled biomolecules [19]. Figure 10.3 illustrates the schematic representation of thin film as biosensors.

10.8 Application of Thin Film in Electronics

Recent advances in thin-film electronics on flexible substrates have shown great promise. Nanocomposite sheets including cellulose nanofibers and ZnS nanoparticles have been successfully produced, revealing their potential for incorporation into high-frequency electronics.

Additive printing techniques have been investigated for creating stack-structured thin-film transistors (TFTs) on flexible substrates employing a variety of functional materials, including silicon, organics, and inorganics. Thermoelectric films and conductive elastic materials were used to create self-powered flexible temperature–pressure bimodal sensors that can detect external tactile stimuli as well as precisely sense temperature [20] (Fig. 10.4).

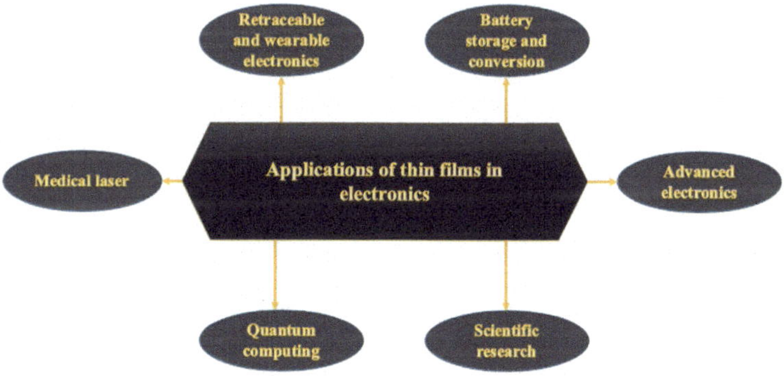

Fig. 10.4 Application of thin films in electronics

10.8.1 Retractable and Wearable Electronics

Recent advances in thin-film electronics on flexible substrates have allowed the creation of wearable gadgets. Thin-film transistors, detectors, and parts for storing energy can be made from flexible materials like as plastic, cloth, or metal foils. This enables the development of linear, portable, breathable electronic devices for medical monitoring, human–machine interaction, and intelligent clothing purposes.

10.8.2 Biomedical Thin Films

Thin films will continue to play an important role in medical applications. Emerging trends incorporate smart bioactive thin films that trigger biological reactions and promote the regeneration of tissue in implantation and engineering substrates. Thin-film drug delivery systems use stimuli-responsive polymers to offer regulated and localized medication dispersion.

- **Battery Storage and Conversion**

Thin-film techniques are being investigated for contemporary battery backup and transmission devices.

- **Thin-film batteries** have more energy density, adaptability, as well as security than traditional batteries.
- **Thin-film solar cells** use substances, such as perovskites, to obtain outstanding performance at low costs. Thin-film fuel cells and thermoelectric devices for mobile and reusable batteries.

10.8.3 Quantum and Neuromorphic Computing

Thin films will be critical in realizing realistic quantum computing devices and neuromorphic (brain-inspired) computer systems. Superconducting thin sheets for quantum bits and detectors is required for further progress in thin film applications.

Memristive thin-film devices can simulate postsynaptic behavior for neuromorphic prosthetic computing.

- **Medical lasers**

Medical lasers, which are becoming more common for non-intrusive treatments, are made with thin films. The global market for medical laser systems is predicted to increase from $5116 million in 2016 to $12,586 million by 2023, at a CAGR of 13.6%. Medical lasers can be utilized to eliminate connective tissue, alleviate pain, reduce irritation and edema, and improve oxygenation.

- **Advanced Electronics**

Thin-film applications will drive substantial breakthroughs in the electronics industry.

- **Semiconductor Technology**

Thin films can increase the effectiveness of semiconductor devices by increasing conductivity and insulation, resulting in improved electronic components.

- **Flexible Electronics**

Thin-film technologies will play a critical role in the creation of adaptable and inexpensive electronic devices, such as wearables and displays. Thin films will be used to fabricate qubits and other components for quantum computers, allowing for quicker and more powerful computing.

10.9 Environmental Applications

Thin films are intended to make contributions to environmental friendliness.

- **Scientific Research and Development**

Thin-film technologies will remain a focus of investigation, contributing toward novel applications.

10.9.1 Thin-Film Photocatalysis for Environmental Remediation

Thin-film photocatalysts have attracted interest in addressing the post-separation issue with powder photocatalysts. Metal-oxide and metal sulfide thin-film photocatalysts can break down organic contaminants into less dangerous compounds. The most commonly tested pollutants include MB, RhB, and Mo dye. The thin-film method can boost immobilized innovation and perhaps disintegrate contaminants on a large scale. Thin-film photocatalysts suffer effectiveness challenges due to their low surface area and reactive sites, as photocatalysis is a phenomenon of the surface. For thin-film photocatalysts to be extensively utilized in different sectors, further research and adjustments are needed. Adding a tiny amount of surfactant can enhance the surface area of thin films. Using different substrates and modifying the electrical structure can considerably enhance photocatalytic activity. Water contamination poses a significant concern to the world, prompting increased investigation and enthusiasm for developing filtration technology. Metal-oxide and metal sulfide thin films may efficiently eliminate organic contaminants from wastewater [21].

10.10 Application of Thin Films in Medical Fields

10.10.1 Thin Films for Tissue Engineering Application

Many vital innovations rely heavily on thin films. Thin films can be sprayed to various surfaces depending on their physicochemical qualities, including moisture absorption, reactivity, conductivity, and corrosion resistance. Thin films can be applied on numerous surfaces using various methods, such as casting in solvent, spraying, wrapping, covalent conjugation, surface-initiated polymerization, and self-assembly (e.g., electrostatic interactions, and hydrogen bonding).

Biomaterials of various topologies and characteristics, both natural and synthetic, were used to create these thin films. Thin films' ability to manipulate physicochemical properties makes them suitable for tissue engineering applications. Thin-film components, procedures, and circumstances must be nontoxic and not disrupt cellular functioning. Thin films are increasingly being used for osseointegration. We specialize in dentistry and musculoskeletal devices, biodegradable scaffolds, and biomimetic materials for tissues engineering and biomedical applications [22]. Figure 10.5 generally describes the applications of thin film in tissue engineering.

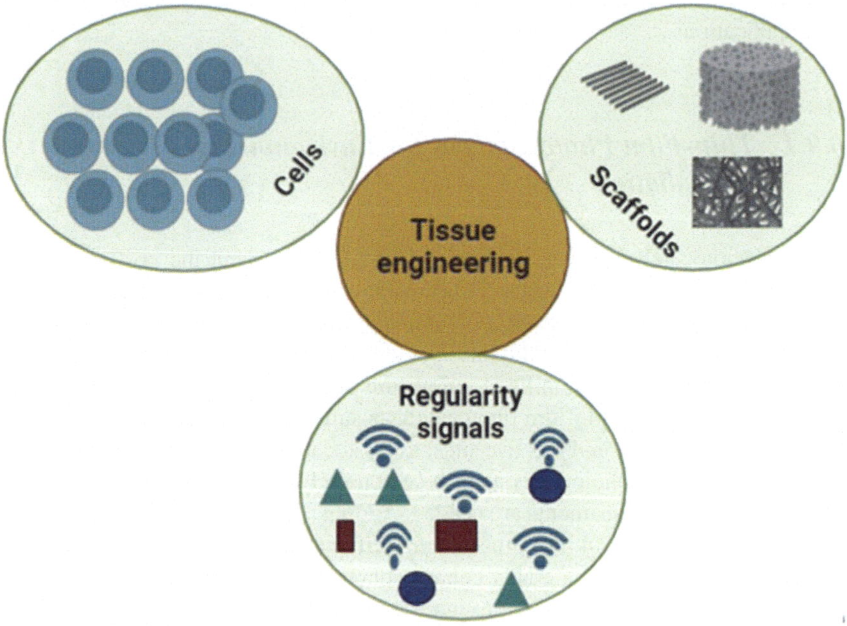

Fig. 10.5 Applications of thin film in tissue engineering

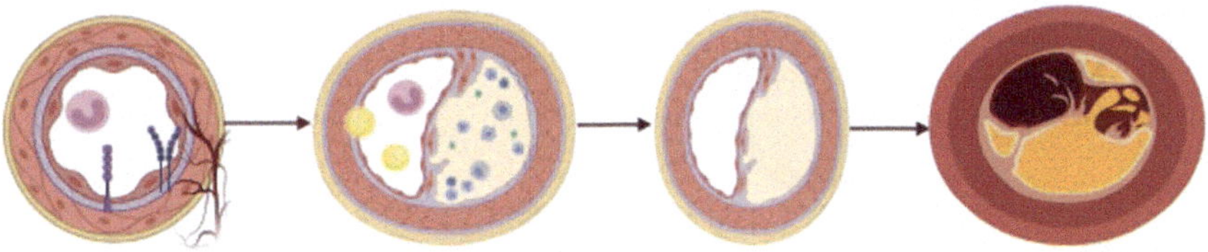

Fig. 10.6 Role of thin film in cardiovascular disease

10.10.2 Thin-Film Development in Cardiovascular Diseases

Heart disease has emerged as the leading cause of death in humans, and numerous ways for studying it have been created. Among these, heart-on-a-chip has arisen over the past few decades as a way of creating disease (or normal) models in vitro and is seen as an exciting avenue for studying heart disease. Heart-on-a-chip has several advantages over other approaches, including its great mobility, throughput, and capacity to simulate microenvironments in vivo. It has demonstrated substantial promise in disease etiology research and medication screening. After analyzing the blueprint of the existing heart-on-a-chip, we recommended that a highly integrated heart-on-a-chip comprises four elements: microfluidic chips, cells/micro tissues, microactuators to build the microenvironment, and microsensors for results readout [23]. Figure 10.6 illustrates the role of thin film in cardiovascular disease.

10.10.3 Biomedical Applications

Thin films are combined with a variety of biomedical technologies:

1. Nanostructured thin films improve osseointegration and antimicrobial characteristics in orthopedic implants.
2. Stimulus-responsive thin films for regulated medication delivery and tissue engineering.
3. Thin-film biosensors for immediate tracking of biological parameters and pathological biomarkers.
4. Thin films are used in drug delivery systems.

Figure 10.7 describes the role of the thin film in drug delivery systems in the field of biomedical applications.

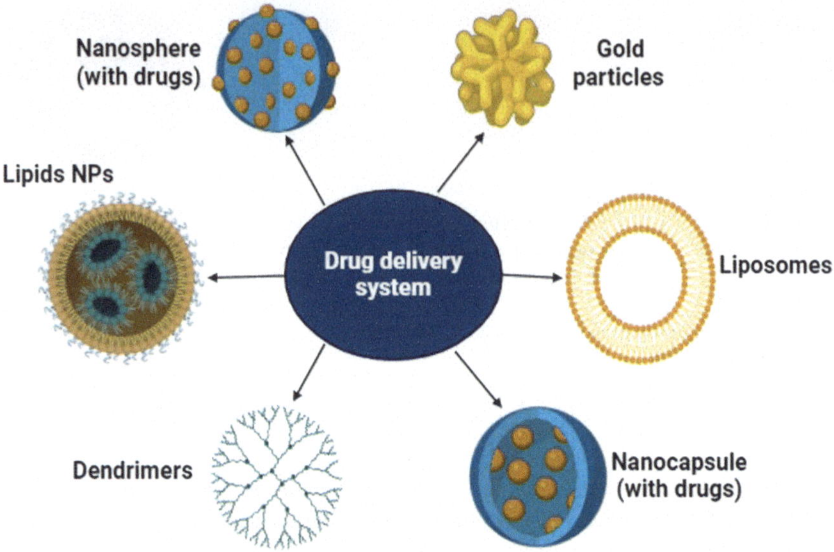

Fig. 10.7 Thin film in drug delivery system

10.10.4 Bio-inspired Lightweight Films for Inactive Midday Radiant Refrigeration

Passive radiative midday cooling is a newer technique that contributes to carbon–neutral thermal management. The approach is based on optically tailored materials with specific solar and mid-infrared absorption and emission capabilities. To reduce climate change, environmentally friendly cooling materials or coatings must cover large areas due to their low emissive power of 100 Wm^{-2} during the day. As a result, biocompatible materials are urgently needed to create acceptable coverings that have no negative environmental influence. It is demonstrated how chitosan films of various thicknesses can be made from mildly acidic water-based solutions [24].

10.11 The Mutual Compatibility of Thin Films in Thermodynamic Cooling Mechanisms

The suitability of low-dimensional thermoelectric materials, such as thin films and nanowires, for use in thermoelectric coolers is investigated. According to first-order thermoelectric theory, the non-dimensional determines the cold and hot junction temperatures in a thermoelectric circuit. The conventional theory's forecasts for performance have been implemented in thermoelectric cooler systems, indicating that they may be miniaturized without compromising effectiveness and that system functionality is mostly determined by (Z_eT_a). A non-dimensional thermoelectric system representation for a cooler is built, and standard thermoelectric cooler efficiency measures are shown, along with estimates based on conventional framework. Performance is measured as an indicator of thermoelectric element length under appropriate system circumstances. Cooler performance may suffer dramatically when miniaturized, especially if the cooling elements are implemented on the size of many recently suggested thermoelectric thin films and nanostructured materials. According to system theory, performance is determined by three non-dimensional parameters: effective thermoelectric figure of merit (Z_eT_a), heat transfer ability, and heat rejection to the ambient environment. Low-dimensional materials like thermoelectric elements may need to be reconsidered due to their impact on cooler effectiveness, which is influenced by both material qualities and relative scale of system thermal conductance. High-performance coolers based on thermoelectric effects must be developed [25].

10.12 Application of Thin Films in Jewelry

An object of jewelry consists of a substrate on which thin sheets of dielectric, metal, and semiconductor materials are placed in an aesthetically attractive pattern. One protective layer is formed by the substrate, while the second is made up of a plastic covering. The pattern can be viewed through either the substrate or the plastic coating. In one version, the substrate is glazed ceramic, while in another, it is nonglare glass. In each embodiment, a light opaque layer is placed on one of the protective layers to reflect light to the observer. A jewelry mount is attached to one of the layers. The current invention pertains generally to jewelry items and more specifically to jewelry articles made from thin films of metals and dielectrics [26].

10.13 Future Emerging Trends and Problems/Challenges

- **Incredible Creative Potential**

Thin films can be employed solely for ornamental purposes, permitting intricate patterns and sensations that typical plating procedures cannot match.

- **Visual Impacts**

Certain thin films can produce optical effects, such as color changes based on viewing angle. This attribute is very enticing for luxury jewelry since it provides a dynamic visual aspect.

- **Customized Designs and Fabrics**

Thin-film technology allows developers to incorporate a variety of textures and patterns, resulting in distinctive and imaginative jewelry items that stand out in the crowded marketplace.

10.14 Application of Thin Films in Forensic Science

Thin films are increasingly being utilized to improve fingerprint identification and visualization, offering substantial benefits over older methods:

- **Metal Thin Films**

Recent advances have demonstrated that metal thin films can be put on surfaces to improve the appearance of hidden fingerprints. These films improve the distinction between the fingerprint ridges and their surroundings, rendering them clearer to analyze. These films are created using techniques, such as sputtering, which allows for exact regulation of width and material characteristics.

- **Surface Modification**

Thin films can change the surface characteristics of substrates, allowing fingerprint remains to adhere more effectively. This change can improve fingerprint retrieval from a variety of surfaces, including non-porous materials, such as glass and plastic [27].

10.15 Application of Thin Film in Memory Storage Devices

Thin films serve an important part in the creation and evolution of memory storage systems. Here are some significant uses of thin films in memory devices:

- **Magnetic Storage**

Magnetic thin films are critical elements in magnetic storage systems like hard disk drives (HDDs). They are employed in magnetic tape to save data by magnetizing particular parts of a thin film.

- **Thin-Film Memory**

Sperry Rand created thin-film memory, a pioneering competitor to magnetic-core memory, in the 1960s. It used vacuum evaporation to produce thin layers of perm alloy (an iron-nickel alloy) on glass plates. This enabled extremely swift access speeds in the range of 670 ns, but was costly to manufacture.

- **Non-volatile Memory**

Thin films can be utilized to make permanent memories memory devices that keep data even when the electricity is switched off. Non-volatile memory arrays, for example, can be built using thin-film transistors (TFTs) constructed of materials, such as cadmium selenide.

- **Optical Storage**

Optical storage media, such as CDs and DVDs, are made from thin films. The information is preserved by using a laser that melts tiny cracks in a thin dye coating. The data are gathered by reflecting a laser off the thin sheet and measuring variations in reflectivity.

- **Solid-State Drives (SSDs)**

Solid-state drives maintain information using flash memory, which is based on thin-film transistors. Thin films enable miniaturization of memory cells, which results in better storage densities in SSDs contrasted to traditional hard disk drives [28].

- **Material Science**

Continued research into thin-film deposition techniques, such as atomic layer deposition (ALD) and chemical vapor deposition (CVD), will promote advancements in material characteristics and functionalities.

- **Nanotechnology**

The combination of thin films with nanotechnology will allow the invention of novel materials and techniques with higher efficiency in numerous applications, such as medicine and electronics.

10.16 Material Limitations

10.16.1 Compliance Concerns

Because thin films are utilized in a variety of applications, it can be difficult to ensure consistency with varied substrates and environments. For example, the adherence of thin films to diverse materials might vary, resulting in breakdown or malfunction in purposes, such as biomedical implants or electronic devices.

10.16.2 Mechanical Characteristics

Thin films' mechanical characteristics may differ from those of the bulk materials to which they are applied. This disparity can cause brittleness, wear, or failure under stress, especially in applications that need extreme resilience, such as laparoscopic tool finishes or safeguards in electronics.

10.16.3 Manufacturing Challenges Related to Thin Films

- **Adaptability**

While thin-film deposition techniques, such as atomic layer deposition (ALD) and chemical vapor deposition (CVD), are useful in the testing facility, scaling these procedures to mass production can be difficult and expensive. Maintaining uniformity as well as consistency across vast areas is still a considerable task.

- **Production Costs**

High-quality thin-film production requires costly materials and techniques. As the need for thin films grows, finding economical options while retaining quality will be critical.

10.16.4 Environmental Concerns

- **Material Toxicity**

Certain materials used in thin-film manufacture may be hazardous to the environment and human health. For example, certain metal oxides or chemicals used in deposition procedures can be harmful. To minimize environmental impact, these materials must be handled and disposed of safely.

- **Sustainability**

The production methods for thin films can involve large amounts of energy and materials. The industry will increasingly focus on establishing sustainable procedures and products that have a low environmental footprint.

10.16.5 Technological Constraints

- **Performance Variability**

Deposition circumstances, such as temperature, pressure, and chemical composition, might have a significant impact on thin-film performance. Variations in these values might cause uneven film qualities, compromising the dependability of devices that use these films.

- **Connectivity with Preexisting Techniques**

As thin films are integrated into existing technologies, reliability with older systems might be a challenge. Demonstrating that new thin-film technologies interact

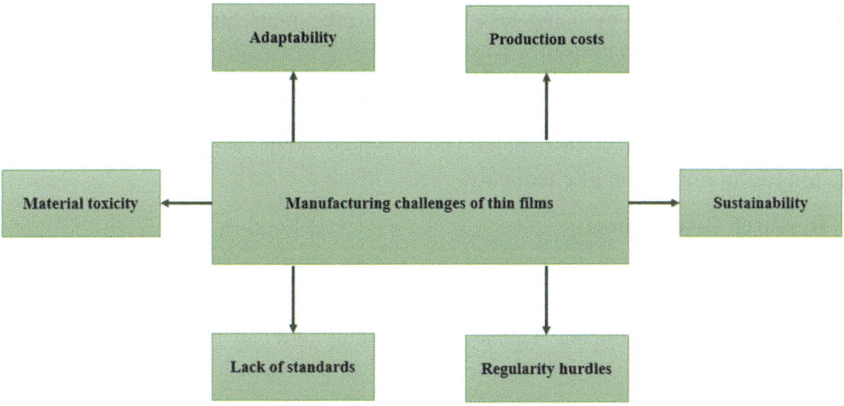

Fig. 10.8 Manufacturing challenges related to thin films

effectively with existing devices and techniques is critical for wider adoption. Figure 10.8 generally describes the manufacturing challenges related to thin film.

10.16.6 Regulatory and Standardization Issues

- **Lack of Standards**

As thin films are applied in critical areas, such as health care and electronics, the absence of standardized testing and quality assurance protocols can lead to variability in product performance and safety. Establishing comprehensive standards will be necessary to ensure reliability and safety.

- **Regulatory Hurdles**

In the medical field, for instance, thin-film coatings used in implants and devices must undergo rigorous testing and approval processes. Navigating these regulatory pathways can be time consuming and costly, potentially delaying the introduction of innovative thin-film solution.

10.17 Green Thin-Film Deposition and Sustainability

- **Green CVD (Chemical Vapor Deposition)**

Green CVD (chemical vapor deposition) promotes sustainable practices. Conventional chemical vapor deposition (CVD) procedures frequently use toxic ingredients and need a significant amount of energy. The "Green CVD" idea strives to minimize adverse consequences by using less hazardous precursors and supplying them in stoichiometric ratios to minimize wastage.

- **Decreasing Use of Energy During the Sedimentation Process**

By the United Nations Sustainable Development Goals, we are concentrating on the long-term utilization of chemicals.

- **Research Instructions**

The Green CVD movement fosters research into alternate resources and techniques for maintaining or improving the properties of deposited films while minimizing environmental impact. This includes looking into new precursor designs and reactor designs to increase effectiveness and lower emissions.

10.18 Applications for Solar Energy

Thin-film photovoltaics are rapidly being used in solar energy projects. These methods are more ecological than standard crystalline solar cells as they utilize fewer resources and essential components for fabrication. Production techniques can be healthier for employees and the surroundings since they frequently do not use harmful chemicals, such as hydrofluoric acid.

- **Emerging Materials**

Improvements in materials like perovskites and gallium arsenide thin films are extending the limits of solar energy performance. For example, gallium arsenide-based thin films have reached a conversion rate close to 28.8%, whereas perovskite films are being investigated for their inexpensiveness and prospective efficiency of around 20%.

- **Storage of Energy and Efficiency**

10.19 Thin-Film Batteries

The use of thin-film manufacturing in battery construction is resulting in improved energy storage options. Thin-film batteries may be manufactured lighter and more compact, thus rendering them ideal for use in microelectronics and renewable energy applications. These batteries can improve the performance of computerized manufacturing operations and connectivity of devices.

- **Smart Energy Systems**

Thin films can help to create smarter energy systems by allowing microchips to monitor and optimize energy use in real time, increasing ultimate effectiveness as well as sustainability in energy production and consumption.

10.20 Application of Advanced Thin Films

Advanced thin films are transforming several contemporary disciplines with their novel architecture, complex manufacturing strategies, and numerous potential uses. These films, which usually vary in width from nanometers to micrometers, are designed to have distinctive characteristics adapted to particular applications. The process of designing requires careful monitoring of arrangement, structure, and thickness to obtain the necessary electrical, optical, magnetic, or mechanical properties. Fabrication techniques, such as physical vapor deposition (PVD), chemical vapor deposition (CVD), and atomic layer deposition (ALD), enable the production of excellent thin films with consistent height and great substrate adherence. These approaches make it possible to create films with certain qualities that can be incorporated into a wide range of gadgets. Advanced thin films have a diverse range of usages in industries, such as gadgets, vitality, medical treatment, and ophthalmology [29].

10.21 Incorporations with Other Trends

- **Combination of Thin Films with Other Developing Technologies**

The combination of thin films and other developing technologies is propelling development across the industry. Here are some important ways that thin films are merged with other technologies:

- **Flexibility and Ubiquitous Electronics**

The advancement of flexible thin films permits the creation of reusable electronic gadgets that can adapt to complex substrates. The thin films are being combined with:

- Flexible conductors and semiconductors for stretchy circuits and screens.
- Piezoelectric materials for capturing energy and monitoring in intelligent fabrics.
- Biodegradable coverings for medical gadgets and implantation.

- **The Internet of Things (IoT) and Smart Sensors.**

Thin films play an important role in IoT sensor technology. They are paired with:

1. Nanomaterials improve accuracy as well as selectivity in gas and chemical sensors.
2. Microfluidics in lab-on-a-chip devices and biological sensors.
3. Communication via wireless modules for remote surveillance of IoT networks.

- **Battery Storage and Converting**

1. Thin films enable breakthroughs in energy preservation and transmission techniques.

2. Thin-film batteries and supercapacitors for adaptable and energy-dense storage.
3. Thin-film photovoltaics for lightweight and convenient solar energy production.
4. Thermoelectric thin films for heat loss extraction in energy-saving systems.

10.22 Quantum Computing and Spintronics

The unique features of thin films are being used in new industries, such as quantum computing and spintronics.

1. Quantum computers use super thin films for quantum bits (qubits).
2. Spintronics uses magnetic thin films for spin-based logic and memory devices.
3. Topological insulator thin films enabling dissipation-free transportation of spin-polarized electrons.

10.23 Thin Film in Additive Printing

Thin-film deposition procedures are being coupled with additive production technologies.

1. Thin-film coatings on 3D-printed items enhance surface characteristics and functioning.
2. Thin layer deposition in 3D-printed buildings for electronic components and sensors.
3. Thin film-based inks for directly printing conductive designs and components.

10.24 Conclusion

Thin-film technology has advanced immensely and is used widely in multiple industries due to its unique qualities and possibility. This conclusion drawn from the extensive research on thin films also highlights a few important points. Most technologies, from electronics to alternative energy alternatives, heavily utilize thin films. This has opened up much wider avenues in solid-state physics and chemistry as they can enhance the functionality of bulk materials by providing certain physical and chemical features to the material. Thin films have found their applications in a wide range of instruments, such as piezoelectric devices, MEMS, smart contact lenses, and photovoltaic systems.

References

1. Acosta, E. 2021. Thin films/properties and applications. In *Thin Films*. IntechOpen.

2. Rao, M., and M. Shekhawat. 2013. A brief survey on basic properties of thin films for device application. *International Journal of Modern Physics: Conference Series.*
3. Muralt, P., R.G. Polcawich, and S. Trolier-McKinstry. 2009. Piezoelectric thin films for sensors, actuators, and energy harvesting. *MRS bulletin* 34 (9): 658–664.
4. Quandt, E., and H. Holleck. 1995. Materials development for thin film actuators. *Microsystem Technologies* 1: 178–184.
5. Vitushinsky, R., S. Schmitz, and A. Ludwig. 2008. Bistable thin-film shape memory actuators for applications in tactile displays. *Journal of Microelectromechanical Systems* 18 (1): 186–194.
6. Yu, X., et al. 2006. Carbon nanotube-based transparent thin film acoustic actuators and sensors. *Sensors and Actuators A: Physical* 132 (2): 626–631.
7. Kohl, M., et al. 2004. A novel actuation mechanism on the basis of ferromagnetic SMA thin films. *Sensors and Actuators A: Physical* 114 (2–3): 445–450.
8. Dai, S., et al. 2024. A high mechanical strength, self-healing silicone elastomer for thin film thermal actuator. *Colloids and Surfaces A: Physicochemical and Engineering Aspects* 689: 133506.
9. Li, C., et al. 2023. Miniaturized retractable thin-film sensor for wearable multifunctional respiratory monitoring. *Nano Research* 16 (9): 11846–11854.
10. Imai, A., et al. 2023. Flexible thin-film neural electrodes with improved conformability for ECoG measurements and electrical stimulation. *Advanced Materials Technologies* 8 (21): 2300300.
11. Schock, H. 1996. Thin film photovoltaics. *Applied surface science* 92: 606–616.
12. Yan, N., et al. 2020. Recent progress of thin-film photovoltaics for indoor application. *Chinese Chemical Letters* 31 (3): 643–653.
13. Hegedus, S. 2006. Thin film solar modules: The low cost, high throughput and versatile alternative to Si wafers. *Progress in Photovoltaics: Research and Applications* 14 (5): 393–411.
14. Kareem, M., et al. 2024. A review of thin films used in smart contact lenses. *Advanced Engineering Materials* 26 (4): 2300363.
15. Chemseddine, A., and M. Fearheiley. 1994. Improved CdS buffer/window layers for thin film solar cells. *Thin Solid Films* 247 (1): 3–7.
16. Kuo, Y. 2013. Thin film transistor technology—Past, present, and future. *The Electrochemical Society Interface* 22 (1): 55.
17. Chopra, K., P. Paulson, and V. Dutta. 2004. Thin-film solar cells: An overview. *Progress in Photovoltaics: Research and applications* 12 (2–3): 69–92.
18. Ma, J., X. Ye, and B. Jin. 2011. Structure and application of polarizer film for thin-film-transistor liquid crystal displays. *Displays* 32 (2): 49–57.
19. Mittal, P., S. Yadav, and S. Negi. 2021. Advancements for organic thin film transistors: Structures, materials, performance parameters, influencing factors, models, fabrication, reliability and applications. *Materials Science in Semiconductor Processing* 133: 105975.
20. Catania, F., et al. 2022. Thin-film electronics on active substrates: Review of materials, technologies and applications. *Journal of Physics D: Applied Physics* 55 (32): 323002.
21. Pedanekar, R., S. Shaikh, and K. Rajpure. 2020. Thin film photocatalysis for environmental remediation: A status review. *Current Applied Physics* 20 (8): 931–952.
22. Mozafari, M., et al. 2016. Thin films for tissue engineering applications. In *Thin Film Coatings for Biomaterials and Biomedical Applications*, 167–195. Elsevier.
23. Yang, Q., et al. 2021. Fabrication and Biomedical Applications of Heart-on-a-chip. *International Journal of Bioprinting* 7(3).
24. Lauster, T., et al. 2023. From Chitosan to Chitin: bio-inspired thin films for passive daytime radiative cooling. *Advanced Science* 10 (11): 2206616.
25. Miner, A. 2007. The compatibility of thin films and nanostructures in thermoelectric cooling systems. *Journal of Heat Transfer*
26. Coupry, C., and D. Brissaud. 1996. Applications in art, jewelry and forensic science. In *Raman Microscopy*, 421–453. Elsevier.
27. Neole, N.G. 2024. *Analytical Techniques in Forensic Science: Spectroscopy and Chromatography*, in *Cases on Forensic and Criminological Science for Criminal Detection and Avoidance*, 188–240. IGI Global.

28. Singh, A.K. 2024. *Experimental Techniques for the Characterization of Magnetic Thin Films.*
29. Rubin, S., et al. 2023. The world of advanced thin films: design, fabrication, and applications. *Fusion of Multidisciplinary Research, An International Journal* 4 (1): 393–406.

The manufacturer's authorised representative in the EU is Springer Nature Customer Service Centre GmbH, Europaplatz 3, 69115 Heidelberg, Germany. If you have any concerns regarding our products, please contact ProductSafety@springernature.com

Printed and bound by CPI Group (UK) Ltd, Croydon, CR0 4YY
01/10/2026
02225934-0010